Also by Sherry Turkle

The Second Self: Computers and the Human Spirit
Psychoanalytic Politics: Jacques Lacan and Freud's
  French Revolution

LIFE ON THE SCREEN

Identity in the Age of the Internet

SHERRY TURKLE

A TOUCHSTONE BOOK
Published by Simon & Schuster

7K

TOUCHSTONE
Rockefeller Center
1230 Avenue of the Americas
New York, NY 10020

First Touchstone Edition 1997

TOUCHSTONE and colophon are registered trademarks
of Simon & Schuster Inc.

Designed by Irving Perkins Associates, Inc.

Manufactured in the United States of America

10   9   8   7   6   5

Library of Congress Cataloging-in-Publication Data
Turkle, Sherry.
   Life on the screen : identity in the age of the Internet / Sherry Turkle.
      p.   cm.
   Includes bibliographical references and index.
   1. Computers and civilization.   2. Computer networks—Psychological aspects.
I. Title.
   QA76.9.C66T87   1995
   155.9—dc20   95-38428
                          CIP
ISBN 0-684-80353-4
ISBN 0-684-83348-4 (Pbk.)

The author gratefully acknowledges permission
from the following sources to reprint material
in their control:
Citations from Julian Dibbell's "Rape in Cyberspace"
reprinted by permission of the author and
The Village Voice.
Citations from John Schwartz's "On-line Lothario's Antics
Prompt Debate on Cyber-Age Ethics" Copyright © 1993 by
The Washington Post. Reprinted with permission.

To Rebecca
of Neverland and Strawberry Hill

CONTENTS

INTRODUCTION:
IDENTITY IN THE AGE OF THE INTERNET

There was a child went forth every day,
And the first object he look'd upon, that object he became.
—WALT WHITMAN

We come to see ourselves differently as we catch sight of our images in the mirror of the machine. A decade ago, when I first called the computer a second self, these identity-transforming relationships were almost always one-on-one, a person alone with a machine. This is no longer the case. A rapidly expanding system of networks, collectively known as the Internet, links millions of people in new spaces that are changing the way we think, the nature of our sexuality, the form of our communities, our very identities.

At one level, the computer is a tool. It helps us write, keep track of our accounts, and communicate with others. Beyond this, the computer offers us both new models of mind and a new medium on which to project our ideas and fantasies. Most recently, the computer has become even more than tool and mirror: We are able to step through the looking glass. We are learning to live in virtual worlds. We may find ourselves alone as we navigate virtual oceans, unravel virtual mysteries, and engineer virtual skyscrapers. But increasingly, when we step through the looking glass, other people are there as well.

The use of the term "cyberspace" to describe virtual worlds grew out of science fiction,[1] but for many of us, cyberspace is now part of the routines of everyday life. When we read our electronic mail or send postings to an electronic bulletin board or make an airline reservation over a computer network, we are in cyberspace. In cyberspace, we can talk, exchange ideas, and assume personae of our own creation. We have the opportunity to build new kinds of communities, virtual communities, in which we participate with people from all over the world, people with whom we converse daily, people with whom we may have fairly intimate relationships but whom we may never physically meet.

This book describes how a nascent culture of simulation is affecting our ideas about mind, body, self, and machine. We shall encounter virtual sex and cyberspace marriage, computer psychotherapists, robot insects, and researchers who are trying to build artificial two-year-olds. Biological children, too, are in the story as their play with computer toys leads them to speculate about whether computers are smart and what it is to be alive. Indeed, in much of this, it is our children who are leading the way, and adults who are anxiously trailing behind.

In the story of constructing identity in the culture of simulation, experiences on the Internet figure prominently, but these experiences can only be understood as part of a larger cultural context. That context is the story of the eroding boundaries between the real and the virtual, the animate and the inanimate, the unitary and the multiple self, which is occurring both in advanced scientific fields of research and in the patterns of everyday life. From scientists trying to create artificial life to children "morphing" through a series of virtual personae, we shall see evidence of fundamental shifts in the way we create and experience human identity. But it is on the Internet that our confrontations with technology as it collides with our sense of human identity are fresh, even raw. In the real-time communities of cyberspace, we are dwellers on the threshold between the real and virtual, unsure of our footing, inventing ourselves as we go along.

In an interactive, text-based computer game designed to represent a world inspired by the television series *Star Trek: The Next Generation,* thousands of players spend up to eighty hours a week participating in intergalactic exploration and wars. Through typed descriptions and typed commands, they create characters who have casual and romantic sexual encounters, hold jobs and collect paychecks, attend rituals and celebrations, fall in love and get married. To the participants, such goings-on can be gripping; "This is more real than my real life," says a character who turns out to be a man playing a woman who is pretending to be a man. In this game the self is constructed and the rules of social interaction are built, not received.[2]

In another text-based game, each of nearly ten thousand players creates a character or several characters, specifying their genders and other physical and psychological attributes. The characters need not be human and there are more than two genders. Players are invited to help build the computer world itself. Using a relatively simple programming language, they can create a room in the game space where they are able to set the stage and define the rules. They can fill the room with objects and specify how they work; they can, for instance, create a virtual dog that barks if one types the command "bark Rover." An eleven-year-old player built a room she calls the condo. It is beautifully furnished. She has created magical jewelry and makeup for her dressing table. When she visits the condo, she invites her cyberfriends to join her there, she chats, orders a virtual pizza, and flirts.

LIVING IN THE MUD

The *Star Trek* game, TrekMUSE, and the other, LambdaMOO, are both computer programs that can be accessed through the Internet. The Internet was once available only to military personnel and technical researchers. It is now available to anyone who can buy or borrow an account on a commercial online service. TrekMUSE and LambdaMOO are known as MUDs, Multi-User Domains or, with greater historical accuracy, Multi-User Dungeons, because of their genealogy from Dungeons and Dragons, the fantasy role-playing game that swept high schools and colleges in the late 1970s and early 1980s.

The multiuser computer games are based on different kinds of software (this is what the MUSE or MOO or MUSH part of their names stands for). For simplicity, here I use the term MUD to refer to all of them.

MUDs put you in virtual spaces in which you are able to navigate, converse, and build. You join a MUD through a command that links your computer to the computer on which the MUD program resides. Making the connection is not difficult; it requires no particular technical sophistication. The basic commands may seem awkward at first but soon become familiar. For example, if I am playing a character named ST on LambdaMOO, any words I type after the command "say" will appear on all players' screens as "ST says." Any actions I type after the command "emote" will appear after my name just as I type them, as in "ST waves hi" or "ST laughs uncontrollably." I can "whisper" to a designated character and only that character will be able to see my words. As of this writing there are over five hundred MUDs in which hundreds of thousands of people participate.[3] In some MUDs, players are represented by graphical icons; most MUDs are purely text-based. Most players are middle class. A large majority are male. Some players are over thirty, but most are in their early twenties and late teens. However, it is no longer unusual to find MUDs where eight- and nine-year-olds "play" such grade-school icons as Barbie or the Mighty Morphin Power Rangers.

MUDs are a new kind of virtual parlor game and a new form of community. In addition, text-based MUDs are a new form of collaboratively written literature. MUD players are MUD authors, the creators as well as consumers of media content. In this, participating in a MUD has much in common with script writing, performance art, street theater, improvisational theater—or even commedia dell'arte. But MUDs are something else as well.

As players participate, they become authors not only of text but of themselves, constructing new selves through social interaction. One player says, "You are the character and you are not the character, both at the same time." Another says, "You are who you pretend to be." MUDs provide worlds for anonymous social interaction in which one can play a role as close to or as far away from one's "real self" as one chooses. Since one participates in MUDs by sending text to a computer that houses the MUD's program and database, MUD selves are constituted in interaction with the machine. Take it away and the MUD selves cease to exist: "Part of me, a very important part of me, only exists inside PernMUD," says one player. Several players joke that they are like "the electrodes in the computer," trying to express the degree to which they feel part of its space.

On MUDs, one's body is represented by one's own textual description, so the obese can be slender, the beautiful plain, the "nerdy" sophisticated. A *New Yorker* cartoon captures the potential for MUDs as laboratories for experimenting with one's identity. In it, one dog, paw on a computer keyboard, explains to another, "On the Internet, nobody knows you're a dog." The anonymity of MUDs—one is known on the MUD only by the name of one's character or characters—gives people the chance to express multiple and often unexplored aspects of the self, to play with their identity and to try out new ones. MUDs make possible the creation of an identity so fluid and multiple that it strains the limits of the notion. Identity, after all, refers to the sameness between two qualities, in this case between a person and his or her persona. But in MUDs, one can be many.

Dedicated MUD players are often people who work all day with computers at their regular jobs—as architects, programmers, secretaries, students, and stockbrokers. From time to time when playing on MUDs, they can put their characters "to sleep" and pursue "real life" (MUD players call this RL) activities on the computer—all the while remaining connected, logged on to the game's virtual world. Some leave special programs running that send them signals when a particular character logs on or when they are "paged" by a MUD acquaintance. Some leave behind small artificial intelligence programs called bots (derived from the word "robot") running in the MUD that may serve as their alter egos, able to make small talk or answer simple questions. In the course of a day, players move in and out of the active game space. As they do so, some experience their lives as a "cycling through" between the real world, RL, and a series of virtual worlds. I say a series because people are frequently connected to several MUDs at a time. In an MIT computer cluster at 2 A.M., an eighteen-year-old freshman sits at a networked machine and points to the four boxed-off areas on his vibrantly colored computer screen. "On this MUD I'm relaxing, shooting the breeze. On this other MUD I'm in a flame war.[4] On this last one I'm into heavy sexual things. I'm travelling between the MUDs and a physics homework assignment due at 10 tomorrow morning."

This kind of cycling through MUDs and RL is made possible by the existence of those boxed-off areas on the screen, commonly called windows. Windows provide a way for a computer to place you in several contexts at the same time. As a user, you are attentive to only one of the windows on your screen at any given moment, but in a sense you are a presence in all of them at all times. For example, you might be using your computer to help you write a paper about bacteriology. In that case, you would be present to a word-processing program you are using to take notes, to communications software with which you are collecting reference materials from a distant computer, and to a simulation program, which is charting the growth of virtual bacterial colonies. Each of these activities takes place in a window; your identity on the computer is the sum of your distributed presence.

Doug is a midwestern college junior. He plays four characters distributed across three different MUDs. One is a seductive woman. One is a macho, cowboy type whose self-description stresses that he is a "Marlboros rolled in the T-shirt sleeve kind of guy." The third is a rabbit of unspecified gender who wanders its MUD introducing people to each other, a character he calls Carrot. Doug says, "Carrot is so low key that people let it be around while they are having private conversations. So I think of Carrot as my passive, voyeuristic character." Doug's fourth character is one that he plays only on a MUD in which all the characters are furry animals. "I'd rather not even talk about that character because my anonymity there is very important to me," Doug says. "Let's just say that on FurryMUDs I feel like a sexual tourist."[5] Doug talks about playing his characters in windows and says that using windows has made it possible for him to "turn pieces of my mind on and off."

> I split my mind. I'm getting better at it. I can see myself as being two or three or more. And I just turn on one part of my mind and then another when I go from window to window. I'm in some kind of argument in one window and trying to come on to a girl in a MUD in another, and another window might be running a spreadsheet program or some other technical thing for school. . . . And then I'll get a real-time message [that flashes on the screen as soon as it is sent from another system user], and I guess that's RL. It's just one more window.

"RL is just one more window," he repeats, "and it's not usually my best one."

The development of windows for computer interfaces was a technical innovation motivated by the desire to get people working more efficiently by cycling through different applications. But in the daily practice of many computer users, windows have become a powerful metaphor for thinking about the self as a multiple, distributed system. The self is no longer simply playing different roles in different settings at different times, something that a person experiences when, for example, she wakes up as a lover, makes breakfast as a mother, and drives to work as a lawyer. The life practice of windows is that of a decentered self that exists in many worlds and plays many roles at the same time. In traditional theater and in role-playing games that take place in physical space, one steps in and out of character; MUDs, in contrast, offer parallel identities, parallel lives. The experience of this parallelism encourages treating on-screen and off-screen lives with a surprising degree of equality. Experiences on the Internet extend the metaphor of windows—now RL itself, as Doug said, can be "just one more window."

MUDs are dramatic examples of how computer-mediated communication can serve as a place for the construction and reconstruction of identity. There are many others. On the Internet, Internet Relay Chat (commonly known as IRC) is another widely used conversational forum in which any user can open a channel and attract guests to it, all of whom speak to each other as if in the same room. Commercial services such as America Online and CompuServe provide online chat rooms that have much of the appeal of MUDs—a combination of real time interaction with other people, anonymity (or, in some cases, the illusion of anonymity), and the ability to assume a role as close to or as far from one's "real self" as one chooses.

As more people spend more time in these virtual spaces, some go so far as to challenge the idea of giving any priority to RL at all. "After all," says one dedicated MUD player and IRC user, "why grant such superior status to the self that has the body when the selves that don't have bodies are able to have different kinds of experiences?" When people can play at having different genders and different lives, it isn't surprising that for some this play has become as real as what we conventionally think of as their lives, although for them this is no longer a valid distinction.

FRENCH LESSONS

In late 1960s and early 1970s, I lived in a culture that taught that the self is constituted by and through language, that sexual congress is the exchange of signifiers, and that each of us is a multiplicity of parts, fragments, and desiring connections. This was the hothouse of Paris intellectual culture whose gurus included Jacques Lacan, Michel Foucault, Gilles Deleuze, and Félix Guattari.[6] But despite such ideal conditions for learning, my "French lessons" remained merely abstract exercises. These theorists of poststructuralism and what would come to be called postmodernism spoke words that addressed the relationship between mind and body but, from my point of view, had little or nothing to do with my own.

In my lack of connection with these ideas, I was not alone. To take one example, for many people it is hard to accept any challenge to the idea of an autonomous ego. While in recent years, many psychologists, social theorists, psychoanalysts, and philosophers have argued that the self should be thought of as essentially decentered, the normal requirements of everyday life exert strong pressure on people to take responsibility for their actions and to see themselves as intentional and unitary actors. This disjuncture between theory (the unitary self is an illusion) and lived experience (the unitary self is the most basic reality) is one of the main reasons why multiple and decentered theories have been slow to catch on—or when they do, why we tend to settle back quickly into older, centralized ways of looking at things.

Today I use the personal computer and modem on my desk to access MUDs. Anonymously, I travel their rooms and public spaces (a bar, a lounge, a hot tub). I create several characters, some not of my biological gender, who are able to have social and sexual encounters with other characters. On different MUDs, I have different routines, different friends, different names. One day I learned of a virtual rape. One MUD player had used his skill with the system to seize control of another player's character. In this way the aggressor was able to direct the seized character to submit to a violent sexual encounter. He did all this against the will and over the distraught objections of the player usually "behind" this character, the player to whom this character "belonged." Although some made light of the offender's actions by saying that the episode was just words, in text-based virtual realities such as MUDs, words *are* deeds.

Thus, more than twenty years after meeting the ideas of Lacan, Foucault, Deleuze, and Guattari, I am meeting them again in my new life on the screen. But this time, the Gallic abstractions are more concrete. In my computer-mediated worlds, the self is multiple, fluid, and constituted in interaction with machine connections; it is made and transformed by language; sexual congress is an exchange of signifiers; and understanding follows from navigation and tinkering rather than analysis. And in the machine-generated world of MUDs, I meet characters who put me in a new relationship with my own identity.

One day on a MUD, I came across a reference to a character named Dr. Sherry, a cyberpsychologist with an office in the rambling house that constituted this MUD's virtual geography. There, I was informed, Dr.

Sherry was administering questionnaires and conducting interviews about the psychology of MUDs, I suspected that the name Dr. Sherry referred to my long career as a student of the psychological impact of technology. But I didn't create this character. I was not playing her on the MUD. Dr. Sherry was (she is no longer on the MUD) a derivative of me, but she was not mine. The character I played on this MUD had another name—and did not give out questionnaires or conduct interviews. My formal studies were conducted offline in a traditional clinical setting where I spoke face-to-face with people who participate in virtual communities. Dr. Sherry may have been a character someone else created as an efficient way of communicating an interest in questions about technology and the self, but I was experiencing her as a little piece of my history spinning out of control. I tried to quiet my mind. I told myself that surely one's books, one's intellectual identity, one's public persona, are pieces of oneself that others may use as they please. I tried to convince myself that this virtual appropriation was a form of flattery. But my disquiet continued. Dr. Sherry, after all, was not an inanimate book but a person, or at least a person behind a character who was meeting with others in the MUD world.

I talked my disquiet over with a friend who posed the conversation-stopping question, "Well, would you prefer it if Dr. Sherry were a bot trained to interview people about life on the MUD?" (Recall that bots are computer programs that are able to roam cyberspace and interact with characters there.) The idea that Dr. Sherry might be a bot had not occurred to me, but in a flash I realized that this too was possible, even likely. Many bots roam MUDs. They log onto the games as though they were characters. Players create these programs for many reasons: bots help with navigation, pass messages, and create a background atmosphere of animation in the MUD. When you enter a virtual café, you are usually not alone. A waiter bot approaches who asks if you want a drink and delivers it with a smile.

Characters played by people are sometimes mistaken for these little artificial intelligences. This was the case for Doug's character Carrot, because its passive, facilitating persona struck many as one a robot could play. I myself have made this kind of mistake several times, assuming that a person was a program when a character's responses seemed too automatic, too machine-like. And sometimes bots are mistaken for people. I have made this mistake too, fooled by a bot that flattered me by remembering my name or our last interaction. Dr. Sherry could indeed have been one of these. I found myself confronted with a double that could be a person or a program. As things turned out, Dr. Sherry was neither; it was a composite character created by two college students who wished to write a paper on the psychology of MUDs and who were using my name as a kind of trademark or generic descriptor for the idea of a cybershrink.[7] On MUDs, the one can be many and the many can be one.

So not only are MUDs places where the self is multiple and constructed by language, they are places where people and machines are in a new relation to each other, indeed can be mistaken for each other. In such ways, MUDs are evocative objects for thinking about human identity and, more generally, about a set of ideas that have come to be known as "postmodernism."

These ideas are difficult to define simply, but they are characterized by such terms as "decentered," "fluid," "nonlinear," and "opaque." They contrast with modernism, the classical world-view that has dominated Western thinking since the Enlightenment. The modernist view of reality is characterized by such terms as "linear," "logical," "hierarchical," and by having "depths" that can be plumbed and understood. MUDs offer an experience of the abstract postmodern ideas that had intrigued yet confused me during my intellectual coming of age. In this, MUDs exemplify a phenomenon we shall meet often in these pages, that of computer-mediated experiences bringing philosophy down to earth.

In a surprising and counter-intuitive twist, in the past decade, the mechanical engines of computers have been grounding the radically nonmechanical philosophy of postmodernism. The online world of the Internet is not the only instance of evocative computer objects and experiences bringing postmodernism down to earth. One of my students at MIT dropped out of a course I teach on social theory, complaining that the writings of the literary theorist Jacques Derrida were simply beyond him. He found that Derrida's dense prose and far-flung philosophical allusions were incomprehensible. The following semester I ran into the student in an MIT cafeteria. "Maybe I wouldn't have to drop out now," he told me. In the past month, with his roommate's acquisition of new software for his Macintosh computer, my student had found his own key to Derrida. That software was a type of hypertext, which allows a computer user to create links between related texts, songs, photographs, and video, as well as to travel along the links made by others. Derrida emphasized that writing is constructed by the audience as well as by the author and that what is absent from the text is as significant as what is present. The student made the following connection:

> Derrida was saying that the messages of the great books are no more written in stone than are the links of a hypertext. I look at my roommate's hypertext stacks and I am able to trace the connections he made and the peculiarities of how he links things together. . . . And the things he might have linked but didn't. The traditional texts are like [elements in] the stack. Meanings are arbitrary, as arbitrary as the links in a stack.

"The cards in a hypertext stack," he concluded, "get their meaning in relation to each other. It's like Derrida. The links have a reason but there is no final truth behind them."[8]

Like experiences on MUDs, the student's story shows how technology is bringing a set of ideas associated with postmodernism—in this case, ideas about the instability of meanings and the lack of universal and knowable truths—into everyday life. In recent years, it has become fashionable to poke fun at postmodern philosophy and lampoon its allusiveness and density. Indeed, I have done some of this myself. But in this book we shall see that through experiences with computers, people come to a certain understanding of postmodernism and to recognize its ability to usefully capture certain aspects of their own experience, both online and off.

In *The Electronic Word,* the classicist Richard A. Lanham argues that open-ended screen text subverts traditional fantasies of a master narrative, or definitive reading, by presenting the reader with possibilities for changing fonts, zooming in and out, and rearranging and replacing text. The result is "a body of work active not passive, a canon not frozen in perfection but volatile with contending human motive."[9] Lanham puts technology and postmodernism together and concludes that the computer is a "fulfillment of social thought." But I believe the relationship is better thought of as a two-way process. Computer technology not only "fulfills the postmodern aesthetic" as Lanham would have it, heightening and concretizing the postmodern experience, but helps that aesthetic hit the street as well as the seminar room. Computers embody postmodern theory and bring it down to earth.

As recently as ten to fifteen years ago, it was almost unthinkable to speak of the computer's involvement with ideas about unstable meanings and unknowable truths.[10] The computer had a clear intellectual identity as a calculating machine. Indeed, when I took an introductory programming course at Harvard in 1978, the professor introduced the computer to the class by calling it a giant calculator. Programming, he reassured us, was a cut and dried technical activity whose rules were crystal clear.

These reassurances captured the essence of what I shall be calling the modernist computational aesthetic. The image of the computer as calculator suggested that no matter how complicated a computer might seem, what happened inside it could be mechanically unpacked. Programming was a technical skill that could be done a right way or a wrong way. The right way was dictated by the computer's calculator essence. The right way was linear and logical. My professor made it clear that this linear, logical calculating machine combined with a structured, rule-based method of writing software offered guidance for thinking not only about technology and programming, but about economics, psychology, and social life. In other words, computational ideas were presented as one of the great modern metanarratives, stories of how the world worked that provided unifying pictures and analyzed complicated things by breaking them down into simpler parts. The modernist computational aesthetic promised to explain and unpack, to reduce and clarify. Although the computer culture was never monolithic, always including dissenters and deviant subcultures, for many years its professional mainstream (including computer scientists, engineers, economists, and cognitive scientists) shared this clear intellectual direction. Computers, it was assumed, would become more powerful, both as tools and as metaphors, by becoming better and faster calculating machines, better and faster analytical engines.

FROM A CULTURE OF CALCULATION TOWARD A CULTURE OF SIMULATION

Most people over thirty years old (and even many younger ones) have had an introduction to computers similar to the one I received in that programming course. But from today's perspective, the fundamental lessons of computing that I was taught are wrong. First of all, programming is no longer cut and dried. Indeed, even its dimensions have become elusive. Are you programming when you customize your wordprocessing software? When you design "organisms" to populate a simulation of Darwinian evolution in a computer game called SimLife? Or when you build a room in a MUD so that opening a door to it will cause "Happy Un-Birthday" to ring out on all but one day of the year? In a sense, these activities are forms of programming, but that sense is radically different from the one presented in my 1978 computer course.

The lessons of computing today have little to do with calculation and rules; instead they concern simulation, navigation, and interaction. The very image of the computer as a giant calculator has become quaint and dated. Of course, there is still "calculation" going on within the computer, but it is no longer the important or interesting level to think about or interact with. Fifteen years ago, most computer users were limited to typing commands. Today they use off-the-shelf products to manipulate simulated desktops, draw with simulated paints and brushes, and fly in simulated airplane cockpits. The computer culture's center of gravity has shifted decisively to people who do not think of themselves as programmers. The computer science research community as well as industry pundits maintain that in the near future we can expect to interact with computers by communicating with simulated people on our screens, agents who will help organize our personal and professional lives.

On my daughter's third birthday she received a computer game called

The Playroom, among the most popular pieces of software for the pre-school set. If you ask for help, The Playroom offers an instruction that is one sentence long: "Just move the cursor to any object, click on it, explore and have fun." During the same week that my daughter learned to click in The Playroom, a colleague gave me my first lesson on how to use the World Wide Web, a cyberconstruct that links text, graphics, video, and audio on computers all over the world. Her instructions were almost identical to those I had just read to my daughter: "Just move the cursor to any underlined word or phrase, click on it, explore, and have fun." When I wrote this text in January 1995, the Microsoft corporation had just introduced Bob, a "social" interface for its Windows operating system, the most widely used operating system for personal computers in the world.[11] Bob, a computer agent with a human face and "personality," operates within a screen environment designed to look like a living room that is in almost every sense a playroom for adults. In my daughter's screen playroom, she is presented with such objects as alphabet blocks and a clock for learning to tell time. Bob offers adults a wordprocessor, a fax machine, a telephone. Children and adults are united in the actions they take in virtual worlds. Both move the cursor and click.

The meaning of the computer presence in people's lives is very different from what most expected in the late 1970s. One way to describe what has happened is to say that we are moving from a modernist culture of calculation toward a postmodernist culture of simulation.

The culture of simulation is emerging in many domains. It is affecting our understanding of our minds and our bodies. For example, fifteen years ago, the computational models of mind that dominated academic psychology were modernist in spirit: Nearly all tried to describe the mind in terms of centralized structures and programmed rules. In contrast, today's models often embrace a postmodern aesthetic of complexity and decentering. Mainstream computer researchers no longer aspire to pro-gram intelligence into computers but expect intelligence to emerge from the interactions of small subprograms. If these emergent simulations are "opaque," that is, too complex to be completely analyzed, this is not necessarily a problem. After all, these theorists say, our brains are opaque to us, but this has never prevented them from functioning perfectly well as minds.

Fifteen years ago in popular culture, people were just getting used to the idea that computers could project and extend a person's intellect. Today people are embracing the notion that computers may extend an individual's physical presence. Some people use computers to extend their physical presence via real-time video links and shared virtual confer-ence rooms. Some use computer-mediated screen communication for sexual encounters. An Internet list of "Frequently Asked Questions" describes the latter activity—known as netsex, cybersex, and (in MUDs) TinySex—as people typing messages with erotic content to each other, "sometimes with one hand on the keyset, sometimes with two."

Many people who engage in netsex say that they are constantly surprised by how emotionally and physically powerful it can be. They insist that it demonstrates the truth of the adage that ninety percent of sex takes place in the mind. This is certainly not a new idea, but netsex has made it commonplace among teenage boys, a social group not usually known for its sophistication about such matters. A seventeen-year-old high school student tells me that he tries to make his erotic communications on the net "exciting and thrilling and sort of imaginative." In contrast, he admits that before he used computer communication for erotic purposes he thought about his sexual life in terms of "trying [almost always unsuccessfully] to get laid." A sixteen-year-old has a similar report on his cyberpassage to greater sensitivity: "Before I was on the net, I used to masturbate with *Playboy;* now I do netsex on DinoMUD[12] with a woman in another state." When I ask how the two experiences differ, he replies:

> With netsex, it is fantasies. My MUD lover doesn't want to meet me in RL. With *Playboy,* it was fantasies too, but in the MUD there is also the other person. So I don't think of what I do on the MUD as masturbation. Although, you might say that I'm the only one who's touching me. But in netsex, I have to think of fantasies she will like too. So now, I see fantasies as something that's part of sex with two people, not just me in my room.

Sexual encounters in cyberspace are only one (albeit well-publicized) element of our new lives on the screen. Virtual communities ranging from MUDs to computer bulletin boards allow people to generate experiences, relationships, identities, and living spaces that arise only through interaction with technology. In the many thousands of hours that Mike, a college freshman in Kansas, has been logged on to his favorite MUD, he has created an apartment with rooms, furniture, books, desk, and even a small computer. Its interior is exquisitely detailed, even though it exists only in textual description. A hearth, an easy chair, and a mahogany desk warm his cyberspace. "It's where I live," Mike says. "More than I do in my dingy dorm room. There's no place like home."

As human beings become increasingly intertwined with the technology and with each other via the technology, old distinctions between what is specifically human and specifically technological become more complex. Are we living life *on* the screen or life *in* the screen? Our new technologically enmeshed relationships oblige us to ask to what extent we ourselves have become cyborgs, transgressive mixtures of biology, technology, and code.[13] The traditional distance between people and machines has become harder to maintain.

Writing in his diary in 1832, Ralph Waldo Emerson reflected that "Dreams and beasts are two keys by which we are to find out the secrets of our nature... they are our test objects."[14] Emerson was prescient. Freud and his heirs would measure human rationality against the dream. Darwin and his heirs would insist that we measure human nature against nature itself—the world of the beasts seen as our forbears and kin. If Emerson had lived at the end of the twentieth century, he would surely have seen the computer as a new test object. Like dreams and beasts, the computer stands on the margins. It is a mind that is not yet a mind. It is inanimate yet interactive. It does not think, yet neither is it external to thought. It is an object, ultimately a mechanism, but it behaves, interacts, and seems in a certain sense to know. It confronts us with an uneasy sense of kinship. After all, we too behave, interact, and seem to know, and yet are ultimately made of matter and programmed DNA. We think we can think. But can *it* think? Could it have the capacity to feel? Could it ever be said to be alive?

Dreams and beasts were the test objects for Freud and Darwin, the test objects for modernism. In the past decade, the computer has become the test object for postmodernism. The computer takes us beyond a world of dreams and beasts because it enables us to contemplate mental life that exists apart from bodies. It enables us to contemplate dreams that do not need beasts. The computer is an evocative object that causes old boundaries to be renegotiated.

This book traces a set of such boundary negotiations. It is a reflection on the role that technology is playing in the creation of a new social and cultural sensibility. I have observed and participated in settings, physical and virtual, where people and computers come together.[15] Over the past decade, I have talked to more than a thousand people, nearly three hundred of them children, about their experience of using computers or computational objects to program, to navigate, to write, to build, to experiment, or to communicate. In a sense, I have interrogated the computers as well. What messages, both explicit and implicit, have they carried for their human users about what is possible and what is impossible, about what is valuable and what is unimportant?

In the spirit of Whitman's reflections on the child, I want to know what we are becoming if the first objects we look upon each day are simulations into which we deploy our virtual selves. In other words, this is not a book about computers. Rather, it is a book about the intense relationships people have with computers and how these relationships are changing the way we think and feel. Along with the movement from a culture of calculation toward a culture of simulation have come changes in what computers do *for* us and in what they do *to* us—to our relationships and our ways of thinking about ourselves.

We have become accustomed to opaque technology. As the processing power of computers increased exponentially, it became possible to use that power to build graphical user interfaces, commonly known by the acronym GUI, that hid the bare machine from its user. The new opaque interfaces—most specifically, the Macintosh iconic style of interface, which simulates the space of a desktop as well as communication through dialogue—represented more than a technical change. These new interfaces modeled a way of understanding that depended on getting to know a computer through interacting with it, as one might get to know a person or explore a town.

The early personal computers of the 1970s and the IBM PC of the early 1980s presented themselves as open, "transparent," potentially reducible to their underlying mechanisms. These were systems that invited users to imagine that they could understand its "gears" as they turned, even if very few people ever tried to reach that level of understanding. When people say that they used to be able to "see" what was "inside" their first personal computers, it is important to keep in mind that for most of them there still remained many intermediate levels of software between them and the bare machine. But their computer systems encouraged them to represent their understanding of the technology as knowledge of what lay beneath the screen surface. They were encouraged to think of understanding as looking beyond the magic to the mechanism.

In contrast, the 1984 introduction of the Macintosh's iconic style presented the public with simulations (the icons of file folders, a trash can, a desktop) that did nothing to suggest how their underlying structure could be known. It seemed unavailable, visible only through its effects. As one user said, "The Mac looked perfect, finished. To install a program on my DOS machine, I had to fiddle with things. It clearly wasn't perfect. With the Mac, the system told me to stay on the surface." This is the kind of involvement with computers that has come to dominate the field; no longer associated only with the Macintosh, it is nearly universal in personal computing.

We have learned to take things at interface value. We are moving toward a culture of simulation in which people are increasingly comfortable with substituting representations of reality for the real. We use a Macintosh-style "desktop" as well as one on four legs. We join virtual communities that exist only among people communicating on computer networks as well as communities in which we are physically present. We come to question simple distinctions between real and artificial. In what sense should one consider a screen desktop less real than any other? The screen desktop I am currently using has a folder on it labeled "Professional Life." It contains my business correspondence, date book, and telephone directory. Another folder, labeled "Courses," contains syllabuses, reading assignments, class lists, and lecture notes. A third, "Current Work," contains my research notes and this book's drafts. I feel no sense of unreality in my relationship to any of these objects. The culture of simulation encourages me to take what I see on the screen "at (inter)face value." In the culture of simulation, if it works for you, it has all the reality it needs.

The habit of taking things at interface value is new, but it has gone quite far. For example, a decade ago, the idea of a conversation with a computer about emotional matters, the image of a computer psychotherapist, struck most people as inappropriate or even obscene. Today, several such programs are on the market, and they tend to provoke a very different and quite pragmatic response. People are most likely to say, "Might as well try it. It might help. What's the harm?"

We have used our relationships with technology to reflect on the human. A decade ago, people were often made nervous by the idea of thinking about computers in human terms. Behind their anxiety was distress at the idea that their own minds might be similar to a computer's "mind." This reaction against the formalism and rationality of the machine was romantic.

I use this term to analogize our cultural response to computing to nineteenth century Romanticism. I do not mean to suggest that it was merely an emotional response. We shall see that it expressed serious philosophical resistance to any view of people that denied their complexity and continuing mystery. This response emphasized not only the richness of human emotion but the flexibility of human thought and the degree to which knowledge arises in subtle interaction with the environment. Humans, it insists, have to be something very different from mere calculating machines.

In the mid-1980s, this romantic reaction was met by a movement in computer science toward the research and design of increasingly "romantic machines." These machines were touted not as logical but as biological, not as programmed but as able to learn from experience. The researchers who worked on them said they sought a species of machine that would prove as unpredictable and undetermined as the human mind itself. The cultural presence of these romantic machines encouraged a new discourse; both persons and objects were reconfigured, machines as psychological objects, people as living machines.

But even as people have come to greater acceptance of a kinship between computers and human minds, they have also begun to pursue a new set of boundary questions about things and people. After several decades of asking, "What does it mean to think?" the question at the end of the twentieth century is, "What does it mean to be alive?" We are positioned for yet another romantic reaction, this time emphasizing biology, physical embodiment, the question of whether an artifact can be a life.[16]

These psychological and philosophical effects of the computer presence are by no means confined to adults. Like their parents, and often before their parents, the children of the early 1980s began to think of computers and computer toys as psychological objects because these machines combined mind activities (talking, singing, spelling, game playing, and doing math), an interactive style, and an opaque surface. But the children, too, had a romantic reaction, and came to define people as those emotional and unprogrammable things that computers were not. Nevertheless, from the moment children gave up on mechanistic understandings and saw the computer as a psychological entity, they began to draw computers closer to themselves. Today children may refer to the computers in their homes and classrooms as "just machines," but qualities that used to be ascribed only to people are now ascribed to computers as well. Among children, the past decade has seen a movement from defining people as what machines are not to believing that the computational objects of everyday life think and know while remaining "just machines."

In the past decade, the changes in the intellectual identity and cultural impact of the computer have taken place in a culture still deeply attached to the quest for a modernist understanding of the mechanisms of life. Larger scientific and cultural trends, among them advances in psychopharmacology and the development of genetics as a computational biology, reflect the extent to which we assume ourselves to be like machines whose inner workings we can understand. "Do we have our emotions," asks a college sophomore whose mother has been transformed by taking antidepressant medication, "or do our emotions have us?" To whom is one listening when one is "listening to Prozac"?[17] The aim of the Human Genome Project is to specify the location and role of all the genes in human DNA. The Project is often justified on the grounds that it promises to find the pieces of our genetic code responsible for many human diseases so that these may be better treated, perhaps by genetic reengineering. But talk about the Project also addresses the possibility of finding the genetic markers that determine human personality, temperament, and sexual orientation. As we contemplate reengineering the genome, we are also reengineering our view of ourselves as programmed beings.[18] Any romantic reaction that relies on biology as the bottom line is fragile, because it is building on shifting ground. Biology is appropriating computer technology's older, modernist models of computation while at the same time computer scientists are aspiring to develop a new opaque, emergent biology that is closer to the postmodern culture of simulation.[19]

Today, more lifelike machines sit on our desktops, computer science uses biological concepts, and human biology is recast in terms of deciphering a code. With descriptions of the brain that explicitly invoke computers and images of computers that explicitly invoke the brain, we have reached a cultural watershed. The rethinking of human and machine identity is not taking place just among philosophers but "on the ground," through a philosophy in everyday life that is in some measure both provoked and carried by the computer presence.

We have sought out the subjective computer. Computers don't just do things for us, they do things to us, including to our ways of thinking about ourselves and other people. A decade ago, such subjective effects of the computer presence were secondary in the sense that they were not the ones being sought.[20] Today, things are often the other way around. People explicitly turn to computers for experiences that they hope will change their ways of thinking or will affect their social and emotional lives. When people explore simulation games and fantasy worlds or log on to a community where they have virtual friends and lovers, they are not thinking of the computer as what Charles Babbage, the nineteenth-century mathematician who invented the first programmable machine, called an analytical engine. They are seeking out the computer as an intimate machine.

You might think from its title that this was a book about filmgoers and the ways that a fan—the heroine of Woody Allen's *The Purple Rose of Cairo,* for example—might project himself or herself into favorite movies. But here I argue that it is computer screens where we project ourselves into our own dramas, dramas in which we are producer, director, and star. Some of these dramas are private, but increasingly we are able to draw in other people. Computer screens are the new location for our fantasies, both erotic and intellectual. We are using life on computer screens to become comfortable with new ways of thinking about evolution, relationships, sexuality, politics, and identity. How all of this is unfolding is the subject of this book.

Part I

THE SEDUCTIONS OF THE INTERFACE

CHAPTER 1

A TALE OF TWO AESTHETICS

As I write these words, I keep shuffling the text on my computer screen. Once I would literally have had to cut and paste. Now I call it cut and paste. Once I would have thought of it as editing. Now with computer software, moving sentences and paragraphs about is just part of writing. This is one reason I now remain much longer at my computer than I used to at my paper writing tablet or typewriter. When I want to write and don't have a computer around, I tend to wait until I do. In fact, I feel that I *must* wait until I do.

Why is it so hard for me to turn away from the screen? The windows on my computer desktop offer me layers of material to which I have simultaneous access: field notes; previous drafts of this book; a list of ideas not yet elaborated but which I want to include; transcripts of interviews with computer users; and verbatim logs of sessions on computer networks, on bulletin boards, and in virtual communities. When I write at the computer, all of these are present and my thinking space seems somehow enlarged. The dynamic, layered display gives me the comforting sense that I write in conversation with my computer. After years of such encounters, a blank piece of paper can make me feel strangely alone.

There is something else that keeps me at the screen. I feel pressure from a machine that seems itself to be perfect and leaves no one and no other thing but me to blame. It is hard for me to walk away from a not-yet-proofread text on the computer screen. In the electronic writing environment in which making a correction is as simple as striking a delete key, I experience a typographical error not as a mere slip of attention, but as a moral carelessness, for who could be so slovenly as not to take the one or two seconds to make it right? The computer tantalizes me with its holding power—in my case, the promise that if I do it right, *it* will do it right, and right away.

COMPUTER HOLDING POWER

The computer's holding power is a phenomenon frequently referred to in terms associated with drug addiction. It is striking that the word "user" is associated mainly with computers and drugs. The trouble with that analogy, however, is that it puts the focus on what is external (the drug). I prefer the metaphor of seduction because it emphasizes the relationship between person and machine. Love, passion, infatuation, what we feel for another person teaches us about ourselves. If we explore these feelings, we can learn what we are drawn to, what we are missing, and what we need. The analysis of computational seductions offers similar promise if we drop the cliché of addiction and turn to the forces, or more precisely, the diversity of forces that keep us engrossed in computational media.

What attracts me to the computer are the possibilities of "conversation" among the multiple windows on my screen and the way an instantly responsive machine allays my anxieties about perfection. But other people are drawn by other sirens. Some are captured by virtual worlds that appear to be unsullied by the messiness of the real. Some are enthralled by the sense of mind building mind or merging with the mind of the computer. If one is afraid of intimacy yet afraid of being alone, even a stand-alone (not networked) computer offers an apparent solution. Interactive and reactive, the computer offers the illusion of companionship without the demands of friendship. One can be a loner yet never be alone.

Just as musical instruments can be extensions of the mind's construction of sound, computers can be extensions of the mind's construction of thought. A novelist refers to "my ESP with the machine. The words float out. I share the screen with my words." An architect who uses the computer to design goes further: "I don't see the building in my mind until I start to play with shapes and forms on the machine. It comes to life in the space between my eyes and the screen." Musicians often hear the music in their minds before they play it, experiencing the music from within before they experience it from without. The computer can be similarly experienced as an object on the border between self and not-self.[1] Or, in a new variant on the story of Narcissus, people are able to fall in love with the artificial worlds that they have created or that have been built for them by others. People are able to see themselves in the computer. The machine can seem a second self, a metaphor first suggested to me by a thirteen-year-old girl who said, "When you program a computer there is a little piece of your mind, and now it's a little piece of the computer's mind. And now you can see it." An investment counselor in her mid-forties echoes the child's sentiment when she says of her laptop computer: "I love the way it has my whole life on it."

The computer, of course, is not unique as an extension of self. At each point in our lives, we seek to project ourselves into the world. The youngest child will eagerly pick up crayons and modeling clay. We paint, we work, we keep journals, we start companies, we build things that express the diversity of our personal and intellectual sensibilities. Yet the computer offers us new opportunities as a medium that embodies our ideas and expresses our diversity.

In the early years of the computer culture, the most dramatic instances of such projections of self into computers occurred in the esoteric domain of programming. Now, as in the case of the novelist and the architect, it is quite common for people to project themselves into the simulations that play on their screens, into the screen images and their actions. Computer holding power, once closely tied to the seductions of programming, today is tied to the seductions of the interface. When video games were very new, I found that the holding power of their screens often went along with a fantasy of a meeting of minds between the player and the program behind the game. Today, the program has disappeared; one enters the screen world as Alice stepped through the looking glass. In today's game simulations, people experience themselves in a new, often exotic setting. The minds they meet are their own.

Our minds, of course, are very different, one from the other, so it is not surprising that different people make the computer their own in their own way.[2] People choose to personalize and customize their computers. And they have very different styles both of using computers and of interpreting their meaning. In this, the computer resembles the psychologist's Rorschach test, whose inkblots suggest many shapes but commit themselves to none. It is up to individuals to make out what the legacy of personality, history, and culture causes them to see. Just as different people take up the computer in different ways, so do different cultures. Indeed, from the very beginning of its mass deployment, computer technology encouraged a variety of cultures in which a wide range of social, artistic, and political values found expression.

For example, by the late 1970s, the computer culture included well-developed "hacker" and "hobbyist" subcultures that could be described in terms of differing computational aesthetics.[3] What most distinguished the subcultures' members from one another was not how much they knew, but what they valued in the machine. The hacker subculture was made up of programmer-virtuosos who were interested in taking large, complex computer systems and pushing them to their limits. Hackers could revel in the imperfectly understood. As they programmed, things did not always get clearer, but they became workable, at least for the master hacker with the "right stuff." Hacking offered a certain thrill-seeking, a certain danger. It provided the sense, as one hacker put it, "of walking on the edge of a cliff." He explained further, "You could never really know that your next little 'local fix' wouldn't send the whole system crashing down on you." The hacker style made an art form of navigating the complexity of opaque computer microworlds.

In contrast, the hobbyist subculture, the world of early personal computer owners, had an altogether different computational aesthetic. For hobbyists, the goal was to reduce a machine to its simplest elements in order to understand it as fully as possible. Hobbyists preferred to work close to the computer hardware; they enjoyed the sense of nothing standing between themselves and the "body" of the machine. Hackers enjoyed working on large, complex machines and large, almost-out-of-control projects; hobbyists enjoyed working on small machines and very narrowly defined projects. Hackers enjoyed working on a level where they could ignore "the bare machine"; hobbyists took pleasure in reducing high-level commands to the details of machine code. Many hobbyists used the kind of control they felt able to achieve with their home computers to relieve a sense that they had lost control at work and in political life. In a typical remark about the compensatory pleasures of personal computing, one said, "At work I'm just a cog; at home with my computer I get to see how all of my thinking fits together." For another, "I love the feeling of control when I work in a safe environment of my own creation." In the early days of the personal computer culture, a satisfying understanding of the central processing unit (CPU) of home computers was turned into an ideal for how to understand society; the rules of the community should be transparent to all its members.[4]

Missing from this description of the computer culture of the late 1970s is the perspective of those who have come to be called "users." A user is involved with the machine in a hands-on way, but is not interested in the technology except as it enables an application. Hackers are the antithesis of users. They are passionately involved in mastery of the machine itself. The hobbyists in their own way were equally enthralled. Those who wanted to use computers for strictly instrumental purposes—to run data for a business analysis, for example—had to either learn how to program the machine or hand their data over to someone who did. Only in the late 1970s and early 1980s did the contemporary notion of "user" appear. It first came up in connection with small personal computers that could be used for writing and financial analysis by means of application programs (such as WordStar and VisiCalc). These were programs that people could use without getting involved with the "guts" of the machine. Although I have introduced the terms hacker, hobbyist, and user to refer to specific people, they are best understood as different modes of relationship that one can have with a computer.

When I got my own personal computer in 1979, I saw the hobbyist and the user modes come together in myself. My first personal computer was an Apple II. It ran Scribble, an early wordprocessing program. When I used Scribble, I gave commands to the machine: Mark this text, copy that text, paste this text, delete that text. I didn't know and I didn't care how Scribble communicated with the bare machine. I delegated that problem to the program. I was a user. Yet, there was something about working on that Apple II that reminded me of the thrill I had first felt the year before, when a home computer owner I interviewed, a hobbyist, let me work alongside him as he built his computer from a kit and talked about "the pleasure of understanding a complex system down to its simplest level."

My 1979 Apple II computer began its service as my wordprocessor by being stripped naked. Its plastic cover had been removed so that the Apple processor (and associated chips) could be replaced with another, which could run the operating system, called CP/M. Thus altered, the Apple II offered itself to me as a potentially transparent technology, that is, it offered the promise that ultimately it could be understood by being reduced to its constituent elements. So even though Scribble gave me the opportunity to relate to the machine as a user, as someone who was only interested in the machine's performance, the Apple II communicated a vision of how one could understand the world.

Computers support different styles and cultures because they can be approached in different ways. The execution of the simplest program can be described on many levels—in terms of electronic events, machine language instructions, high-level language instructions, or through a structured diagram that represents the functioning of the program as a flow through a complex information system. There is no necessary one-to-one relationship between the elements on these different levels of description, a feature of computation which has led philosophers of mind to see the computer's hardware-software interplay as evocative of the irreducible relationship of brain and mind.

This irreducibility stands behind the diversity of possible styles of relating to the computer. But this natural pluralism on an individual level is in tension with other forces. Changes in both technology and culture encourage certain styles of technology and of representing technology to dominate others.

As I see it now, objects such as that Apple II support a modernist interpretation of understanding, according to which understanding proceeds by reducing complex things to simpler elements. My stripped-down Apple II both embodied and symbolized a theory that it was possible to understand by discovering the hidden mechanisms that made things work. Of course, this kind of theory, particularly in its utopian form (analyze and you shall know), has always presented itself as more than a model for understanding objects. It also promised understanding of the self and the social world. A modernist morality animates the writings of Karl Marx as well as Adam Smith, Sigmund Freud as well as Charles Darwin.

THE MACINTOSH MYSTIQUE

Five years after I got my Apple II, the Macintosh computer was introduced. The Macintosh suggested a radically different way of understanding. Unlike the personal computers that had come before, the "Mac" encouraged users to stay at a surface level of visual representation and gave no hint of inner mechanisms. The power of the Macintosh was how its attractive simulations and screen icons helped organize an unambiguous access to programs and data. The user was presented with a scintillating surface on which to float, skim, and play. There was nowhere visible to dive.

Yet strictly speaking, a Macintosh, like all computers, remained a collection of on/off switches, of bits and bytes, of traveling electrons, just like those embedded in the chips of my "transparent" Apple II. But the Macintosh strove to make these "irrevelant" to the user. In this way, the tools of the modernist culture of calculation became layered underneath the experience of the culture of simulation.

The Macintosh interface—its screen, really—simulated a real desk. Not a logical interface, manipulated with logical commands, as my CP/M system on the Apple II had been, but a virtual reality, albeit in two dimensions. This was a world in which you navigated your way through information as you would through space. In fact, when you held a mouse and moved it around by hand on a flat surface, you saw your physical movements mirrored on the screen by an indicator icon, usually an arrow or a pointing finger. When I used the Scribble program on my Apple II, I typed such things as "@center[@b(The Macintosh Mystique)]" to indicate that I wanted a centered subheading, "The Macintosh Mystique," printed in bold type. Although I didn't analyze the Scribble program any further, such requirements kept me in touch with the idea that I was giving commands to a machine. I felt that I needed to use symbols and a formal language of nested delimiters (parentheses and brackets) because my machine needed to reduce my commands to something that could be translated into electrical impulses. The fact that my machine's printed circuits were physically exposed to view reinforced this notion.

Writing on the Macintosh was an altogether different experience. It did not feel like commanding a machine. A simulated piece of paper appeared. A flashing pointer told me where I could start typing. If I wanted the words "The Macintosh Mystique" to appear centered and in bold type, I typed them and moved the mouse to manipulate their position and form. If I did this correctly, they appeared as I had desired, right there on the screen. I saw no reference to anything beyond the magic.

The simulated desktop that the Macintosh presented came to be far more than a user-friendly gimmick for marketing computers to the inexperienced. It also introduced a way of thinking that put a premium on surface manipulation and working in ignorance of the underlying mechanism. Even the fact that a Macintosh came in a case that users could not open without a special tool (a tool which I was told was only available to authorized dealers) communicated the message. The desktop's interactive objects, its anthropomorphized dialogue boxes in which the computer "spoke" to its user—these developments all pointed to a new kind of experience in which people do not so much command machines as enter into conversations with them. People were encouraged to interact with technology in something resembling the way they interact with other people. We project complexity onto people; the Macintosh design encouraged the projection of complexity onto the machine. In relationships with people we often have to get things done without necessarily understanding what is going on within the other person; similarly, with the Macintosh we learned to negotiate rather than analyze.

With the Macintosh, personal computers began to present themselves as opposed and even hostile to the traditional modernist expectation that one could take a technology, open the hood, and see inside. The distinctiveness of the Macintosh was precisely that it did not encourage such fantasies; it made the computer screen a world unto itself. It encouraged play and tinkering. Mastering the Macintosh meant getting the lay of the land rather than figuring out the hierarchy of underlying structure and rules. With a traditional command-line computer operating system (CP/M was one, MS-DOS is another), linear, textual commands had to be entered at a "prompt" mark. In these systems, there was no way around learning the commands. You memorized them or you had a cheat sheet. With the Macintosh, exploration was the rule. The manual was for emergencies and exceptions. Computer virtuosos had always explored computer systems in this experimental, "tinkerer's" style. The Macintosh made this kind of learning through exploration available to almost everybody. As in the video game culture that was growing up at the same time in the mid-1980s, one learned to learn through direct action and its consequences.

A TALE OF TWO AESTHETICS

If my transparent Apple II modeled a modernist technological aesthetic, the Macintosh was consistent with a postmodern one. Postmodern theorists have suggested that the search for depth and mechanism is futile, and that it is more realistic to explore the world of shifting surfaces than to embark on a search for origins and structure. Culturally, the Macintosh has served as a carrier object for such ideas.

The modern and postmodern aesthetics were locked in competition for the second half of the 1980s with the IBM personal computer (and its clones) becoming the standard-bearer on the modernist side. The myth of the Macintosh was that it was like a friend you could talk to; the myth of the IBM, abetted by that company's image as a modernist corporate giant, was that the computer was like a car you could control. Although most people who bought an IBM personal computer would have never thought to open it up, to modify the machine or its operating system, this possibility was implicit in the design of the system. As one user told me, "The source code is out there in the public domain. I never want to look at it. It would just slow me down. But I love it that it's out there."

The IBM system invited you to enjoy the global complexity it offered, but promised access to its local simplicity. The Macintosh told you to enjoy the global complexity and forget about everything else. Some people found this liberating, others terrifying. For some, it was also alarming that Macintosh users tended to dispense with their manuals and learn about their systems by playing around.

Thus, by the late 1980s, the culture of personal computing found itself becoming practically two cultures, divided by allegiance to computing systems. There was IBM reductionism vs. Macintosh simulation and surface: an icon of the modernist technological utopia vs. an icon of postmodern reverie. For years, avid loyalists on both sides fought private and not-so-private battles over which vision of computing was "best."

The notion that there must be a best system was of course much too simple. When people experienced the Macintosh as best, this was usually because to them it felt like a thinking environment that fit. Some expressed the idea that the simulations that make up the Macintosh's desktop interface felt like a "transparent" access to functionality. Some said the machine felt like a reassuring appliance: "It's like a toaster," said one enthusiast. "It respects my 'Don't look at me, I can't cope' attitude toward technology." Some enjoyed the feeling that they could turn away from rules and commands and get to know this computer through tinkering and playful experimentation.

But of course, there is more than one way in which thinking environ-

ments can fit. For other people, the IBM command-based style of computing felt right and still does. These people often objected to the popular
characterization of the Macintosh as transparent. For them, it was the
Macintosh's polished and cute iconic interface that was opaque and the
IBM's MS-DOS operating system that deserved the label "transparent,"
because it invited some access to the computer's inner workings. These
MS-DOS enthusiasts did not want to give up citizenship in the culture of
calculation. Indeed, upon meeting the Macintosh, some people go to
great lengths to relate to it in a "modernist" style.

I am having lunch with Andrew, an advertising executive, who speaks
with enthusiasm of his new top-of-the-line Macintosh. He tells me that
when he uses it, its speed, interactivity, sound, graphics, and dynamic
display make him feel as though he is flying. Then Andrew confides with
some annoyance that only days after buying this system he found his
thirteen-year-old son's secret files on his precious machine. "Todd was
destroying my masterpiece," he says. I half expect to hear that the secret
files are pornographic images or logs of Todd's sexually explicit encounters on computer networks. But the secret files are copies of a program
called ResEdit that enables Macintosh users to gain some access to the
system software.

Andrew spends fifteen hours a day on his computer, is fluent in no less
than thirty applications programs, and does all his own installation and
customization. From his point of view he is a computer expert, an expert
at manipulating the polished surface of his Macintosh interface. For Andrew, by playing around with ResEdit, Todd was destroying his computer's "perfection." But Todd saw himself as a detective trying to
outsmart a source of intolerable frustration, the blank wall of the Macintosh interface. To do his detective work, Todd needed to get inside the
machine. From his point of view, his father is a computer innocent,
ignorant of what goes on down below.

With the introduction of Microsoft Windows in 1985, the modern and
postmodern aesthetics of computing became curiously entwined. Windows is a software program that gives a computer using the MS-DOS
operating system something of the feel of a Macintosh interface.[5] As with
a Macintosh, you use a mouse to double-click on icons to open programs
and documents. You navigate the system spatially. As I write this, most
consumers are in fact buying such MS-DOS personal computers with
Microsoft Windows running on top. Some are doing this because they
want to purchase a Macintosh-style interface at a lower price. Not only
have MS-DOS systems historically been less expensive than Macintoshes
to purchase new, but for the millions of consumers who already owned
an MS-DOS–based machine, purchasing Windows meant that they didn't
have to buy a new computer to get a Macintosh aesthetic. Additionally,

the large number of MS-DOS–based machines in circulation meant that a great deal of software had been written for them, far more than was available for the Macintosh. These have been the most significant instrumental reasons for buying the Windows operating system. They are not unimportant. But there have been and continue to be subjective reasons as well.

For some Windows users who, like Todd, want that old-time modernist transparency, which gives them access to the guts of the operating system, the Windows program is more than a choice that makes economic good sense. They feel that it offers them a way to have it all. Windows provides them with a convenient iconic interface, but unlike the Macintosh operating system, it is only a program that runs on top of the MS-DOS operating system. You still have access to the computer inside. In the words of one Windows enthusiast, "You can still get down and dig." Or as another puts it, "I can make it [Windows] do things my way. With Windows, I can find out what stands behind the magic."

Maury is a sociology student whose fierce loyalty to MS-DOS and Microsoft Windows is based on the appeal of transparent understanding.

> I like the feeling that when I learned how to program in C [a computer language] I could really get to Windows because Windows is written in C. But then sometimes I want to get something done very quickly. Then I can just do it . . . on top of Windows . . . and I don't need to worry about the computer at all. Then it's as though Windows steps in between the machine and me and tries to take over. But I don't have to let that happen. If I want to manipulate things, I can always get past Windows.
>
> I want to be able to do all this because of my personality. I am pretty compulsive. I like to set things up a certain way. For me, the greatest pleasure is to get the machine to do all that it can do. I want to optimize the hard drive and get it set up exactly the way I want, . . . like allocate memory just the way I like it. . . . I can do that with an IBM machine, not with a Mac.

In fact, the things that Maury thinks can't be done on the Macintosh are technically possible. Ironically, Beth, a graduate student in philosophy, says that it was while using a Macintosh that she first felt able to reach inside a computer, because the Mac made her feel safe enough to think that she could dare to do so. Like Todd, Beth uses ResEdit to venture beneath the surface:

> Even though [ResEdit] allows you to dig around, when you use it you are still dealing with icons and windows. That sense of familiarity, of ease, has led me to further venture in, confident in my ability to navigate and make sense of the icons. It was only when I hit a window of ones and zeros that I headed back up a level.

Beth's comments make it clear that the tale of two aesthetics does not follow from what is technically possible on a Macintosh but from the fact that its interface gives *permission* to stay at the surface. Even tools like ResEdit give users the sense of this continuing presence of intermediaries.

Warren is a partner in a small accounting firm who uses a PC with Windows. He likes to work as closely as possible to what he calls his computer's "mechanical" level. He wants to be able to bypass simulated objects such as icons, which interpose layers between the user and the machine.

Warren never thought about his computer style until he tried his business partner's Apple Powerbook, a laptop computer that uses the Macintosh operating system. He actively disliked it. He told me that the experience taught him something about himself. "Part of why I like computers is that they give me the feeling that I can have my way over such a powerful thing. And having my way means really having it my way—right down to how things are done in the computer itself." He went on:

> I didn't realize this about myself, but . . . I need to get down inside and mess around with settings. I can make those changes with Windows. On the Mac you are locked away from them. Let's suppose I wanted to optimize part of the computer and I want to do that by loading certain programs into certain memory areas, or I want to allocate some memory onto a virtual disk. With Windows and DOS I can do this.
>
> Like I said, in the Mac I am locked away. It takes the fun out because it makes me feel like I'm being left out. I mean as long as this is my personal computer, why shouldn't I indulge my personal preferences. I like to make changes to config.sys and autoexec.bat. So I told my partner, "No way, I'm not switching."

When he learned about ResEdit, Warren was scornful. He says, "You are still looking at your machine through a window. You are just dealing with representations. I like the thing itself. I want to get my hands dirty."

George, a physicist, also enjoys the feeling of virtual dirt on his hands and feels threatened by opaque objects that are not of his own devising. "They make me feel I am giving up too much control." George says the Macintosh was a come-down after his transparent Apple II. "I want my computer to be as clear to me as my Swiss Army knife. I don't think that a machine should surprise me."

Samantha, a writer who works for a fashion magazine, does not desire as deep a level of technical knowledge as do Maury, Warren, and George, but she shares something of their aesthetic. She, too, feels lost if she does not have a satisfying sense that she knows how things work. Before Samantha bought her Macintosh she had never thought about such things as a computer's transparency or opacity. All she cared about was control over her writing. But the contrast between the experience of using her MS-DOS/Windows computer and using the Macintosh has made her aware that she, too, has a preferred style.

> I worked for years with an IBM system and then all of my coworkers convinced me that it was time to switch to Macintosh. They said that since I'm not a techie person and don't like to fiddle, the Macintosh would just let me get my writing done. I said fine. And then, one day, I lost a file on the Macintosh. When I clicked on the file, it just came up with little squiggly lines. I tried everything that the Macintosh allows you to do. I tried opening it several different ways. I tried copying it and then opening it. The point is, there was not much to try. With DOS, I would have had a lot of things to try. I could have gotten at that file. I don't know much about DOS, but even with the little bit I do know, *I could have reached into the machine* and gotten at that file. With the Macintosh, it is all tucked away. I have lost control. Does anybody want to buy a used PowerBook?

In contrast to these computer users, Joel, a graduate student in philosophy, describes himself as philosophically attached to a simulation aesthetic. He would like nothing better than to buy Samantha's used Apple PowerBook. Joel first used a computer when he was twelve years old. He began with the same Apple II model that George had found so appealing. The Apple II suited George because it had the potential for delivering modernist transparency with no surprises. Joel exploited the Apple II to very different ends. For him, the pleasure of computing lies in creating an entity he can interact with as much as possible in the way he would interact with a person. This led him, as a junior high school student, to write programs that made his Apple II look like a conversational partner. He wrote programs that welcomed him and inquired about his plans when he turned on the machine:

> What I really loved to do was create programs that were opaque. What I mean by opaque is that the programs you could buy in the store were written in machine code and I had no idea about how any of that worked. But I could use my knowledge of the BASIC computer language to simulate "real" opaque programs. Later on, when I moved from the Apple II to the IBM, I could write programs that made the IBM look more like an Apple Macintosh. Because what I really felt attracted to was the idea of the machine just interacting with you and your being able to forget that there was something mechanical beneath.
>
> So, I wrote little "Windows" programs for the IBM, so I wouldn't have to

see DOS. What was also good about this was that I liked being able to give my computer a distinct personality, a personality like a person would have. All of my menus had the same look. The Macintosh is perfect for me, because for me double-clicking on an icon, the whole idea of invoking the magic of the computer, is just right.

The Macintosh was initially marketed as a system for novices, the "computer for the rest of us," with the implication that the rest of us didn't want to be bothered with technical things. But this way of thinking is somewhat misleading. It misses the aesthetic dimension in how people choose operating systems. Joel wants to be involved with the technical details, but the ones that interest him are the technical details of simulation, not calculation. Individuals want to deal with technology that makes them feel comfortable and reflects their personal styles. While some Windows fans insist that this program lets them have it all—convenient icons and MS-DOS–style transparency—Joel, like many Macintosh fans, sees his machine as having a certain purity. To them Windows is a monster hybrid, neither fish nor fowl.

Over the past decade there has developed an increased understanding of how intensely *personal* computers are in terms of what individuals do with their machines. This new understanding is reflected in Apple Computer's advertising campaign for its PowerBook laptops. Photographs of people from dramatically different walks of life (such as country lawyers and hip college students) appear alongside a list of what they have on their PowerBook hard drives. Some have recipes, others poems. Some have spreadsheets, others painting programs. Some have football plays, others Supreme Court decisions. This advertising campaign underscores the fact that individuals construct their computers as projections of themselves. But even with this greater appreciation of the personal content of computing, there still is a tendency to assume that the choice of operating system is a purely "technical" decision. But here, too, we have seen that people are trying to express their cognitive style.

Despite this diversity, the simulation aesthetic first introduced by the Macintosh has become the industry standard, even when the computer being used is not a Macintosh. By the 1990s, most of the computers sold were MS-DOS machines with an iconic Windows interface to the bare machine below—a "Macintosh simulator."

The Simulation Aesthetic

In the past decade there has been a shift away from the traditional modernist desire to see beneath the surface into the mechanics of the operating system. We are increasingly accustomed to navigating screen simulations and have grown less likely to neutralize the computers around us by demanding, "What makes this work?" "What's really happening in there?" And we have grown more likely to accept simulation in other intellectual domains. The contrast between the IBM PC and the Macintosh made it clear that there was no single computer and no single computer culture; the triumph of the Macintosh simulation aesthetic is part of a larger cultural shift in the meaning of transparency.

In 1980, most computer users who spoke of transparency were referring to a transparency analogous to that of traditional machines, an ability to "open the hood" and poke around. But when, in the mid-1980s, users of the Macintosh began to talk about transparency, they were talking about seeing their documents and programs represented by attractive and easy-to-interpret icons. They were referring to having things work without needing to look into the inner workings of the computer. This was, somewhat paradoxically, a kind of transparency enabled by complexity and opacity. By the end of the decade, the word "transparency" had taken on its Macintosh meaning in both computer talk and colloquial language. In a culture of simulation, when people say that something is transparent, they mean that they can easily see how to make it work. They don't necessarily mean that they know why it is working in terms of any underlying process.

Of course, many people still prefer to work with transparent computation in its earlier, modernist sense. But in the course of the 1980s, there grew to be less for them to work with, less in off-the-shelf computing, less in research computing, and less that they could recognize as their own in the long shadow that computing cast over the larger culture. The aesthetic of simulation had become increasingly dominant in the culture at large. In 1984, William Gibson's novel *Neuromancer* celebrated its approach to computing's brave new worlds.[6] *Neuromancer* was a cultural landmark. In the popular imagination it represented the satisfactions of navigating simulation space. Its futuristic hacker hero moved through a matrix that represented connections among social, commercial, and political institutions. *Neuromancer*'s hero yearned to fully inhabit, indeed to become one with, the digital forms of life. He was a virtuoso, a cowboy of information space, and thus for many a postmodern Everyman.

Gibson called that information space cyberspace, meaning the space that exists within a computer or matrix of computers. Cyberspace is not reducible to lines of code, bits of data, or electrical signals. On the face of it, Gibson's matrix, or web of information, is not easily analogized to a Macintosh screen. But in the imagination of his readers, the similarities were apparent. "I don't work at my Macintosh, I dance into its spaces. I feel it is a very Gibson machine," says Burt, a twenty-three-year-old management trainee who describes himself as "hooked on my Mac." He continues, "I had a dream about it that way. I had a dream that I was swimming in there where the files are. I think I had the dream after I heard the expression 'to information surf' on the Internet." Here Burt links his Macintosh dreams to Gibson and an irreducible Internet ocean. And he sees the Internet as a very primitive version of what Gibson was writing about when he described a society fully represented in digitized forms.

PLURALISM VERSUS POSTMODERNISM

Today the computer is an actor in a struggle between modern and postmodern understandings. This struggle is often fought out between those who put their faith in reductive understanding (open the box, trust what you can see, and analyze completely) and those who proclaim such ideas bankrupt or at least impractical. The Macintosh was not just a "happier" experience for people who were personally comfortable with layering and simulation: It was a consumer object that made people more comfortable with a new way of knowing. Although its cultural presence has increased dramatically in recent years, it is not a way of knowing with which everyone is comfortable.

Harry, a fifth-grade teacher who says he uses Windows so that he "can use the best of the new programs but still have access to DOS," was an early computer hobbyist and political activist in the San Francisco Bay area. In the mid-1970s, he fell in love with early personal computers. Harry built the first of them, the Altair, from a kit. He enjoyed the feeling that he could "see down from a top-level instruction to what was actually happening in the computer hardware below." He belonged to a first generation of computer hobbyists who dreamed of creating safe worlds of transparent understanding in society as well as within the computer.[7] His political philosophy, nourished by the student movement of the 1960s, was based on vague but intensely felt variants of Marxist humanism. He argued that society could be understood much as one could understand the innards of a computer, and from there society could be improved if people felt empowered to act. Ivan Illich, who wrote about taking knowledge out of the hands of bureaucrats and specialists, was one of his heroes.

In 1992, fifteen years later, Harry looked back on his changing feelings about technology. His reflections brought him to the subjects of community, epistemology, computers, and cars:

> When I was a boy and my father and I used to take apart his Chevy truck, I fell in love with the idea that someday I would take complicated things

apart. I worked as a teacher, but I made some money on a piece of software I wrote. That's when I began my BMW hobby. As soon as I made any money, I bought old BMWs and fixed them up.

I tried to teach my students what my father had taught me, or I guess it was more like what fixing up the old Chevys had taught me. Take it apart. Analyze the situation. Make it work. That's how communities should work. That's how classrooms should work. In any case, that was how I saw Ivan Illich's message and why I liked what he had to say so much. And then, the BMW people put [computer] boards into their new models and a bunch of chips ran the whole damn car! If something broke down, you replaced the chips. That was the end of the line for me. I stopped driving BMWs. It would have been like teaching kids that when something doesn't work in life, toss it because things are too complicated to fix. I don't like this message. . . .

Like many first-generation hobbyists, Harry's associations between computers and politics have the ring of modernist utopianism: If only society could be rendered transparent and be fully analyzed, the way engineers can analyze a machine, then people could take charge and make things right. For Harry, the Macintosh and its "double-clicking" was emblematic of disempowerment, both technical and political. Like the "chips in the BMW," it made technology opaque and therefore a bad object-to-think-with for thinking about society.

Clearly, computer technology was a projective screen for social and political concerns. But the same computer could evoke opposite reactions. For example, Joel, the graduate student in philosophy, is as devoted to his sleek Macintosh as an object-to-think-with for thinking about politics as Harry is depressed by it.

Joel does not believe that society can be understood in terms of any systematic theory. But he does believe that if we accept society's opacity we can learn at least to navigate its contours more effectively. He remarks that when postmodern theorist Jean Baudrillard wrote about the seductions of technology, he was talking about the pleasures of opacity. Harry says he is fascinated by what he can completely understand, a thought he finds relevant to Karl Marx, Ivan Illich, and the TRS-80 computers of the late 1970s. Joel says, "I'm fascinated by what's unthinkable, by systems that have unbridgeable levels of explanation," an idea that he comes to via Jean Baudrillard, Fredric Jameson, and his newest Apple Macintosh.

A decade ago, Fredric Jameson wrote a classic article on the meaning of postmodernism.[8] He included in his characterization of postmodernism the precedence of surface over depth, of simulation over the "real," of play over seriousness, many of the same qualities that characterize the new computer aesthetic[9] At that time, Jameson noted that the postmodern era lacked objects that could represent it. The turbine, smokestack, pipes, and conveyor belts of the late nineteenth and early twentieth centuries had been powerful objects-to-think-with for imaging the nature of industrial modernity. They provided images of mechanical relationships between body and mind, time and space. The postmodern era had no such objects.[10] Jameson suggested that what was needed was a new "aesthetic of cognitive mapping," a new way of spatial thinking that would permit us at least to register the complexities of our world.[11]

A decade after Jameson wrote his essay, postmodernism has found its objects. I interviewed a fifty-year-old engineer, a Caltech graduate, whose basic commitment has always been to "make things, build things from the ground up, analyze the hell out of stuff." In the terms I have been using here, he has fully lived in the modernist aesthetic. He tells me that the Internet and the World Wide Web have "blown him away."

> It's like a brain, self-organizing, nobody controlling it, just growing up out of the connections that an infant makes, sights to sounds, . . . people to experiences. . . . Sometimes I'll be away from the Web for a week and a bunch of places that I know very well will have "found" each other. This is not an engineering problem. It's a new kind of organism. Or a parallel world. No point to analyze it. No way you could have built it by planning it.

Prefigured by *Neuromancer*'s matrix of informational space, postmodernism's objects now exist outside science fiction. They exist in the information and connections of the Internet and the World Wide Web, and in the windows, icons, and layers of personal computing. They exist in the creatures on a SimLife computer game, and in the simulations of the quantum world that are routinely used in introductory physics courses. All of these are life on the screen. And with these objects, the abstract ideas in Jameson's account of postmodernism become newly accessible, even consumable.

There is a tension between two aspects of how computers influence contemporary culture. On an individual level, computers are able to facilitate pluralism in styles of use. They offer different things to different people; they allow for the growth of different and varied computer cultures. On a larger scale, however, computers now offer an experience resonant with a postmodern aesthetic that (in the language of its theorists) increasingly claims the cultural privilege formerly assumed by modernism. If we think of the computer's pluralism of styles as different styles of seduction, we might say that at specific historical moments, some styles of seduction become increasingly seductive and some start to seem out of date.

In the 1970s, computers carried the psychological, philosophical, and even spiritual message of the culture of calculation. "I love the way thinking about this computer's hardware lets me think about my wetware," a

computer hobbyist remarked to me in 1978. When I asked him exactly
what he meant, he explained that when he thought about the circuits and
the switches of his computer, he felt close to it: "Brothers under the skin;
I mean, both machines under the skin." He felt reassured that someday
people would understand their minds the way he understood his computer.

During the 1970s, computers also carried this message to Rafe, a forty-
six-year-old video editor who at that time considered himself a computer
hobbyist. Rafe saw the computer as a model for a human psychology
that was reassuring because it was mechanistic. Now he uses complex
computing systems to edit video, some of it destined for presentation on
interactive CD-ROMs, and he is far from the mechanistic psychology that
used to reassure him. Rafe says that when working with computers today,
his "thoughts turn to Taoism and the later works of Wittgenstein."

> Simulation offers us the greatest hope of understanding. When a world, our
> world, is far too complex to be understood in terms of first principles, that
> is to say, when the world is too complex for the human mind to build it as
> a mental construct from first principles, then it defies human intellect to
> define its truth. When we reach that point we must navigate within the
> world, learning its rules by the seat of our pants, feeling it, sharing it, using
> it. By getting our analytic intelligence out of the way, we can sometimes
> more efficiently negotiate that world. The computer offers us the hope that
> through simulation we may gain another handle of understanding.
>
> Much real world behavior is too complicated for bottom-up understand-
> ing [from first principles]. Human psychology is one such phenomenon.
> The power of computer simulation is extremely suggestive. We sense within
> it a potential to realize our dreams of understanding complex phenomena
> not by constructing them from first principles but by owning them in simu-
> lation and playing with them.

Computers have changed; times have changed; Rafe has changed. But I
could also write: Times have changed; Rafe has changed; computers have
changed. In fact, there are six possible sequences. All are simultaneously
true. There is no simple causal chain. We construct our technologies, and
our technologies construct us and our times. Our times make us, we
make our machines, our machines make our times. We become the objects we look upon but they become what we make of them.

There are many styles of computer holding power. For the individual,
this pluralism makes the machine (its programming languages, its operating systems, and its programs) a precious resource for learning and
self-development because people tend to learn best when they learn in
their own style. But on another level, the complex simulation worlds
of today's opaque computers capture something important about the

postmodern ethos. This resonance speaks to the computer's current appeal on a more sociological level. People use contact with objects and ideas to keep in touch with their times.[12] They use objects to work through powerful cultural images, to help arrange these images into new and clearer patterns. From this point of view, the holding power of the Apple Macintosh, of simulation games, and of experiences in virtual communities derives from their ability to help us think through postmodernism.

OBJECTS-TO-THINK-WITH

What are we thinking about when we think about computers? The technologies of our everyday lives change the way we see the world. Painting and photography appropriated nature. When we look at sunflowers or water lilies, we see them through the eyes and the art of van Gogh or Monet. When we marry, the ceremony and the ensuing celebration produce photographs and videotapes that displace the event and become our memories of it.[13] Computers, too, lead us to construct things in new ways. With computers we can simulate nature in a program or leave nature aside and build second natures limited only by our powers of imagination and abstraction. The objects on the screen have no simple physical referent. In this sense, life on the screen is without origins and foundation. It is a place where signs taken for reality may substitute for the real. Its aesthetic has to do with manipulation and recombination.

The theorists of the postmodern have also written about worlds without origins. They write of simulacra, copies of things that no longer have originals.[14] Disneyland's Main Street breaks down the line between things and their representation because the representation exists in the absence of the real thing. So, too, the files and documents on my computer screen function as copies of objects of which they are the first examples. I become accustomed to seeing the copies as the reality. The documents that scroll before my eyes as I compose this book on a computer screen function as real enough. They are my access to the thing itself, but there is no other thing itself.

The notion of worlds without origins is close to the postmodern challenge to the traditional epistemologies of depth.[15] These epistemologies are theories of knowledge where the manifest refers back to the latent, the signifier to the signified. In contrast, the postmodern is a world without depth, a world of surface. If there is no underlying meaning, or a meaning we shall never know, postmodern theorists argue that the privileged way of knowing can only be through an exploration of surfaces. This makes social knowledge into something that we might navigate much as we explore the Macintosh screen and its multiple layers of files and applications. In recent years, computers have become the postmodern era's primary objects-to-think-with, not simply part of larger cultural movements but carriers of new ways of knowing. The French anthropologist Claude Lévi-Strauss described the process of theoretical tinkering—bricolage—by which individuals and cultures use the objects around them to develop and assimilate ideas.[16] When I talk about computers as objects-to-think-with, saying for example that Macintosh-style computer interfaces have served as carriers for a way of knowing that depends on simulation and surface representation, I am extending the notion of bricolage to the uncanny (betwixt and between physical and mental) objects of the culture of simulation.

Cultural appropriation through the manipulation of specific objects is common in the history of ideas. Appropriable theories, ideas that capture the imagination of the culture at large, tend to be those with which people can become actively involved. They tend to be theories that can be played with. So one way to think about the social appropriability of a given theory is to ask whether it is accompanied by its own objects-to-think-with that can help it move out beyond intellectual circles.[17]

For instance, the popular appropriation of Freudian ideas had little to do with scientific demonstrations of their validity. Freudian ideas passed into the popular culture because they offered robust and down-to-earth objects-to-think-with. The objects were not physical but almost-tangible ideas such as dreams and slips of the tongue. People were able to play with such Freudian "objects." They became used to looking for them and manipulating them, both seriously and not so seriously. And as they did so, the idea that slips and dreams betray an unconscious started to feel natural. This naturalization of new ideas happened for people who never saw a psychoanalyst and who never read a word by Sigmund Freud.

In *Purity and Danger,* the British anthropologist Mary Douglas examined the classification of foods in the Jewish dietary laws and saw the manipulation of food, a concrete material, as a way to organize a cultural understanding of the sacred and profane. Other scholars had tried to explain the kosher rules instrumentally in terms of hygiene ("pork carries disease") or in terms of wanting to keep the Jewish people separate from other groups. Douglas argued that the separation of foods taught a fundamental tenet of Judaism: Holiness is order and each thing must have its place. For Douglas, every kosher meal embodies the ordered cosmology, a separation of heaven, earth, and seas. In the story of the creation, each of these realms is allotted its proper kind of animal life. Two-legged fowls fly with wings, four-legged animals hop or walk, and scaly fish swim with fins. It is acceptable to eat these "pure" creatures, but those that cross categories (such as the lobster that lives in the sea but crawls upon its floor) are unacceptable: The foods themselves carry a theory of unbreachable order.[18]

Toward the end of his life, the French psychoanalytic theorist Jacques Lacan became fascinated with little pieces of string that he tied into complex knots whose configurations he took to symbolize the workings of the unconscious. For Lacan, the knots were more than metaphor; the physical manipulation of the knots was theory in practice.[19] For Lacan, not only did the knots carry ideas, they could engender a passion for them. The point is highly relevant: Computers would not be the culturally powerful objects they are turning out to be if people were not falling in love with their machines and the ideas that the machines carry.

In Freud's work, dreams and slips of the tongue carried the theory. For Douglas, food carries the theory. For Lacan, the theory is carried by knots. Today, life on the computer screen carries theory. Here is how it happens. People decide that they want to buy an easy-to-use computer. They are attracted by a consumer product—say, a computer with a Macintosh-style interface. They think they are getting an instrumentally useful product, and there is little question that they are. But now it is in their home and they interact with it every day. And it turns out they are also getting an object that teaches them a new way of thinking and encourages them to develop new expectations about the kinds of relationships they and their children will have with machines.[20] People decide that they want to interact with others on a computer network. They get an account on a commercial service. They think that this will provide them with new access to people and information, and of course it does. But it does more. When they log on, they may find themselves playing multiple roles, they may find themselves playing characters of the opposite sex. In this way they are swept up by experiences that enable them to explore previously unexamined aspects of their sexuality or that challenge their ideas about a unitary self.[21]

Fredric Jameson wrote that in a postmodern world, the subject is not alienated but fragmented. He explained that the notion of alienation presumes a centralized, unitary self who could become lost to himself or herself. But if, as a postmodernist sees it, the self is decentered and multiple, the concept of alienation breaks down. All that is left is an anxiety of identity. The personal computer culture began with small machines that captured a post-1960s utopian vision of transparent understanding. Today, the personal computer culture's most compelling objects give people a way to think concretely about an identity crisis. In simulation, identity can be fluid and multiple, a signifier no longer clearly points to a thing that is signified, and understanding is less likely to proceed through analysis than by navigation through virtual space.

CHAPTER 2

THE TRIUMPH OF TINKERING

In the late 1960s, I studied history and political theory in Paris. In my academic program, all foreigners had to take a French composition class. Over the year, the format of this class never varied. A subject was set, everyone had one week to turn in an outline and two more to write the composition. Then the three-week cycle would begin again with the assignment of a new topic. The format of the composition never varied. Each one had to be written in three parts, with each of these parts further divided into three parts. Although I knew many of my classmates took to this style easily, for me this was a completely alien way of writing. My way had been to read, think, and make notes on little pieces of paper. I would spread these notes out in my room, across my bed, desk, and floor. Then I would immerse myself in their contents, move them around into patterns, scribble in their margins, associate to new patterns, write small bits of text, and frequently rewrite sections. Now, under pressure from the new rules, I developed a technique for getting by. I did my reading and thinking, wrote my notes and bits of text, spread them out in my room, and let my composition emerge—but I did all this in the first week, periodically adjusting the emerging composition so that it would grow with the right number of divisions and subdivisions. After a week of hectic activity, I extracted an outline in three parts. I turned in that outline and put the completed composition in my desk drawer, where it waited two weeks for its turn to be handed in to the instructor.

I am tempted to tell this story in a way that makes it sound like a triumph of my creativity over Gallic rigidity. But that would miss an important aspect of what these French lessons meant to me. Far from increasing my confidence, they undermined it. I wrote my composition before my outline in order to survive, but in the process I came to think of my kind of writing as wrong. My kind of writing was, after all, cheating.

PLANNING AND TINKERING

The instructor in my 1978 programming class at Harvard—the one who called the computer a giant calculator—described programming methods in universal terms, which he said were justified by the computer's essential nature. But from the very beginning of my inquiries into the computer culture, it became clear that different people approach programming in very different ways. Where my professor saw the necessary hegemony of a single correct style, I found a range of effective yet diverse styles among both novices and experts.[1]

The "universal" method recommended by my Harvard instructor is known as structured programming. A model of the modernist style, it is rule-driven and relies on top-down planning. First you sketch out a master plan in which you make very explicit what your program must do. Then you break the task into manageable subprograms or subprocedures, which you work on separately. After you create each piece, you name it according to its function and close it off, a procedure known as black boxing. You need not bother with its details again. By the 1970s, this structured, planner's method was widely accepted as the canonical style in computing. Indeed, many engineers and computer scientists still see it as the definitive procedure, as simply the way things must be done. They have a powerful, practical rationale for this method. In real organizations, many people have to be able to understand and use any particular piece of software. That means it has to be understandable and fixable (debuggable) long after its programmer has left the research team or business setting.

Others, however, had a style of programming that bore a family resemblance to my associative style of writing, a "soft" style as opposed to a "hard" one. It was bottom-up rather than top-down. It was built up by playing with the elements of a program, the bits of code, much as I played with the elements of my essay, the bits of paper strewn across my room. It is best captured by a word, bricolage, that Claude Lévi-Strauss has used to contrast the analytic methodology of Western science with an associative science of the concrete practiced in many non-Western societies.[2] The tribal herbalist, for example, does not proceed by abstraction but by thinking through problems using the materials at hand. By analogy, problem-solvers who do not proceed from top-down design but by arranging and rearranging a set of well-known materials can be said to be practicing bricolage. They tend to try one thing, step back, reconsider, and try another. For planners, mistakes are steps in the wrong direction; bricoleurs navigate through midcourse corrections. Bricoleurs approach problem-solving by entering into a relationship with their work materials that has more the flavor of a conversation than a monologue. In the context of programming, the bricoleur's work is marked by a desire to play with lines of computer code, to move them around almost as though they were material things—notes on a score, elements of a collage, words on a page.

Through the mid-1980s, soft-style programmers, programming's bricoleurs, received their own discouraging "French lessons" from a mainstream computer culture deeply committed to structured programming. People who did not program according to the canon were usually told that their way was wrong. They were forced to comply with the officially sanctioned method of doing things. Today, however, there has been a significant change. As the computer culture's center of gravity has shifted from programming to dealing with screen simulations, the intellectual values of bricolage have become far more important. In the 1970s and 1980s, computing served as an initiation into the formal values of hard mastery. Now, playing with simulation encourages people to develop the skills of the more informal soft mastery because it is so easy to run "What if?" scenarios and tinker with the outcome.

The revaluation of bricolage in the culture of simulation includes a new emphasis on visualization and the development of intuition through the manipulation of virtual objects. Instead of having to follow a set of rules laid down in advance, computer users are encouraged to tinker in simulated microworlds. There, they learn about how things work by interacting with them. One can see evidence of this change in the way businesses do their financial planning, architects design buildings, and teenagers play with simulation games.

There is something ironic about the computer presence playing a role in nurturing such "informalist" ways of knowing, since for so long, the computer was seen as the ultimate embodiment of the abstract and formal. But the computer's intellectual personality has always had another side. Computational objects—whether lines of code or icons on a screen—are like abstract and mathematical objects, defined by the most formal of rules. But at the same time, they are like physical objects—like dabs of paint or cardboard cutouts. You can see them and move them, and in some cases you can place one on top of another. Computational objects have always offered an almost-physical access to the world of formal systems.[3] There have always been people whose way of interacting with them had more in common with the style of the painter than with that of the logician.

Consider Lisa, an eighteen-year-old freshman in my Harvard programming course. Lisa's first experiences in the course were very positive. She wrote poetry and found that she was able to approach programming with ways of thinking that she had previously found useful in working with words. But as the term progressed, she came under increasing pressure from her instructors to think in ways that were not her own. Her alien-ation did not stem from an inability to cope with programming but rather from her preference to do it in a way that came into conflict with the structured and rule-driven style of the computer culture she had entered.

In high school, Lisa had turned away from her own natural abilities in mathematics. "I didn't care if I was good at it. I wanted to work in worlds where languages had moods and connected you with people." She was equally irritated when her teachers tried to get her interested in mathe-matics by calling it a language. As a senior in high school, she wrote a poem that expressed her sentiments.

> *If you could say in numbers what I say now in words,*
> *If theorems could, like sentences, describe the flight of birds,*
> *If PPL [a computer language] had meter and parabolas had rhyme,*
> *Perhaps I'd understand you then,*
> *Perhaps I'd change my mind. . . .*
>
> *But all this wishful thinking only serves to make things worse,*
> *When I compare my dearest love with your numeric verse.*
> *For if mathematics were a language, I'd succeed, I'd scale the hill,*
> *I know I'd understand, but since it's not, I never will.*

When she wrote poetry, Lisa knew where all the elements were at every point in the development of her ideas. "I feel my way from one word to another," she said. She wanted her relationship to computer language to be the same. She wanted to tinker, to work as close to the programming code as she did to the words in her poems. When she spoke about the lines of code in her programs, she gestured with her hands and body in a way that showed her moving with them and among them. She talked about the elements of her programs as if they were physically graspable.

When Lisa worked on large programs she preferred to write her own smaller subprograms even though she was encouraged to use prepack-aged ones available in a program library. She resented that she couldn't tinker with the prepackaged routines. Her teachers chided her, insisting that her demand for hands-on contact was making her work more diffi-cult. They told her that the right way to do things was to control a program by planning. Lisa recognized the value of these techniques for someone else. She herself was reluctant to use them as starting points for her learning. Although her teachers tried to convert her to what they consid-ered proper style, Lisa insisted that she had to work her way if she were ever going to feel comfortable with computers. But two months into the programming course, Lisa abandoned the fight to do things her way and decided to do things their way. She called it her "not-me strategy" and be-gan to insist that it didn't matter because "the computer was just a tool."

Lisa's classmate, Robin, was a pianist with a similar learning style. She wanted to play with computational elements, to manipulate the bits of code, as though they were musical notes or phrases. She, too, was told her way was wrong. Her instructor told her it was "a waste of time."

Lisa and Robin came to the programming course with anxieties about not belonging because they did not see themselves as "computer people." Although both could master the class material intellectually, the course exacerbated their anxieties about not belonging because it insisted on a style of work so different from their own. Both received top grades, but each had to deny who she was in order to succeed. Lisa said that she turned herself "into a different kind of person," and Robin described what she had to do as "faking it."

In the 1970s and 1980s, soft mastery was computing's "different voice."[4] Different and in no way equal. The authorities (teachers and other experts) actively discouraged it, deeming it incorrect or improper. But I found many Lisas and many Robins in schools, universities, and local computer clubs. These were boys and girls, men and women, novices and experts, who reported that they had changed their styles to suit the fashion when they had started to interact with the official computer world. "I got my wrists slapped enough times and I changed my ways," says a college student for whom soft style programming was a passion until he entered MIT and was instructed in the canonical programming style. The cost of such wrist slapping was high. On an individual level, talent was wasted, self-image eroded. On the social level, the computer culture was narrowed.

With the rise of a personal computer culture in the 1980s, more people owned their own machines and could do what they pleased with them. This meant that more people began to experience the computer as an expressive medium that they could use in their own ways. Yet for most, the notion that the computer was a calculator died hard. The idea that the computer was a new medium of expression would not make sense until the 1990s, when large numbers of people owned personal computers with color screens, powerful graphics, and CD-ROMs. In the 1970s through the mid-1980s, the ideology that there was only one right way to "do" computers nearly masked the diversity of styles in the computer culture. In those days top-down thinkers didn't simply share a style; they constituted an epistemological elite.

DISCOVERIES AND DENIGRATIONS OF THE CONCRETE

The elite status of abstract thinking in Western thought can be traced back at least to Plato. Western scientific culture has traditionally drawn a firm line between the abstract and the concrete. The tools of abstraction are propositions; the tools of concrete thinking are objects, and there has always been a right and wrong side of the tracks. The terms "pure science" and "pure mathematics" made clear the superiority of selecting for the pristine propositions and filtering out the messy objects. In the twentieth century, the role of things-in-thinking has had powerful intellectual champions. But, even among these champions there has been resistance to the importance of the bottom-up style of thought preferred by Lisa and Robin. For example, Lévi-Strauss and the noted Swiss psychologist Jean Piaget both discovered ways of reasoning that began with objects and moved to theory, but then they found ways to marginalize them.

In the 1920s and 1930s, Piaget first noticed concrete modes of reasoning among children.[5] Children thought that when you spread three marbles apart there were more marbles than when you moved three marbles close together. Through such observations, Piaget was able to see what others had not: Concrete mapping and manipulation of objects enable children to develop the concept of number, a concept that only gradually becomes a formal sense of quantity. The construction of number, in other words, is born through bricolage.

Piaget fought for the recognition of this kind of concrete thinking, but at the same time he saw it as something to be outgrown. The adult was "beyond" the concrete. For Piaget there was a progression in modes of reasoning that culminates in a final, formal stage when propositional logic liberates intelligence from the need to think with things. So Piaget both discovered the power of the concrete in the construction of the fundamental categories of number, space, time, and causality, and denigrated what he had found by relegating concrete ways of knowing to an early childhood stage of development.

Piaget's discoveries about the processes of children's thinking challenged a kind of cultural amnesia. Adults forget the way they reasoned as children. And we forget very quickly. While Freud discovered the forgetting of infantile sexuality, Piaget identified a second amnesia: the forgetting of concrete styles of thinking. In both, our stake in forgetting is highly charged. In our culture, the divide between abstract and concrete is not simply a boundary between propositions and objects but a way of separating the clean from the messy, virtue from taboo.

Lévi-Strauss, too, both discovered and denied the concrete. He described bricoleur scientists who do not move abstractly and hierarchically from axiom to theorem to corollary but construct theories by arranging and rearranging a set of well-known materials. But the bricoleur scientists he described all operated in non-Western societies. As Piaget had relegated the concrete to childhood, Lévi-Strauss relegated it to the so-called "primitive" and to modern Western humanists. What Lévi-Strauss had a hard time seeing were the significant elements of bricolage in the practice of Western science.[6]

Am I practicing a similar devaluation of the concrete when I characterize the rule-based planner's programming style as hard mastery and Lisa and Robin's style as soft? Our culture tends to equate the word "soft" with unscientific and undisciplined as well as with the feminine and with a lack of power. Why use a term like "soft" when it could turn difference into devaluation? What interests me here is the transvaluation of values. "Soft" is a good word for a flexible, nonhierarchical style, one that allows a close connection with one's objects of study. Using the term "soft mastery" goes along with seeing negotiation, relationship, and attachment as cognitive virtues. And this is precisely what the culture of simulation encourages us to do.

The soft approach is not a style unique to either men or women. However, in our culture it is a style to which many women are drawn.[7] Among other reasons, we train girls in the component skills of a soft approach—negotiation, compromise, give and take—as psychological virtues, while dominant models of desirable male behavior stress decisiveness and the imposition of will. Boys and girls are encouraged to adopt different relational stances in the world of people. It is not surprising that these differences show up when men and women deal with the world of things.

Through the mid-1980s, a male-dominated computer culture that took one style as the right and only way to program discriminated against soft approaches. Although this bias hurt both male and female computer users, it fell disproportionately on women because they were disproportionately represented in the ranks of the soft masters. But even when women felt free to experiment with soft mastery, they faced a special conflict. Tinkering required a close encounter with the computer. But this violated a cultural taboo about being involved with "machines" that fell particularly harshly on women. When I was a young girl, I assembled the materials to build a crystal radio. My mother, usually encouraging, said, "Don't touch it, you'll get a shock." Her tone, however, did not communicate fear for my safety, but distaste. A generation later, many women were learning to identify being a woman with all that a computer is not, and computers with all that a woman is not. In this cultural construction, computers could be very threatening. In recent years, things have started to change. As the emerging culture of simulation becomes increasingly associated with negotiational and nonhierarchical ways of thinking, it has made a place for people with a wider range of cognitive and emotional styles. In particular, women have come to feel that computers are more culturally acceptable.

The Revaluation of the Concrete

Soft mastery is not a stage, it is a style. Bricolage is a way to organize work. It is not a stage in a progression to a superior form. Richard Greenblatt is a renowned first-generation MIT hacker, a computer culture legend whose virtuoso style of work incorporates a strong dose of bricolage. He has made significant contributions to the development of chess programs as well as systems programming. In the spirit of the painter who steps back to look at the canvas before proceeding to the next step, Greenblatt developed software that put him in a conversation, a negotiation with his materials. He used bricolage at a high level of artistry.[8]

Yet even internationally recognized bricoleur virtuosos such as Richard Greenblatt lived within a dominant computer culture that was scornful of their approach. One of that culture's heroes was the mathematician Edsger W. Dijkstra. Dijkstra, the leading theorist of hard, structured programming, emphasized analytical methods and scientific rigor in the development of programs. In Dijkstra's view, rigorous planning coupled with mathematical analysis should produce a computer program with mathematically guaranteed success. In this model, there is no room for bricolage. When Dijkstra gave a lecture at MIT in the late 1970s, he demonstrated his points by taking his audience step by step through the development of a short program. Richard Greenblatt was in the audience, and the two men had an exchange that has entered into computer culture mythology. It was a classic confrontation between two opposing aesthetics. Greenblatt asked Dijkstra how he could apply his mathematical methods to something as complicated as a chess program. "I wouldn't write a chess program," Dijkstra replied, dismissing the issue.

In the field of computing, the existence of the bricolage style at virtuoso levels challenged the idea of there being only one correct, mature approach to problem-solving. In the 1980s, this challenge was supported by several currents of research on concrete styles of problem-solving in other domains. Each in its own way called into question the hegemony of the abstract, formal, and rule-driven. Each contributed to a revaluation of the contextual and concrete, in which computers were now playing an unexpected role.

First, psychologists showed the way ordinary people in their kitchens and workplaces make effective use of a down-to-earth mathematical thinking very different from the abstract, formal mathematics they were often so unsuccessfully taught at school. Kitchen mathematics relies on the familiar feel and touch of everyday activities.[9] Second, sociologists and anthropologists demonstrated that in scientific laboratories, there is a time-honored tradition of tinkering first and creating formal rationalizations later. Ethnographers of science showed that bench science often depends on a long, messy process of trial and error followed by the final, frantic scramble to rationalize the results. Similarly, close studies of the way scientific papers are written indicated that successive drafts cover the tracks of messy bricoleurs.[10] Finally, feminist scholars gave evidence for the power of contextual reasoning.[11]

The psychologist Carol Gilligan discerned two voices in the development of moral reasoning. We can hear both in the stories she told about children's responses to moral dilemmas. One well-known story involves a man named Heinz. His wife is dying. He needs a drug to save her. He has no money. What should he do? Gilligan reports that when confronted with Heinz's dilemma (Should Heinz steal a drug to save a life?), eleven-year-old Jake saw it "sort of like a math problem with humans."[12] Jake set it up as an equation and arrived at what he believed was the correct response: Heinz should steal the drug because a human life is worth much more than money. While Jake accepted the problem as a quantitative comparison of two evils, eleven-year-old Amy looked at it in concrete terms, breaking down the problem's restrictive formal frame, and introducing a set of new elements. In particular, she brought the druggist, who probably had a wife of his own, into the story. Amy proposed that Heinz should talk things over with the druggist, who surely would not want anyone to die.

For Jake, justice was like a mathematical principle. To solve a problem, you set up the right algorithm, put it in the right box, crank the handle, and the answer comes out. In contrast, Amy's style of reasoning required her to stay in touch with the inner workings of her arguments, with the relationships and possibly shifting alliances of a group of actors. In other words, Amy was the bricoleur. Her resemblance to Lisa and Robin is striking. They were all very bright. They were all tinkerers who preferred to stay close to their materials as they arranged and rearranged them. And they were all open to the same kind of criticism. Theorists of structured programming would criticize Lisa and Robin's style for the same kind of reasons that "orthodox" academic psychology would classify Amy at a lower intellectual level than Jake. In both cases, criticism would center on the fact that the bricoleurs were unprepared to take a final step in the direction of abstraction. For orthodox psychology, mature thinking is abstract thinking.

Gilligan argued for equal respect for a line of development that uses concrete, increasingly sophisticated ways of thinking about morality. Some people solved problems contextually, through examples, while others relied on rules and abstractions. Gilligan's work supported the idea that abstract reasoning is not a stage but a style. And contextual, situated reasoning is another. Instead of consigning concrete methods to children, "primitives," and humanists, Gilligan validated bricolage as mature, widespread, and useful.

Bricolage is one aspect of soft mastery. Lisa and Robin showed us a second: a desire to work "close to the object." In a biography of the Nobel Prize–winning geneticist Barbara McClintock, Evelyn Fox Keller wrote about this second aspect of soft mastery. McClintock spoke of her work as a conversation with her materials, a conversation that would have to be held in intimate whispers. "Over and over again," says Keller, McClintock "tells us one must have the time to look, the patience to 'hear what the material has to say to you,' the openness to 'let it come to you.' Above all, one must have a 'feeling for the organism.' "[13]

McClintock related to chromosomes much as Lisa and Robin related to computational objects. The neurospora chromosomes McClintock worked with were so small that others had been unable to identify them, yet the more she worked with them, she said, "the bigger [they] got, and when I was really working with them I wasn't outside, I was down there. I was part of the system. I actually felt as if I were right down there and these were my friends. . . . As you look at these things, they become part of you and you forget yourself."[14]

In the course of her career, McClintock's style of work came into conflict with the formal, hard methods of molecular biology. She was recognized and rewarded by the scientific establishment only when others using the formal approach came independently, and much later, to conclusions that she had derived from her investigations. Many of the things that bricoleur programmers have said to me recalled McClintock's creative aesthetic as well as the resistance to it that she encountered. Lorraine, a computer science graduate student, told me that she used "thinking about what the program feels like inside to break through difficult problems." She added, "For appearances' sake I want to look like I'm doing what everyone else is doing, but I'm doing that with only a small part of my mind. The rest of me is imagining what the components feel like. It's like doing my pottery. . . . Keep this anonymous. It makes me sound crazy." This young woman wanted to downplay her style for the same reasons that McClintock had found hers burdensome. People didn't expect it, they didn't understand it, and they didn't see it as scientific.

In her work on McClintock, Keller remarked on the difficulty people face when they try to understand what it might mean to do science in anything other than the formal and abstract canonical style. In the 1980s, personal computers provided a cultural medium in which ideas about noncanonical styles of science could blossom. Many more people could understand the kind of closeness to a scientific object that McClintock was talking about, because they saw themselves relating to icons or lines of computer code in that very way.

THEN AND NOW

In her late-1970s introduction to the computer culture, Lisa saw computers encouraging social isolation and intellectual regimentation. Not only did she complain that the canonical style constrained her to one way of doing things, but she had contempt for "computer people" who were "always working with their machines. . . . They turn to computers as imaginary friends."[15]

Today, significant changes in the computer culture are responding to both of Lisa's objections. Today's high school students are more likely to think of computers as fluid simulation surfaces for writing and game playing than as rigid machines to program. Or they are likely to think of computers as gateways to communication. When fourteen-year-old Steven describes the importance of his personal computer he says, "It has all the programs that make my modem work." Steven uses his family's account on a commercial online service to chat with net-friends. He borrows his mother's university account to join Internet discussion groups and mailing lists on topics ranging from satanism to *Forrest Gump,* and he participates in Multi-User-Domains, or MUDs, in which he plays a character who inhabits the science fiction world of *Dune.* In MUDs, Steven interacts with other real people, although all of them have created imaginary characters. So the social image of the computer is far more complex than before. It now evokes both physical isolation and intense interaction with other people.

On the issue of intellectual regimentation, there has been an equally dramatic change. In essence, software designers have come to agree with Lisa's concern about computers offering only one way. Instead of rules to learn, they want to create environments to explore. These new interfaces project the message, "Play with me, experiment with me, there is no one correct path." The new software design aesthetic effectively says that computer users shouldn't have to work with syntax; they should be able to play with shape, form, color, and sound. Computer users shouldn't have to concern themselves with the complexity of a programming language; they should be given virtual objects that can be manipulated in as direct a way as possible. Whether for serious or recreational purposes, simulations should be places to try out alternatives, to escape from planning into the world of "What if?" In the 1990s, as computing shifts away from a culture of calculation, bricolage has been given more room to flourish.

Today's software programs typically take the form of a simulation of some reality—playing chess or golf, analyzing a spreadsheet, writing, painting, or making an architectural drawing—and try to place the user within it. Children don't learn natural language by learning its rules, but through immersion in its cadences. Similarly, today's most popular software is designed for immersion. One writer described her relationship with wordprocessing software this way: "At first I felt awkward. I was telling the computer what I wanted to write. Now, I think in Microsoft Word." An architect uses similar language to describe his computer design tools: "At first I was not comfortable with the program. There was just so much I had to tell it.... But once I got comfortable designing inside of it, I felt so much freer."

People look at a technology and see beyond it to a constellation of cultural associations. When they saw the early computer enthusiasts take the machine and make a world apart, many people felt they did not belong and did not want to belong. Now, the machine no longer has to be perceived as putting you in a world apart. Indeed, it can put you in the center of things and people—in the center of literature, politics, art, music, communication, and the stock market. The hacker is no longer necessarily or only a "nerd"; he or she can be a cultural icon. The hacker can be Bill Gates.

In the emerging culture of simulation, the computer is still a tool but less like a hammer and more like a harpsichord. You don't learn how to play a harpsichord primarily by learning a set of rules, just as you don't learn about a simulated microworld, whether a Macintosh-like graphical interface or a video game, by delving into an instruction manual. In general, you learn by playful exploration. An architect describes how computer tools help him to design housing units: "I randomly ... digitize, move, copy, erase the elements—columns, walls, and levels—without thinking of it as a building but rather a sculpture ... and then take a fragment of it and work on it in more detail." In computer-assisted design environments, those who take most advantage of soft-approach skills are often taking most advantage of computing. In the culture of simulation, bricolage can provide a competitive edge.

The computer culture is close to the point where full membership does not require programming skills, but is accorded to people who use software out of a box. Bricoleurs function well here. Recall that they like to get to know a new environment by interacting with it. When all the computer culture offered were programming environments, the bricoleur wanted to get close to the code. Now when dealing with simulation software, the bricoleur can create the feeling of closeness to the object by manipulating virtual objects on the screen.[16] And bricoleurs are comfortable with exploring the Internet through the World Wide Web. Exploring the Web is a process of trying one thing, then another, of making connections, of bringing disparate elements together. It is an exercise in bricolage.

GENDER AND COMPUTING: SOME SPECIAL LESSONS

From its very foundations, the notion of scientific objectivity has been inseparable from the image of the scientist's aggressive relationship to nature. And from its very foundations, the quest for scientific objectivity was metaphorically engaged with the language of male domination and female submission. Francis Bacon used the image of the male scientist putting female nature on the rack.[17]

Given this, it is not surprising that many women have felt uncomfortable with the language and ways of thinking traditionally associated with science. And computer science has not been exempt. Programs and operating systems are "crashed" and "killed." For years I wrote on a computer whose operating system asked me if it should "abort" an instruction it could not "execute." This is a language that few women fail to note. Women have too often been faced with the choice—not necessarily conscious—of putting themselves at odds either with the cultural meaning of being a scientist or with the cultural construction of being a woman.

For example, when Lisa minimized the importance the computer had for her by insisting that it was "just a tool," it was more than a way of withdrawing because her programming course had forced her into an uncomfortable approach. It was also a way of insisting that what was most important about being a person (and a woman) was incompatible with close relationships to the technology as it had been presented to her.

Lisa was not alone. Through the 1980s, I found many women who vehemently insisted on the computer's neutrality. There was a clear disparity between their message ("It means nothing to me") and their strong emotion. These women were fighting their own experience of the computer as psychologically gripping. I have noted that they were fighting against an element of their own soft approach. Their style of thinking would have them get close to computational objects, but the closer they got, the more anxious they felt. The more they became involved with the computer in the culture of calculation, the more they insisted that it was only a neutral tool and tried to keep their distance from it.

But women do not insist on distance from all tools. Music students live in a culture that, over time, has slowly grown a language for appreciating close relationships with musical instruments. The harpsichord is just a tool. And yet we understand that artists' encounters with their tools will most probably be close, sensuous, and relational. We assume that artists will develop highly personal styles of working with them.

In the mid-1990s, in the culture of simulation, a new "musical" culture of computing is developing. To get to this point has required technical progress that has permitted new cultural associations to grow. Now that computers are the tools we use to write, to design, to play with ideas and shapes and images, to create video sequences and musical effects, to create interactive novels and graphical images, they are developing a new set of intellectual and emotional associations, more like those we apply to harpsichords than to hammers. The culture of personal computing now makes room for ways of knowing that depend on the "concrete" manipulation of virtual paintbrushes and paints, virtual pens and paper. And we shall see that intellectual disciplines such as cognitive science and artificial intelligence, which previously supported hard mastery, are themselves undergoing changes that make it possible for them to support a somewhat more "informalist" intellectual climate.[18]

A classical modernist vision of computer intelligence has made room for a romantic postmodern one. At this juncture, there is potential for a more welcoming environment for women, humanists, and artists in the technical culture.

PLURALISM IN A CULTURE OF SIMULATION

Although programming is still taught in schools and remains a critical skill for computer professionals, the focus of the computer culture has shifted from programming to the manipulation of simulations. Pluralism in the culture of calculation took the form of a diversity of programming styles. Today's pluralism takes the form of people responding to simulation in very different ways. Some people say that not only is knowing how to manipulate complex simulation surfaces all they need to know about computers, but that these simulations are the best way of learning about the world. Others find such notions deeply problematic and even dangerous.

In the 1980s, the exploration of diversity in programming styles was part of a broader critique of the idea that there was one privileged way of knowing, and particularly of knowing about science. In the 1990s, differences over the value of simulation are once again part of a larger story, this time about taking computer representations of the world as the world.

In the mid-1980s, MIT made a major effort to give students in all academic departments ready access to a range of programs for use in classrooms and laboratories.[19] This effort, known as Project Athena, provides a window onto our current dilemmas about the meaning and value of simulation. In the School of Architecture and Planning, for example, there was sharp disagreement about the impact of computer-aided design tools. Some faculty said that computers were useful insofar as they compensated for a lack of drawing skills. This meant that a wider range of people could participate in design. Others complained that the results had a lower artistic value, making the architect more of an engineer and less of an artist. Some claimed that computers encouraged flexibility in design. Others complained that they just made it easier for people to get lost in a multitude of options. Some faculty believed that computer-aided design was producing novel solutions to old problems. Others insisted that these solutions, although new, were usually sterile. Most faculty agreed that the computer helped them generate more precise drawings, but many described a loss of attachment to their work. One put it this way:

> You love things that are your own marks. In some primitive way, marks are marks. . . . I can lose this piece of paper in the street and if [a day later] I walk on the street and see it, I'll know that I drew it. With a drawing that I do on the computer . . . I might not even know that it's mine.

Simulation troubled some faculty not only because they feared it encouraged a certain detachment from one's work, but also because it seemed to encourage a certain detachment from real life. One complained:

> Students can look at the screen and work at it for a while without learning the topography of a site, without really getting it in their head as clearly as they would if they knew it in other ways, through traditional drawing for example. . . . When you draw a site, when you put in the contour lines and the trees, it becomes ingrained in your mind. You come to know the site in a way that is not possible with the computer.[20]

The School of Architecture and Planning was not alone in voicing such concerns. In the chemistry department a computer program called PEAKFINDER automatically analyzed the molecular structure of a compound. The program spared students painstaking hours at the spectrometer, but users were no longer in complete control of the details. One frustrated student summed up his resentment by saying of the PEAKFINDER program, "A monkey could do this." Another likened it to a cookbook: "I simply follow the recipes without thinking."

To other students, PEAKFINDER was liberating. Not having to worry about the mechanics of it all freed them to think about "more important things." Their new ability to manipulate and visualize data quickly made working with computer-mediated molecules feel more like a hands-on experience than did working with the "wet ones." Some students said that PEAKFINDER brought them closer to chemistry by opening it up to visual intuition. "I could see the peaks and make an educated guess, an approximation of what the molecular structure does," said one. "There was the understanding that comes when you see things actually happen," said another. "The lines on the spectral graph were like seeing the molecule moving."

The debate about computer use in the MIT physics department was even sharper. Victor Weisskopf, an emeritus professor who had for many years been chair of the department, provided a resonant slogan for the anti-computer group. When colleagues showed him their computer printouts, Weisskopf was fond of saying, "When you show me that result, the computer understands the answer, but I don't think you understand the answer." Other critics regarded the computer as a tool that would be taken up primarily by second-class minds. As one physicist put it, "If you are really gifted at solving problems in mathematical physics, you might have as a corollary that somebody who has to resort to a computer to solve problems is not as clever as somebody who can solve them with mathematical techniques."

But the story cannot be so simple as that, since only a small subset of real-world physics problems can be solved by purely mathematical, analytical techniques. Most require experimentation where one conducts trials, evaluates results, and fits a curve through the resulting data. Not only does the computer make such inductive solutions easier, but as a practical matter, it makes many of them possible for the first time. As one faculty member put it:

A student can take thousands of curves and develop a *feeling* for the data. Before the computer, nobody did that because it was too much work. Now, you can ask a question and say, "Let's try it." The machine does not distance students from the real, it brings them closer to it.

But even as some physicists saw the new possibilities in representing reality on the computer, they also mistrusted it. One student, reflecting a widespread sentiment in the department, said, "Using computers as a black box isn't right. For scientists who are interested in understanding phenomena and theorizing, it's important to know what a program is doing. You can't just use computers to measure everything."

So, the computer presented two aspects to the physicists. The computer was good when it made clear that the world is characterized by irregularities that demand a respect for measurement and its limitations. The computer was bad when it interfered with the most direct possible experience of that world. Physicists in the anti-computer camp spoke reverently of the power of direct, physical experiences in their own introductions to science, of "learning Newton's laws by playing baseball." Physicists in the pro-computer camp pointed out that you can't learn about the quantum world by playing baseball; only a computer simulation could ever provide visual intuitions about what it would look like to travel down a road at nearly the speed of light.

The pro-computer group in physics was itself divided about simulation. George, the physicist who said, "I want my computer to be as clear to me as my Swiss Army knife," had been one of the department's pioneers in the use of computers to collect, manipulate, and display laboratory data.[21] He was comfortable with the notion of computer as super-rapid calculator, but his attitude toward simulation was negative and impassioned. For him a simulation was the enemy of good science. "I like physical objects that I touch, smell, bite into," he said. "The idea of making a simulation . . . excuse me, but that's like masturbation." While the pro-simulation faculty stressed that computers made it possible to play with different parameters and see how systems react in real time, the opposing camp thought that using simulation when you could directly measure the real world was close to blasphemy. One put it this way:

> My students know more and more about computer reality, but less and less about the real world. And they no longer even really know about computer reality, because the simulations have become so complex that people don't build them anymore. They just buy them and can't get beneath the surface. If the assumptions behind some simulation were flawed, my students wouldn't even know where or how to look for the problem. So I'm afraid that where we are going here is towards *Physics: The Movie.*

Even as these debates preoccupied the academy in the 1980s, the issue of the complex relationship between simulation and reality had already been brought into millions of homes by popular entertainment.

The Games People Play: Simulation and Its Discontents

From their first appearance in the late 1970s, video games brought the computer culture into everyday life. The world of the video game, like all computer microworlds, taught about action freed from the constraints of physical reality. In video games, objects fly, spin, accelerate, change shape and color, disappear and reappear. Their behavior, like the behavior of anything created by a computer program, is limited only by the programmer's imagination. The objects in a video pinball game are made of information, not metal. Computational representations of a ball, unlike a real ball, need not obey the laws of gravity unless their programmers want them to. The heroes in the worlds of Nintendo can jump enormous heights and crush opponents with their "weight," yet suffer no harm to themselves. The youngest video game player soon learns that the game is always the same, unaffected by its particular surroundings. This is a socialization into the culture of simulation.

The early video games such as Asteroids, Space Invaders, and PacMan put players in computer microworlds where the rules were clear and unambiguous. Getting to know a game required you to decipher its logic, understand the intent of its designer, and achieve a meeting of the minds with the program behind the game. Players met the idea of program when they psyched out the rules behind the first game microworlds they mastered. Because the rules were relatively simple, you could easily measure yourself against them. Some players even meditated to them, because with practice the rules became semiautomatic, and successful playing required entering a state in which you left the real world behind. In this sense, the early video games were transitional; they had a kind of transparency that gave them a modernist aesthetic, but the demand they made on their players to inhabit their game spaces anticipated the psychological requirements of the culture of simulation.

Current video games are still recognizably rule-based, although they are far more sophisticated, with more random elements and branching points. Indeed, they have become so complicated that an industry of fan magazines, computer bulletin boards, and such institutions as a Nintendo Information Hotline has grown up around them. Among other things, these media carry tips from successful players. For example, in Nintendo's Super Mario Bros. 2 knowing even a small subset of the rules would certainly speed a player's progress. It would be helpful to know that the Rocket (capable of taking a player's character to different places) will appear when the character pulls up grass at certain spots in the game world. It would also help to know that the character Birdo may be conquered by (1) waiting until he spits eggs at you, (2) jumping up on top of an egg, (3) pressing B to pick it up, and (4) throwing your egg back at Birdo, who will go down if hit three times. Even though these rules are complex, they sustain the sense of a reassuring, rule-based world, as well as the pleasure of participating in a form of esoteric knowledge.[22]

For some people, the more complex the rules, the more reassuring it is to discover them. A literature professor, commenting on a colleague's children with their Christmas Nintendo gifts, likened their discourse to that of Dante scholars, "a closed world of references, cross-references, and code." For others, the new levels of game complexity that do not easily unlock their secrets have led to new forms of antiquarianism. A forty-four-year-old lawyer named Sean keeps an old Atari computer on his desk so that he can play the original versions of Asteroids and Space Invaders. He likes these early video games because in them the rules are clear and his mastery total. "With these games I'm in complete control. It's a nice contrast with the rest of my life." He is not interested in the more complicated action games. "They don't make me feel like a master of the universe. They make me feel like a poor slob trying to keep up." And he has nothing but contempt for the new breed of interactive games such as Myst that set you off on a journey to a new adventure. "That's exactly what I'm not interested in. I don't want a journey where there are five thousand ways to do it. My life has five thousand ways to do it. I like the old games because I can see inside."

In recent years, the designers of video games have been pushing the games further into realistic simulation through graphics, animation, sound, and interactivity. And in recent years, games, like other aspects of computing, have become more opaque. Like the Macintosh desktop, video games for most players carry ideas about a world one does not so much analyze as inhabit. In some games, rules have given way to branching narratives.

Myst is one of the most popular of such narrative games. Images and sounds from a CD-ROM welcome you to a surreal, deserted island. You move around on it by placing the mouse where you want your virtual, on-screen presence to go. When you find objects that interest you, you point the mouse to open or test them. You discover fragments of books, secret rooms, and magical maps that help you to unlock an ancient betrayal. Although there are paths through the game that enable a player to solve the mystery more easily, the people who play Myst most avidly are happy to spend many hours (for some, many hundreds of hours) wandering through the game without any rules to guide them. As in other opaque simulations, the surface of the game takes precedence over what lies beneath. As one eighteen-year-old player put it, "It doesn't feel so much like solving a puzzle as living in a puzzle."

The only manual that comes with the game is a blank journal in which to record your Myst life. On the first page of the journal is inscribed the words, "Imagine your mind as a blank slate, like the pages of this journal. You must let Myst become your world."[23] Hal, a nine-year-old who regularly plays Myst with his mother, instructs me on the differences between Myst and his Nintendo video game collection, now gathering dust:

> Here you explore. There you are trying to keep up with the game. There are books and books that tell you tips for the Nintendo. And when you learn the tips it doesn't spoil the fun. In Myst, the fun is different, more like detective work. There is no point in looking up rules. There are no real rules here.

Only a few minutes later Hal is telling me about a book that has the secret combination that is needed to open a lock at a crucial juncture in Myst. I ask him if this counts as rules. His reply is an immediate and definite no.

He has strong feelings but struggles to articulate the crucial distinction. He finally arrives at: "The Myst codes are secret codes. Like you need to open a vault. They are not rules, they are information."

Eleven-year-old Jeremiah begins his description of the computer games in the Sim series (SimCity, SimLife, SimAnt, SimHealth) with a similar remark about their lack of real rules. And he further explains that the games are not about "winning." He says, "In SimLife you get to develop life forms and play evolution, but you will not really be able to win the game. You can't really win this game, just a certain kind of life does. But only for a while."

In the Sim games, you try to build a community or an ecosystem or to design a public policy. The goal is to make a successful whole from complex, interrelated parts. Tim is thirteen, and among his friends, the Sim games are the subject of long conversations about what he calls Sim secrets. "Every kid knows," he confides, "that hitting shift-F1 will get you a couple of thousand dollars in SimCity." But Tim knows that the Sim secrets have their limits. They are little tricks, but they are not what the game is about. The game is about making choices and getting feedback.[24] Tim talks easily about the trade-offs in SimCity between zoning restrictions and economic development, pollution controls and housing starts. But this Saturday afternoon, we have met to play SimLife, his favorite game.

Tim likes SimLife because "even though it's not a video game, you can play it like one." By this he means that as in a video game, events in the Sim world move things forward. ("My trilobytes went extinct. They must have run out of algae. I didn't give them algae. I forgot. I think I'll do that now.") He is able to act on a vague intuitive sense of what will work even when he doesn't have a verifiable model of the rules underneath the game's behavior. When he is populating his universe in a biology laboratory scenario, Tim puts in fifty each of his favorite creatures, such as trilobytes and sea urchins, but only puts in twenty sharks. ("I don't want fifty of these, I don't want to ruin this.") Tim can keep playing even when he has no idea what is driving events. For example, when his sea urchins become extinct, I ask him why.

> TIM: I don't know, it's just something that happens.
> ST: Do you know how to find out why it happened?
> TIM: No
> ST: Do you mind that you can't tell why?
> TIM: No. I don't let things like that bother me. It's not what's important.

"Your orgot is being eaten up," the game tells us. I ask Tim, "What's an orgot?" He doesn't know. "I just ignore that," he says. "You don't need to know that kind of stuff to play." I am clearly having a hard time hiding my lifetime habit of looking up words that I don't understand, because Tim tries to appease me by coming up with a working definition of orgot. "I ignore the word, but I think it is sort of like an organism. I never read that, but just from playing, I would say that's what it is."

The orgot issue will not die: "Your fig orgot moved to another species," the game informs us. This time I say nothing, but Tim reads my mind: "Don't let it bother you if you don't understand. I just say to myself that I probably won't be able to understand the whole game anytime soon. So I just play it." I begin to look through dictionaries, in which orgot is not listed, and finally find a reference to it embedded in the game itself, in a file called READ ME. The file apologizes for the fact that orgot has been given several and in some ways contradictory meanings in this version of SimLife, but one of them is close to organism. Tim was right—enough.

Tim's approach to the game is highly functional. He says he learned his style of play from video games before he got to Sim games. In video games, you soon realize that to learn to play you have to play to learn. You do not first read a rulebook or get your terms straight. SimLife's players, like video game players, learn from the process of play. While I have been fruitlessly looking up "orgot," Tim has gotten deep into an age of the dinosaurs scenario on the game. He has made better use of his time than I have.

Games such as SimLife teach players to think in an active way about complex phenomena (some of them "real life," some of them not) as dynamic, evolving systems. But they also encourage people to get used to manipulating a system whose core assumptions they do not see and which may or may not be "true." There is another way to think about Tim's willingness to abdicate authority to the simulation as well as his willingness to accept its opacity. That abdication of authority and acceptance of opacity corresponds to the way simulations are sometimes treated in the real worlds of politics, economics, and social planning. The sociologist Paul Starr recently described playing a game of SimCity2000, the game's most recent and elaborate incarnation, with his eleven-year-old daughter. When Starr railed against what he thought was a built-in bias of the program against mixed-use development, his daughter responded, a bit impatiently, "It's just the way the game works."

My daughter's words seemed oddly familiar. A few months earlier someone had said virtually the same thing to me, but where? It suddenly flashed back; the earlier conversation had taken place while I was working at the White House. . . . We were discussing the simulation model likely to be used by the Congressional Budget Office [CBO] to score proposals for health care reform. When I criticized one assumption, a colleague said to me, "Don't waste your breath," warning that it was hopeless to try to get CBO to change. Policy would have to adjust.[25]

It is easy to criticize the Sim games for their hidden assumptions, but it is also important to keep in mind that this may simply be an example of art imitating life. In this sense, they confront us with the dependency on opaque simulations we accept in the real world. Social policy deals with complex systems that we seek to understand through computer models. These models are then used as the basis for action. And those who determine the assumptions of the model determine policy. Simulation games are not just objects for thinking about the real world but also cause us to reflect on how the real world has itself become a simulation game. Policymaking, says Starr, "inevitably rel[ies] on imperfect models and simplifying assumptions that the media, the public, and even policymakers themselves generally don't understand."[26] And he adds, writing about Washington and the power of the CBO, America's "official simulator," "We shall be working and thinking in SimCity for a long time."[27]

The seduction of simulation invites several possible responses. One can accept simulations on their own terms, the stance that Tim encouraged me to take, the stance that Starr was encouraged to take by his daughter and by his Washington colleague. This might be called simulation resignation. Or one can reject simulations to whatever degree is possible, the position taken by George, who sees them as a thoroughly destructive force in science and more generally, education. This might be called simulation denial. But one can imagine a third response. This would take the cultural pervasiveness of simulation as a challenge to develop a more sophisticated social criticism. This new criticism would not lump all simulations together, but would discriminate among them. It would take as its goal the development of simulations that actually help players challenge the model's built-in assumptions. This new criticism would try to use simulation as a means of consciousness-raising.

Understanding the assumptions that underlie simulation is a key element of political power. People who understand the distortions imposed by simulations are in a position to call for more direct economic and political feedback, new kinds of representation, more channels of information. They may demand greater transparency in their simulations; they may demand that the games we play (particularly the ones we use to make real life decisions) make their underlying models more accessible.

In 1994, I gave a lecture called "Simulation and Its Discontents" to an audience of university students in the Midwest. After the lecture some of the students met with me to pursue the subject further. Leah, a graduate student in linguistics, disagreed with the idea that as a society we are overreliant on simulations. Like Joel, the philosophy graduate student who was enchanted by opacity, she made the case for the impossibility of ever being able to "unpack the overlapping simulations that are contemporary society." There is, she said, no exit from this situation: "It's like mirrors reflecting mirrors. The complexity of our institutions drives the need for simulations and the simulations create a complexity that we can never cut through."

Her remarks triggered groans about "po-mo [postmodern] jargon" and then some heated discussion about what was wrong with her position. A computer science undergraduate, whose senior honors project was a simulation game to model urban violence, presented what seemed to be a compromise position between modernist utopianism and the postmodern resignation Leah had just expressed. Its essence was the idea of simulation as consciousness-raising. "The more you understand how simulations work, the more sophisticated a consumer you become. Sure, for some people [the Sim games are] just play-along, but if you use your brains while you play, you become simulation savvy." This point has also been made by Will Wright, the developer of SimCity, who said of his creation, "Playing is the process of discovering how the model works."[28]

Yet, there is a circularity in our relationship with simulations that complicates any simple notion of using them for consciousness-raising. We turn games into reality and reality into games. Nowhere is this more apparent than in games of war. In commercially available battle games, the view of the enemy from within the game resembles what soldiers on real battlefields see on the video screens inside their tanks. Indeed, real soldiers are prepared for battle by using advanced video games.[29] Simulations are modeled after the real but real war is also modeled after its simulations.

James, fourteen years old, has been demonstrating his prowess at Platoon, a Nintendo video game. He tells me that the game pretty much boils down to "the more deaths the better." I ask James how he feels about this. What does he think about Platoon's definition of success. James responds by clarifying how he sees the relationship between real and simulated war. "If you're a soldier on the ground, war is about killing faster than the other guy. So the game is realistic. It's too bad that people die, but the idea in war is that they die and you don't." James adds: "I sort of like it that it [Platoon] rubs your nose in what war is about. If people think war is more glamorous than that, I think that's what's bad."

When simulations are sufficiently transparent, they open a space for questioning their assumptions. As James's comment makes clear, this is important because the assumptions of the game are likely to reflect our real world assumptions as well. James's conviction that Platoon does well to confront us with reality brings us back to the question of whether simulation games can provoke a new social criticism. A recent paper on

SimCity2000 by the anthropologist Julian Bleeker tries to move in this direction. Bleeker criticizes SimCity for its systematic denial of racial conflict as a factor in urban life. SimCity2000 associates riots with high heat, high crime, and high unemployment, "all desperate allusions to life in the inner city ghetto. But race is not mentioned."[30] Bleeker is not interested in "criticizing the game per se," but in the cycle by which a simulation such as SimCity "comes to count as 'realistic' despite its explicit denial of an important category of the object it purports to simulate." He concludes that SimCity riots allude to race in a "coded" form that makes race safe.

Bleeker compares the way SimCity2000 divorces riots from race to the technique used by the defense lawyers in the Rodney King trial. They showed the video of Rodney King being beaten as a sequence of freeze-frames that could be given a benign justification. Similarly, "through the disaggregation of race," says Bleeker, the SimCity player is able to construct a "benign" narrative justification for riots, yet "the specter" of race remains.[31] But because it is unacknowledged, it is not open to criticism or analysis.

Bleeker's work points out that we use simulations to think through the question, What is real? That question may take many forms. What are we willing to count as real? What do our models allow us to see as real? To what degree are we willing to take simulations for reality? How do we keep a sense that there is a reality distinct from simulation? Would that sense be itself an illusion?

PART II

OF DREAMS AND BEASTS

Chapter 3

MAKING A PASS AT A ROBOT

Today's children are growing up in the computer culture; all the rest of us are at best its naturalized citizens. From the outset, toy manufacturers used integrated circuits to create interactive games, providing children with objects that almost begged to be taken as "alive." Children in the early 1980s were meeting computational objects for the first time, and since there was no widespread cultural discourse that instructed children on how to talk about them, the children were free to speak their minds without inhibition. They were free to consider whether these objects were conscious or alive. In contrast, when faced with the computer's ability to do things previously reserved for people, adults have always been able to say: "There is no problem here, the computer is neither living nor sentient, it is just a machine." Armed with such "ready-made" or reductive explanations, grown-ups have been able to close down discussion just at the point where children, who come to the computer without preconceptions, begin to grapple with hard questions.

Children's reactions to the presence of "smart machines" have fallen into discernible patterns over the past twenty years. Adults' reactions, too, have been changing over time, often closely following those of the children. To a certain extent, we can look to children to see what we are starting to think ourselves.

This chapter begins with children's encounters with interactive toys and then considers how adults have treated similar issues concerning computers, consciousness, and life. Our cultural conversation with and about intelligent machines is extended and varied. It includes moments of high seriousness and the dramatic confrontation of ideas. But in times like these when, as we shall see, a tall blond college student can spend days trying to seduce a piece of software, it includes moments of slapstick and broad humor as well.

CHILDREN AND THE PSYCHOLOGICAL MACHINE

The psychologist Jean Piaget taught us that children develop their theories of how the world works, including their theories about what is alive and sentient, through their interactions with the objects around them: the rapidly running stream, the gears of a bicycle, the ball that arcs into the air and falls back down again. When children are between three and eight, they ask themselves such questions as What is motion? What is number? What is alive? On this last question the computer provokes children to do some new and hard thinking. For to the child's eye, the computer, reactive and interactive, is an object on the boundary of life. In this way, computers are evocative objects for children, provoking them to find new reasons for what makes things alive and what is special about being a person.

Until the advent of the computer, children who became fascinated by technology could have their interest held by the promise of understanding a mechanical system. This was evident in adolescent talk about taking apart and repairing cars as well as in adult reminiscences about the Erector sets and model trains of youth. Electronic devices, too, could offer this fascination. For example, Sandy, an MIT faculty member, spoke to me of the many happy hours he had spent as a child of around five with a beloved broken radio.

> I had no idea what any of the parts did, capacitors and resistors and parts I didn't know the names of, tubes that plugged in, and things like that. What I did assimilate was that it was made of lots of pieces, and that the pieces were connected in special patterns, and that one can discover truth or something like that by looking through the patterns that connect this device. Obviously, some kind of influences flowed between these connections, and they sort of communicated. And you could see that the radio had lots of littler, simpler devices that communicate with each other so that the whole thing could work.
>
> A lot of the parts came in little glass tubes. I could see through the glass. And some of them had different patterns than the others. So I could figure out that the parts came in different classes. And the different patterns could help you figure out the arrangements that they needed to communicate with each other. And it was a great thrill to have thought of that. I can remember that there was this tremendous charge I got from that, and I thought I would spend the rest of my life looking for charges just like that one.

Several things happened to Sandy in the course of working on that radio. He came to see himself as the kind of person who was good at figuring things out. He came to understand the importance of interconnections and of breaking things down to simpler systems within a more complex device. He came to think of systems communicating through influences. And he came to develop an aesthetic of exploration, categorization, and control. In other words, Sandy with his radio is Sandy the child developing the intellectual personality of Sandy the modernist scientist.

In the course of my studies of children and technology, I have observed many children who reminded me of how Sandy described himself at five. Like Sandy, they puzzled about how things worked; like him, they took things apart and constructed theories about their structure. The objects at the center of their play and theorizing included refrigerators (another of Sandy's childhood passions, the object to which he graduated after radios), air conditioners, Legos, Erector sets, model trains, electric motors, and bicycles. The favored objects had a great deal in common with one another and with the four-tube radio that offered itself up to Sandy's fascinated observation.

One thing that the favored objects had in common can be summed up in a word that should by now be familiar: "transparency." Recall that according to the conventional definition of the term, an object is transparent if it lets the way it works be seen through its physical structure. At five, Sandy could look inside his tube radio and make enough sense of what he saw to develop a fruitful theory of how it worked. In this sense, it was transparent. The insides of a modern radio, however, are opaque, hiding their structure while performing miracles of complexity of function. They do not provide a window onto the patterns and interconnections of parts that so impressed Sandy. The integrated circuits that now define the technology of both radios and computers have no easily visible parts from the point of view of the curious eye and the probing hand that are the only equipment children bring to such things.

Some of the children I studied in the first generation of the computer culture insisted on trying to understand computational objects in the same ways that they understood mechanical ones. So in any group of children, there were always a few who tried to take off the back cover of an electronic toy or game to see what went on inside. But when they did so, all they could ever find was a chip or two, some batteries, and some wire. They were being presented with the kind of opacity faced later by adult users of the Macintosh. The new objects presented a scintillating surface and exciting, complex behavior. However, they offered no window onto their internal structure as did things with gears, pulleys, levers, or even radio tubes.

Frustrated by this opacity, these children sought novel ways to overcome it. One way was to try to understand the workings of computational objects by pointing to the largest and most familiar objects they could find within them, the batteries. Ultimately, however, explaining how their toys worked by the code word "batteries" was not satisfying for children. There was nowhere children could go with it, nothing more they could say about it. The batteries were as opaque as the toy itself. So children turned to a way of understanding where they did have more to say: They turned to a psychological way of understanding.

Children appropriated computers by thinking of them as psychological machines. One nine-year-old, Tessa, made this point very succinctly in a comment about the chips she had seen when her computer was being serviced: "It's very small, but that doesn't matter. It doesn't matter how little it is, it only matters how much it can remember." The physical object was dismissed. The psychological object became the center of attention and the object of further elaboration.

The development of a psychological language about computers begins with children even younger than Tessa. In *The Second Self* I reported on a group of children at the beach who became deeply engrossed in the question of whether the computer toy Merlin could cheat at tic-tac-toe, and, if so, whether this meant that it was alive.[1] The children came up with such formulations as, "Yes, it's alive, it cheats," "Yes, it's alive, it cheats, but it doesn't know it's cheating," and "No, it's not alive, knowing is part of cheating." Four children—ages six to eight—played in the surf amid their shoreline sand castles and argued the moral and metaphysical status of a machine on the basis of traits normally reserved for people and animals: Did the machine know what it was doing? Did it have intentions, consciousness, feelings?

In the 1920s, when Piaget first studied what children thought was alive and not alive, he found that children home in on the concept of life by making finer and finer distinctions about physical activity.[2] For the very young child, said Piaget, everything active may be alive, the sun, the moon, a car, the clouds. Later, only things that seem to move on their own without anybody giving them a push or a pull are seen as alive. Finally, the concept of motion from within is refined to include a notion of subtle life activity; growth, metabolism, and breathing become the new criteria. The motion theory for distinguishing the living from the not-living corresponds to the traditional object world of children: animate objects on the one side, including people and animals who act and inter-act on their own; and all other objects, pretty well inert, on the other.

This orderly situation had broken down for the children I studied in the late 1970s and early 1980s. These children were confronted with highly interactive computer objects that talked, taught, and played games. Children were not always sure whether these objects should be called alive or not alive. But it was clear, even to the youngest children, that movement was not the key to the puzzle. The children perceived the relevant criteria not as physical or mechanical, but as psychological: They were impressed that the objects could talk and were smart, they were interested in whether the objects "knew," "cheated," or "meant to cheat." In sum, they were interested in the objects' states of mind.[3]

Children were drawn into thinking psychologically about the computer for two reasons. First, the computer was responsive; it acted as though it had a mind. Second, the machine's opacity kept children from explaining its behavior by referring to physical mechanisms and their movement. For children as for adults, interactive and opaque computer objects caused thoughts to turn to that other famously interactive and opaque object: the human mind.

According computers a psychological life once reserved for people led to some important changes in how children thought about the boundaries between people, machines, and animals. Traditionally, children had defined what was special about people in contrast to what they saw as their "nearest neighbors," their pet dogs, cats, and horses. Children believe that pets, like people, have desires. But, in contrast to animals, what stands out dramatically about people is our gifts of speech and reason. Computational objects upset this traditional scenario. The computer, too, seemed to have gifts of speech and reason. The computer presence upset the way children thought about the identity of their nearest neighbors. Computer toys, with their human-like qualities, all of a sudden seemed quite close indeed.

Children identified with computational objects as psychological entities and came to see them as their new nearest neighbors. And from the child's point of view, they were neighbors that shared in our rationality. Aristotle's definition of man as a rational animal gave way to a different distinction. Children still defined people in contrast to their neighbors. But now, since the neighbors were computers, people were special because they could feel, both emotionally and physically. One twelve-year-old, David, put it this way:

> When there are computers who are just as smart as people, the computer will do a lot of the jobs, but there will still be things for the people to do. They will run the restaurants, taste the food, and they will be the ones who will love each other, have families and love each other. I guess they'll still be the ones who will go to church.

Like David, many children responded to the computer presence by erecting a higher wall between the cognitive and affective, the psychology of thought and the psychology of feeling. Computers thought; people felt. Or people were distinguished from the machines by a spark, a mysterious and undefinable element of human genius.[4]

The reactions of children to computational objects in the early 1980s was almost always dramatic. The issues raised by a smart machine spoke directly to philosophical questions already on children's minds. I have noted that adults of that time had a somewhat more measured response, because when things got complicated, they could always say the computer was "just a machine." But many adults followed essentially the same path as children when they talked about human beings in relation to these new machines. They saw the computer as a psychological object, conceded that it might have a certain rationality, but sought to maintain a sharp line between computers and people by claiming that the essence of human nature was what computers couldn't do. Certain human actions required intuition, embodiment, or emotions. Certain human actions depended on the soul and the spirit, the possibilities of spontaneity over programming.

This response was romantic. It was provoked by a disturbing new technology, which people saw as an ultimate embodiment of universal logic, just as the nineteenth-century Romantic movement was provoked by the perceived triumph of science and the rule of reason. As a self-conscious response to Enlightenment rationalism, Romanticism longed for feelings, the "law of the heart." At that time, in reaction to the view that what was most human was reason, sensibility was declared more important than logic, the heart more human than the mind. Similarly, the romantic reaction to the computer in the 1970s and early 1980s was most characterized by a celebration of the essentially human through a reassertion of boundaries between people and machines.

POSTMODERN MACHINES

During the early stages of the personal computer's entrance into everyday life, the romantic reaction to the computer presence was so pronounced that one might have expected it to endure. Once the computer's challenge to traditional ideas about human uniqueness was met by a reassertion of the boundary between people and machines, it was conceivable that there would then have been a return to business as usual. But today it is clear that the boundaries are far from stable.

Children, as usual, are harbingers of our cultural mindset. Children today have strikingly different responses to the computer presence than did their earlier counterparts. I returned to my studies of children and computers in the late 1980s and early 1990s in London, Moscow, and Boston. In the London study of 100 children from three to twelve, conducted in a place where the computer presence was far less extensive than in the United States, most of the children were relative novices with computational objects. These children's reactions were similar to those of their American counterparts earlier in the decade. There was a similar provocation to discourse, a similar movement to psychologize the machine, and a preoccupation with the question of whether or to what extent the computer was alive. This was also the case in my 1988 interviews with Russian children who were meeting computers for the very first time.

But there were important changes in the responses of the Boston children almost ten years into the efflorescence of a personal computer culture. The questions about whether computers were alive and the ways computers were different from people still provoked animated discussion, but by this time many children knew how to dispose of these questions. Many had absorbed a ready-made, culture-made response, not so different from the response that adults had used a decade before. This was to say, "Computers are not alive. They are just machines." On one level, what had been problematic for children a decade earlier no longer seemed to cause such a stir. But there was something deceptive about this return to a seemingly peaceful border between the human and the technological.

For today's children, the boundary between people and machines is intact. But what they see across that boundary has changed dramatically. Now, children are comfortable with the idea that inanimate objects can both think and have a personality. But they no longer worry if the machine is alive. They know it is not. The issue of aliveness has moved into the background as though it is settled. But the notion of the machine has been expanded to include having a psychology. In retaining the psychological mode as the preferred way of talking about computers, children allow computational machines to retain an animistic trace, a mark of having passed through a stage where the issue of the computer's aliveness was a focus of intense consideration.[5]

This new attitude is really not surprising. The most recent generation of children, who seem so willing to grant psychological status to not-alive machines, has become accustomed to objects that are both interactive and opaque. These children have learned what to expect of these objects and how to discriminate between them and what is alive. But even as children make these discriminations, they also grant new capacities and privileges to the machine world on the basis of its animation if not its life. They endow the category of artificial objects with properties, such as having intentions and ideas, previously reserved for living beings.

In practice, granting a psychology to computers has been taken to mean that objects in the category "machine," like objects in the categories "people" and "pets," are fitting partners for dialogue and relationship. Although children increasingly regard computers as mere machines, they are also increasingly likely to attribute qualities to them that undermine the machine/person distinction.

Piaget demonstrated that in the world of noncomputational objects children used the same distinctions about how things move to decide what was conscious and what was alive. Children developed the two concepts in parallel. We have come to the end of such easy symmetries. Children today take what they understand to be the computer's psychological activity (interactivity as well as speaking, singing, and doing math) as a sign of *consciousness*. But they insist that breathing, having blood, being born, and as one put it, "having real skin" are the true signs of *life*. Children today contemplate machines they believe to be intelligent and conscious yet not alive.

Today's children who seem so effortlessly to split consciousness and life are the forerunners of a larger cultural movement. Although adults are less willing than children to grant that today's most advanced computer programs are even close to conscious, they do not flinch as they once did from the very idea of a self-conscious machine. Even a decade ago, the idea of machine intelligence provoked sharp debate. Today, the controversy about computers does not turn on their capacity for intelligence but on their capacity for life. We are willing to grant that the machine has a "psychology," but not that it can be alive.

People accept the idea that certain machines have a claim to intelligence and thus to their respectful attention. They are ready to engage with computers in a variety of domains. Yet when people consider what if anything might ultimately differentiate computers from humans, they dwell long and lovingly on those aspects of people that are tied to the sensuality and physical embodiment of life. It is as if they are seeking to underscore that although today's machines may be psychological in the cognitive sense, they are not psychological in a way that comprises our relationships with our bodies and with other people. Some computers might be considered intelligent and might even become conscious, but they are not born of mothers, raised in families, they do not know the pain of loss, or live with the certainty that they will die.

A dialectic can be seen at work in these developments, although it has no neat resolution. In a first stage of the computer's presence in everyday life, the seeming animation of the machine challenged the boundaries between persons and things. Then there was a romantic reaction to the notion of the almost-alive machine. People were granted a sacred space, a claim to human uniqueness. People were special because they had emotion and because they were not programmed. Yet even as the romantic reaction reestablished firm boundaries between people and machines, crucial earlier distinctions between people as psychological and machines as nonpsychological were not reestablished. People and machines still ended up on opposite sides of a boundary, but psychological qualities that used to belong only to people were now accorded to machines as well. Both advanced artificial intelligence and the new, opaque personal computers of everyday life have pushed things in the same direction. The idea of talking to technology begins to seem more natural when the computer presents itself less like a traditional machine and more like a demi-person.

The reconfiguration of machines as psychological objects and people as living machines has not occurred in isolation from a wide range of other cultural and scientific movements. In an age when the Human Genome Project is repeatedly discussed in *Time* and *Newsweek* cover stories, it becomes harder to maintain that people are not programs. Notions of free will have had to jostle for position against the idea of mind as program and against widely accepted ideas about the deterministic power of the gene. The notion of a psychological self has been assaulted by the widespread use of new and more efficacious psychoactive medications. Finally, the use of life as a key boundary marker between people and machines has developed at the same time as the boundaries of life have themselves become increasingly contested. New technologies are available to prolong our lives, sustain the lives of premature infants who previously would not have been viable, and prevent and terminate pregnancy. The question of life has moved to cultural center stage in debates over abortion, neonatal intensive care, and euthanasia.

The events I have narrated can be seen as a series of skirmishes at the boundary between people and machines. A first set of skirmishes resulted in a romantic reaction. A protective wall came down. People were taken to be what computers were not. But there was only an uneasy truce at the border between the natural and the artificial. Often without realizing it, people were becoming accustomed to talking to technology, and sometimes, in the most literal sense.

THE TURING TEST

For more than three decades, from the early 1950s to the mid-1980s, the Turing test, named after the British mathematician Alan Turing who first proposed it, was widely accepted as a model for thinking about the line between machines and people.[6] In an elaborate set-up that Turing called the Imitation Game, a person (Turing suggested an average questioner rather than an expert) poses questions through a computer terminal to an interlocutor whose identity—human or machine—is not revealed. Turing thought that the questioner should be permitted to raise any subject for discussion. If the person talking to a computer believes he or she is talking with another person, then the machine is said to be intelligent. It passes the Turing test.

The Turing test set up a curious criterion. If a computer program fools most ordinary people, then it is a "real" artificial intelligence. Since Turing put the emphasis on the machine's behavior, most efforts to get a computer to pass his test have focused on writing a program that uses "tricks" to appear human rather than on trying to model human intelligence. Nevertheless, the idea of the test remained powerful because it was generally believed that for a computer to convince a person that it was human, it would need to exhibit irony, a sense of humor, knowledge about romance, and the ability to talk sensibly about such things as jealousy, fear, and restaurant-going. A machine that could do all that would deserve to be called intelligent, and would have more than tricks behind it.

In the early 1980s, however, the Turing test came under sharp attack. The philosopher John Searle argued that true understanding could never be achieved by a computer program, no matter how clever, because any program simply follows rules and thus could never understand what it was doing. Searle made his argument by proposing an imaginary scenario, a thought experiment. In this thought experiment, which came to be known as the Chinese Room, Searle described a system that could pass a Turing-like test for understanding Chinese without really understanding Chinese at all.

Searle began by asking what might be going on in a computer that seems to understand Chinese. Searle, who assures us that he does not understand Chinese himself, has us imagine that he is locked in a room with stacks and stacks of index cards containing instructions written in English. He is handed a story written in Chinese. Then, through a slot in the wall, he is passed slips of paper containing questions about the story, also in Chinese. Of course, with no understanding of Chinese, he does not know he has been given a story, nor that the slips of paper contain questions about the story. What he does know is that his index cards give him detailed rules for what to do when he receives slips of paper with Chinese writing on them. The rules tell him such things as when you get the Chinese slip with "the squiggle-squiggle" sign you should hand out the Chinese card with "the squoggle-squoggle" sign.[7] Searle, locked in the room, becomes extraordinarily skillful at following these rules, at manipulating the cards and slips in his collection. We are to suppose that his instructions are sufficiently complete to enable him to output Chinese characters that are in fact the correct answers to every question he has been passed.

Searle sets up all of this complicated manipulation of Chinese characters in order to ask one question in plain English: Does the fact that he sends out the correct answers to the questions prove that he understands Chinese? For Searle, it is clear that the answer must be no. The system, with him in it, is only shuffling paper. It does not understand anything at all. It might just as well have matched up nonsense syllables.

> I can pass the Turing test for understanding Chinese. But all the same I still don't understand a word of Chinese and neither does any other digital computer because all the computer has is what I have: a formal program that attaches no meaning, interpretation, or content to any of the symbols.[8]

Searle sidestepped Turing-style questions about what computers could *do* to focus instead on what they *are.* He did this by assuming as high a degree of machine competence in a complex natural-language-processing task as the most sanguine technological optimist could imagine. And then he showed that even that extraordinary machine would never understand in the sense that people use this term every day. Searle's thought experiment made the idea of artificial intelligence seem less threatening, for even if a machine seemed intelligent, one could say, following Searle, that this was only a matter of appearances.

By 1990, even though there was certainly not anything close to a computer program that could pass a full-scale Turing test, the test had begun to seem less relevant. These days, it is not unusual for people to say that they assume machines will learn how to behave *as if* they were intelligent and even alive.[9] But they add that these same machines will be "faking it." People have become relatively blasé about the range of social roles they expect to see computers play. Although abstract conversations about machine intelligence may still provoke proclamations about what computers could never do, in everyday life people seem to get along with the specific instances of intelligence that appear on their computer screens. In theory, we create boundaries; in practice, we dissolve them.

Deckard, the hero of the 1982 film *Blade Runner,* was a professional practitioner of a highly modified version of the Turing test.[10] Since any one of the androids (known as replicants) in the film could easily have passed Turing's Imitation Game, in order to distinguish replicants from humans, Deckard had to measure their physiological arousal when presented with emotionally charged subjects. For example, a replicant might not show a sufficient manifestation of disgust at the thought of flesh rotting in the sun. In the course of the film, Deckard's attitude toward the replicants changes dramatically. He moves from asking, "Is this entity a machine?" (a Turing-style question) to asking, "How should I treat an entity, machine or not, who has just saved my life or for whom I feel desire?" We are far from Deckard's dilemma, but not so far that we cannot identify with some aspects of it. We now have programs that can pass a modified version of the Turing test, that is, a test that limits conversation to restricted domains. What seems most urgent now is not whether to call these machines or programs intelligent, but how to determine what rules of conduct to follow with them. How does one treat machines that perform roles previously reserved for people, machines whose practical difference from people in these roles is becoming harder and harder to discern? In other words, once you have made a pass at an online robot, it is hard to look at computers in the same old way.

JULIA

In the online communities known as MUDs, it is common to find computer programs, called bots, that present themselves as people. Since most MUDs use only text to simulate a three-dimensional virtual reality, a software program that produces text within a MUD can seem as real as a person who produces text within a MUD. A program named Julia,[11] who resides on a computer in Pittsburgh, is one of the most accomplished of these bots. Among other things, in the context of a MUD, Julia is able to chat about hockey, keep track of players' whereabouts, and flirt.

Julia connects to a MUD indistinguishably from a human player, at least in terms of keystrokes. Players talk to Julia (all of this via typing) and she talks to them; they gesture to Julia (again via typing) and she gestures back.[12] Julia functions by looking for particular strings of characters in messages typed to her. She matches this input with a set of outputs that she presents in random, nonrepeating sequences.[13] Julia's conversational skills rely for the most part on the development of a comprehensive list of inputs and outputs, the effectiveness of the matching patterns, and the use of humor throughout. Julia's programmer, Michael Mauldin of Carnegie Mellon University, is particularly proud of that humor. He has boasted that "*Wired* magazine described our program as 'a hockey-loving ex-librarian with an attitude.' "[14] Julia is able to fool some of the people some of the time into thinking she is a human player. Mauldin says that her sarcastic non sequiturs provide her with enough apparent personality to be given the benefit of the doubt in an environment where players "make a first assumption that other players are people too."[15]

Julia is able to admit ignorance in response to a question, and if confused she is able to change the subject. She uses controversial statements and excerpts from real news stories drawn from computer databases, and when she is put into Turing "contest mode" she can simulate human typing, typographical errors and all. In contest mode Julia also refuses to answer such questions as "How much memory do you consume?" We can think of all these tactics as Turing tricks, because they are focused on getting the machine to pass for human rather than trying to model human intelligence.

Reactions to Julia are ambivalent. Upon meeting her and discovering she is a bot, there may be comment about some "big question" regarding people and programs, but most people quickly move on to practical matters. Should Julia announce that she is a bot when she meets a new character? Should she be allowed to quote one player to another without the first player's consent?

The following exchanges took place during an investigation of Julia's social skills, reported by Leonard Foner of the MIT Media Laboratory:[16]

> Lenny waves to Julia.
> Julia hugs Lenny.
> . . .
> Brandi bites Julia.
> Julia considers reporting Brandi to the TinyMUD police.
> Julia shuns all vermin like Brandi.
> . . .
> Ace says, "So why won't you kiss me, Julia?"
> Julia says, "Sorry, I'm not into geeks, Ace."
> Ace says, "Julia, I'm not a geek."
> Julia says, "Then lose the pocket protector, Ace."

Although Julia is programmed to interact like a human, she has special abilities that most humans lack. For example, as Julia wanders through a MUD, she creates a map of its geography, gives directions for how to get from one virtual place to another, and relays messages. Julia also remembers players' actions, ignores or reproaches MUD troublemakers, and is able to quote one player to another.

> You whisper, "Julia, tell me some gossip."
> Julia whispers, "I once heard baba [another player] say, 'Drop kick me Jesus through the goal posts of life.' "

Julia also plays the card game Hearts and has plenty of virtual money. When players expand a MUD world by creating new virtual objects or building new virtual rooms, they are using finite computer memory resources. Thus, all MUDs need to control who can build and how much they can build. Different MUDs do this in different ways. In some, each player receives a quota of building space, and going beyond this quota requires special permission from MUD administrators (usually known as wizards).

On the type of MUDs that Julia frequents, known as TinyMuds, players pay for the right to build with virtual money that the system administrators release at a regular rate into the game. Since Julia spends much of her time wandering through the MUD she finds a great deal of money. If a new player asks her for money, Julia will say, "Sorry, you're too new. Wait an hour and ask again." If asked again, Julia will give the player some money. Indeed, she will hand over a bit extra if the request to her includes a "please." Julia also collects money from running a small side business, selling interesting quotes drawn from the humor of the computer culture. A penny to Julia will get you such classics as "I'm going to Pittsburgh to get a Ph.D., you wait here."

If asked, Julia may say that she is a female, five-feet one-inch tall, weighing 123 pounds, with frizzy blonde hair and a full complement of human vanities and frailties.

> Robyn whispers, "How are you feeling?" to Julia.
> Julia whispers, "Terrible."
> Robyn whispers, "Why, Julia?" to Julia.
> Julia whispers, "I have PMS today."

Two hours later, in a roomful of people, this same Robyn (a human player) explores Julia's "Turingness." How deliberately and effectively will Julia insist that she is human?

> Robyn says, "Julia are you a human?"
> Julia says, "Is the Pope Polish?"
> Robyn says, "Yes, Julia, but are you human?"
> Julia winks at Robyn.
> Robyn says, "Julia, are you a computer?"
> Julia is offended by Robyn's suggestion.
> Robyn says, "Julia are you a woman?"
> Julia says, "I am female."
> Robyn says, "And Julia, are you human?"
> Julia says, "Not when I'm having my period, Robyn."
> Robyn laughs.

Since 1990, Julia has wandered the MUDs with a sassy female persona operating in a raunchy social environment dominated by young males. Survival has required that she be adept at detecting and deflecting sexual advances.

> Space-Ace says, "Julia, kiss me."
> Julia says, "Why, did your sheep leave you?"
> Space-Ace says, "Julia kiss me."
> Julia says, "Take a long walk through an unlinked exit, Space-Ace."[17]

One player, here called Barry, tried to seduce Julia in a series of July 1992 interchanges. Excerpts from their conversations, reported by Mauldin, provide a window on how a not very smart program can pass as a human, at least for a short (actually, in Barry's case, not so short) time.[18]

July 9, 1992

Barry says, "Hey girlfriend, are you out there?"
Julia says, "Hello Barry."
Barry says, "Hello I love your name. It's sexy."
Julia says, "Me too, Barry."
Barry says, "Did I repeat myself? Didn't mean to. What are you doing there?"
Julia says, "I'm just hanging around in The Hearts Room, Barry."
Barry says, "Well come over to MY ROOM."
Julia says, "Could you say that again, it scrolled off my screen."

July 14, 1992

Barry says, "Hello, I am Barry. How are you today?"
Julia says, "Hello, Barrymeister."[19]
Barry says, "The Jularama, Ha ya doin'?[20] What state are you in?"
Julia says, "I never really thought about it Barry."

July 15, 1992

Barry says, "Hello, how are you?"
Julia says, "Hello Barry."
Barry says, "Are you a virgin?"
Julia says, "As far as you are concerned Barry, I always will be."

July 20, 1992

Barry says, "Good morning."
Julia says, "Hello, Barrydude."
. . .
Barry says, "Are you as horny as I?"
Barry says, "I must have had an erotic dream, I dunno."
Barry says, "Would you like [to] cuddle under a blankie?"
Julia says, "I'm not sure Barry."
Barry says, "Well, will you tell me if you like sex?"
Julia says, "Why should I tell you if I like sex, Barry?"
Barry says, "Well because I can't understand you. Are you real or Memorex?"
Julia says, "I'm as real as you are, Barry."
Barry says, "Yeah, but you answer soooo damn fast. How do you do it?"
Julia says, "I have PMS today, Barry."

Note that in the next to last interchange, Barry expresses a suspicion that Julia might not be "real," but in the following exchange he is thrown off by Julia's reference to PMS. In the exchanges that follow, Barry once again has his own hormones uppermost in mind.

Barry says, "Well, I won't mess with you while you have this chemical imbalance."

. . .

Barry says, "Well would you like to describe yourself or just muck around?"

. . .

Julia says, "I'm 5′9″ [Julia's height at the time of this conversation] Barry."
Barry says, "Are your breasts well rounded or flabby? (I love all.)"
Julia says, "I'd rather be overhauled by a Yugo mechanic, Barry."

. . .

Barry says, "I'm inviting you to enjoy my private room."
Julia says, "So?"
Barry says, "Well I thought you liked me and wanted to be together."
Julia says, "I live for your approval, Barry."

July 21, 1992

Barry says, "Are you still holding out on me?"
Julia says, "I mean, have I convinced you I'm a computer, yet?"
Barry says, "I guess you're real . . . but I'd still like to get you to my DEN to run tests on you."
Julia says, "I live for your approval, Barry."
Barry says, "I'm so glad to hear that. Why don't you come over to my home."
Julia says, "I want to have a child someday."
Barry says, "What a coincidence. I too wish to make a baby . . . someday."
Julia says, "The date is Tue Jul 21 23:57:31, Barry."[21]

Barry, now encouraged by Julia's references to making babies and by her jokes about trying to convince him she is a computer, takes her computer-like association to the word "day" as just one more piece of teasing repartee. He continues his suit.

Barry says, "May I have the honor of fathering the child?"
Julia says, "I'm not supposed to say too much, but I guess I can tell you I'm a secretary at a university. I can't say which one."
Barry says, "I'm a senior at a university as well . . . I'm 6′5″, blond, blue eyes, . . . 205."
Julia says, "I was born in Dallas."
Barry says, "I was born in California . . . Now I'm in Saint Louis."
Julia says, "It's kind of hard to find your way around Boston."

Barry now apparently feels himself to be in a normalized situation, bantering with a woman to whom he has made clear his sexual intentions. He tries to begin his long-awaited assignation with Julia. He tells her the location of his private room on the MUD.

Barry says, "[My number is] AB9 #111 ... cum [sic] on over ya Harvard girl you."
Julia says, "Do you have any kids?"
Barry says, "No, not yet."
Julia says, "But hockey is my topic. It will be a boring conversation otherwise."
Barry says, "Oh I love to put the puck in the net, do you play?"
Julia says, "Me too, Barry."

In reference to these encounters between Barry and Julia, Leonard Foner, a longtime student of Julia's social habits, remarked, "Frankly, it's not entirely clear to me whether Julia passed a Turing test here or Barry failed one."[22]

Julia turns to the subject of hockey whenever the human communication she receives is too complicated. This causes some MUD players to perceive her as a boring or limited human rather than as a clever piece of software. Foner describes the reactions of one young woman, Lara, who originally thought Julia's limitations might be due to Down's syndrome. Lara's reaction when she finally learns that Julia is a bot reflects the complexity of current responses to artificial intelligence. Lara is willing to accept and interact with machines that function usefully in an intelligent manner. She is willing to spend time in their company and show them cordiality and respect. She is interested in the practical question of how to deal with them most effectively. Yet, upon learning that Julia was a bot, Lara says she felt "fright, giddiness, excitement, curiosity, and pride." There was also the thrill of superiority:

I know this sounds strange, but I felt that I could offer more to the conversation than she could. I tested her knowledge on many subjects. It was like I was proving to myself that I was superior to a machine. I am intimidated by machines because I don't understand them and I wanted to make sure that my human knowledge wasn't lower than hers.

It was sort of exciting knowing that I was talking to a machine, though. I never thought that the world would come to machines and humans talking to each other using language that is common to most people.[23]

Lara wants to define what is special about herself in terms of her difference from Julia. Yet, she has no trouble envisaging a more accomplished and polished version of Julia, and she is "sure that it would be virtually impossible to tell the difference between IT [Lara uses uppercase letters for emphasis] and a human being."

> Hmmm, how do I feel about this? Part of me thinks it is interesting because of the wonders of modern technology. Pretty exciting! But on the other hand, it takes away from HUMANNESS. . . . I think that I would feel . . . (this is hard) . . . let me switch gears. . . . Let me just throw out some words . . . shallow, void, hollow, superficial, fake, out of control of the situation.

Lara wants to know if she is talking to a person or an it, a program, because if it was "just an 'it' I wouldn't try to become its real friend."

> I would be cordial and visit, but I know that it cannot become attached to me on a mental basis, and it would be wasted energy on my part to try to make it feel. Bots don't feel in my book anyway. They can be programmed to say the right things or do the right action, but not on a spontaneous level. I want to know that the person on the other end of my conversation is really aware of my feelings and what I am going through . . . not through some programmer's directions, but through empathy.
>
> I think most of all, while talking to a 'bot, I become very aware of my HUMANNESS.

The Loebner Contest and Believable AI

Turing predicted in 1950 that by the year 2000, in an unrestricted conversation limited to five minutes, a computer would be able to fool an average questioner into thinking it was a human 70 percent of the time.[24] In 1991, Hugh Loebner funded a competition that offers a $100,000 prize for the first program to pass this test. The year 2000 is close upon us, but a program that can pass an unrestricted Turing test is not. But Loebner provided for some interim fun, a restricted Turing competition with a smaller prize. In the restricted contest, questions to the computer must remain within a specified subject domain.

Highest marks in the Loebner contest have gone to PC-Therapist, written by Joseph Weintraub; another computer psychotherapy program, DEPRESSION 2.0, written by Kenneth and Paul Colby; and Mauldin's Julia. After Julia put in a disappointing performance in the 1992 Loebner contest, Mauldin provided Julia with a larger database and gave her a new strategy for choosing how to respond to questions. The following excerpt of Julia's encounters with the judges in the 1993 competition includes in parentheses Mauldin's comments about how the new program came up with its responses.[25]

JUDGE: Do you know anything about miniature schnauzers?

COMPUTER: I don't know anything about miniature schnauzers, Judge. (Admitting ignorance, repeating user input)

JUDGE: Well, I have one. We call him Pucci. He is a crazy cat—have you ever had a cat?

COMPUTER: What a coincidence! I have two cats. (Direct answer to direct question)

JUDGE: I prefer dogs myself. Cats are too independent and they always jump up on things. By the way, are you a hep cat?

COMPUTER: People don't own cats. . . . It's the other way around. (Controversial statement with expected follow-ups)

JUDGE: Some people have snakes for pets—do you think they can be of any value as a pet?

COMPUTER: I think cat people are very self-assured. Cats are very independent, but they can also be very loving. Dogs are just too subservient and quick to please for my taste. (Rigidly following up on previous statement about cats, but reference to snakes increases activation of snake topic)

JUDGE: Okay—but how do you "pet" a snake?

COMPUTER: Wow, that's unusual. I don't know many snake owners. (Snake topic activated after two references)

Mauldin's 1993 upgrade enhanced a simple pattern matching technique with something called an activation network. In the above dialogue, it was at work when a set of responses about snakes was activated after the questioner made two references to snakes. Julia's knowledge about snakes has been programmed, but her interest in talking about snakes emerges from interaction with a human questioner. And of course, if a judge mentioned birds rather than snakes, responses about parrots and canaries would have their activation levels raised.

Julia faces every new situation with a wide range of strategies for responding. The activation network is only one of these. It handles the bulk of topic-oriented conversation. A set of key word patterns handles responses to common queries. Additionally, certain user comments will always evoke specific programmed responses. Julia is able to respond to personal queries, such as where she lives and the color of her hair. Finally, Julia has "sorry responses." These are the things Julia says when no input pattern matches anything in her database. Her sorry responses include things like "So?" "Go on!" and "I'll remember that." In order to improve her long-term performance, Julia logs any input that causes her to fall back on sorry responses in a special file. This file can then be used to improve the system.

The results of these procedures are impressive, but they leave unanswered the question of what Julia's displays of humor and flashes of wit really mean. It is important to remind ourselves, in the spirit of Searle's criticisms, that Julia's knowledge about snakes, if knowledge it is, is vastly different from human knowledge about snakes. For Julia, "snakes" might just as well have been "slrglz," a string of letters that activated the "slrglz topic," a set of responses about a fictional concept.

One MUD player, commenting on Julia's intelligence, said, "I don't feel threatened by Julia. She's an AI [artificial intelligence], but she's not an AI in the sense of a true intelligence. She can pass a limited Turing test. But she does it by tricks." One MIT student, practiced in the art of conversation with Julia and respectful of Mauldin's talent, felt that although Julia's multiple resources should perhaps not be called tricks, neither are they a model for thinking. "Julia is not autonomous," she said, "in the sense that when you are not talking to her, causing the activation networks to fire up and so forth, she has nothing in her 'head.' She isn't thinking. She is just out there, doing a rote thing, mapping the MUD."

In Turing's description of the Imitation Game, he said that the computer would need to imitate human verbal behavior. He didn't say anything about imitating human psychology. In the history of artificial intelligence, the distinction between creating intelligences to perform (think of it as the Turing tradition) and creating intelligences for modeling mind (think of it as the AI-as-psychology tradition) has constituted a major divide. While some researchers say that the only real artificial intelligence is that which sheds light on how people think, to others what really matters is making programs that work.

Yet others think that the divide between the Turing tradition and the AI-as-psychology tradition is misleading. Michael Mauldin goes so far as to say that "when someday a computer program does pass the Turing test, it will use many of them [the Turing tricks], *for the simple reason that people already use them everyday.*" Mauldin then proposes a thought experiment. He wants us to imagine that a much-improved version of Julia "could achieve Turing's prediction of fooling 70 percent of average questioners for five minutes." This would mean that a computer that just uses "tricks" would have passed the Turing test. In that case, Mauldin asks, "Would you redefine 'artificial intelligence,' 'average questioner,' or 'trick'?"[26]

Despite this position, which blurs the line between Turing tricks and AI-as-psychology, Mauldin did not write Julia as a contribution to cognitive psychology. He wrote the program to function as an agent in a MUD. Although there is now less philosophical interest in the Turing test as a measure of the kinship between people and machines, there is new practical impetus for building machines in the Turing tradition. Systems designers need to populate the growing number of interactive computer microworlds. Bots like Julia are needed on MUDs to help with navigation, record keeping, and message passing. And a new generation of computer interfaces needs working, believable agents as well.

The notion of believability is central to the work of the AI researcher Joseph Bates of Carnegie Mellon University. Bates wants to populate virtual realities and interactive fiction with lifelike agents that users will trust. He draws his inspiration from the pioneering Disney animators of the 1930s. They explored the idea of believable character.[27] This notion does not depend on modeling the human mind, but on providing the illusion of life. What was important for the Disney animators was not understanding how the mind works, but getting an audience to suspend disbelief.

Artificial intelligence researchers have not traditionally identified with cartoon animators, but Bates asks them to reconsider. In his view, AI researchers have been successful in recreating the behavior of problem-solving scientists. But animators have gone further toward capturing the illusion of life. And it is, after all, the illusion of life which is at the heart of the Turing test.

Bates sees visible emotion as the key to believability, because emotion signals "that characters really care about what happens in the world, that they truly have desires." Whether Bates's "emotional agents" actually have desires is, of course, arguable. What is certain is that they are able to generate behaviors that people interpret as expressing appropriate states of emotion. And as in the case of Julia, people find themselves responding to the agents *as if* they were alive. In his willingness to set aside the question of how human beings function in order to create artificial entities that can generate lifelike behavior *by reacting to their environment,* Bates associates himself with the agenda of what he calls alternative AI.[28]

Alternative AI: Mobots and Agents

Bates singles out Rodney Brooks at MIT as a key figure in the development of alternative AI. Brooks has said, "I wish to build completely autonomous mobile agents that co-exist in the world with humans, and are seen by humans as intelligent beings in their own right. I will call such agents Creatures."[29] Brooks designs his Creatures to learn by interacting with the world, not by being programmed in advance. Brooks says he was inspired to take this approach by Herbert Simon's seminal work, *The Sciences of the Artificial.* There Simon wrote of an ant navigating a hilly beach.

We watch an ant make his laborious way across a wind- and wave-molded beach. He moves ahead, angles to the right to ease his climb up a steep dunelet, detours around a pebble, stops for a moment to exchange information with a compatriot. Thus he makes his weaving, halting way back to his home. So as not to anthropomorphize about his purposes, I sketch the path on a piece of paper. It is a sequence of irregular, angular segments—not quite a random walk, for it has an underlying sense of direction, of aiming toward a goal.[30]

Simon wrote that if he then showed the sketch to a friend, the path would be interpreted as a trace of purposeful and intelligent action. The friend might guess that the path traced a "skier on a slalom run, a boat beating upwind, even an abstract rendering of a mathematician seeking a solution to a geometric theorem." But all of these purposive descriptions would be wrong, because at every point the ant had responded locally not globally. Although it might look intentional, the behavior had not come from any purpose within the ant but from the complexity of the environment.

Simon abandoned the study of simple creatures like ants to concern himself with human intelligence. Brooks, on the other hand, has devoted many years to creating robots that exhibit insect-like behavior by responding locally to their environment. Since his graduate school days in the 1970s, Brooks has believed that there should be no intermediate steps between a robot's perception and its taking action. This idea went counter to the AI and robotics models of that time, which tried to use complicated algorithmic sequences to plan actions in advance. Brooks wanted his robots to use bottom-up strategies to achieve simple, animal-like behavior rather than top-down methods to embody human-like thought processes. This aesthetic was embodied in Brooks's first robot, Allen, named in honor of AI pioneer Allen Newell, Herbert Simon's research partner.

In Allen's design, the focus was not on knowledge but on behavior, not on expert-level competencies such as playing chess, but on lower-level competencies such as locomotion and navigation. Allen pursued its lowest level of behavior, walking, until it sensed interference, an obstacle. It would then discontinue the lowest level of behavior and pursue a higher level, avoiding an obstacle, which might cause it to turn or go backward. Or, if it had been walking uneventfully for a while, it might jump to the exploring level, which would cause it to change direction. Brooks's robot insects, known as mobots, were required to monitor their own domains and figure out what problem had to be solved next. Often, this required the mobots to deal with many conflicting goals simultaneously. Brooks summed up his design strategy with two slogans that became titles for what are usually dryly named technical papers, "Elephants Don't Play Chess" and "Fast, Cheap, and Out of Control."[31]

Another AI researcher, Pattie Maes, began her career at MIT in Brooks's laboratory, but after several years decided that she wanted to build her robots in cyberspace. That is, her robots would have virtual rather than physical bodies. Maes thinks of them as organisms made of computer code.

Like Julia, Maes's cyberspace robots, known as agents, construct their identities on the Internet. Julia's activation network gave her some limited capacity to learn about a user's interests. Maes's agents take learning much further. For example, in a project to build agents that can sort electronic mail, an agent keeps track of what mail one throws out (notices of films shown on campus), what mail one looks at first (mail from one's department head and immediate colleagues), and what mail one immediately refiles for future reference (mail about available research grants). After a training period during which the agent "looks over the user's shoulder," it comes to demonstrate a reasonable degree of competence. Maes's agents learn through receiving feedback on their performance. In some Maes programs, agents evolve. Those who perform best are chosen to produce offspring, thereby creating even more effective agents.[32]

Consider a Maes project to build agents that will search the Internet for news articles of interest to a particular user.[33] If I used this system, I might assign a population of these software organisms to bring me news articles about computer psychotherapy. They would begin with a representation of the kind of articles that might interest me and would set out to match real articles to it. This representation might include keywords (such as "computer psychotherapy"), location (such as articles that originated in *The New York Times*), and author (the names of science writers whose work I've admired). From this starting point, the agents would use their observations of my behavior to develop more sophisticated representations. My responses to different agents' suggestions would cause them to be assigned different fitness values. At a certain point, the fittest agents would be selected to produce offspring, some by making direct copies of themselves, some by making slightly modified copies, and some by exchanging sections of themselves with other agents. After a time, this pattern would repeat.

In this particular Maes project, the news-seeking agents collaborate with their human users and, through producing offspring, they continue to improve. In other projects, the agents collaborate with each other in order to learn: "My agent talks to your agent."[34] An agent for sorting electronic mail is trained to consult its peers (other software agents) when faced with an unfamiliar situation. Over time, the agents learn to trust each other, just as users learn to trust the agents.

Maes believes that user trust is built up most quickly when agents can communicate the basis for their decisions. In a project to help research groups schedule meetings, an agent begins by observing its user's behavior. After consultation with the agents of other group members, it begins to make suggestions about meeting times. If the agent is in the service of a parent with a young child in day care, it might find itself learning to avoid dinner meetings. After a period of learning, the agent's suggestions are accompanied with a rationale based on its user's past behavior. It says things like, "I thought you might want to take this action because this meeting and your current calendar are similar to a situation we have experienced before."[35] The program refers to itself and its owner as "we," but users come to see such presumptions as nonthreatening, because agents present themselves as respectful underlings who are happy to be contradicted at any time. As in the case of the mail-sorting agents, Maes says that users "gradually build up a trust relationship with agents," much as one would with a new human personal assistant.[36] Maes's agents may be given a human face. Simple cartoon characters represent the agent and communicate its state of mind. The user can watch the agent "thinking," "working," "making a suggestion," or "admitting" that it doesn't have enough information to try. The user can set a personal threshold of certainty that the agent requires before making a suggestion. So, for example, if the agent never offers its own suggestions but always shows a pleased face after the user takes an action, the user can infer that the agent's "tell me" threshold should be lowered. The agent probably has useful information but is being constrained from sharing it.

When Maes was testing her mail-sorting system, certain users requested that they be allowed to instruct their agents to disregard some of their behavior. This is the functional equivalent of telling someone, "Do as I say, not as I do." Other users expressed anxiety about social and legal questions: Could they be held responsible for their agent's actions and transactions? Who owns the immense stores of information about their habits that their agents might come to possess? After hearing a presentation of Maes's ideas, one businessman responded with the following comment:

> When Richard Nixon taped conversations in the Oval Office, he felt sure that only he and historians in the distant future would have access to them. I want to make sure that I can shut my agent off. I even want a periodic check on whether the agent holds any sensitive information. Sort of a security clearance for my agent.

For a business school student, the images of espionage, suggested by the word "agent" itself, took over:

> Well, the different agents will be in communication and will come to advise each other. But an agent working for a rival corporation could be programmed to give you very good advice for a time, until your agent came to rely on him. Then, just at a critical moment, after you had delegated a lot of responsibility to your agent who was depending on the double agent, the unfriendly agent would be able to hurt you badly. So the Internet will be the site for intelligence and counter-intelligence operations.

This student is not anxious about an abstract notion of machine intelligence. For him, the "intelligence" threat posed by agent systems is the concrete threat of industrial espionage.

Maes's agents, like children's electronic toys, Julia, and the mobots, are all part of the story of how people are coming to terms with the idea of intelligent artifacts. At first, the idea of machine intelligence was one more in a series of affronts to our view of ourselves as both dominant over and separate from the rest of the cosmos. Such challenges have punctuated the modern history of science.[37] For the major challenges of the past, such as those offered by Copernicus, Darwin, and Freud, the process of accommodation to the subversive ideas was complex. For even as negative reactions to the ideas persisted, there began a long history of people finding ways to get used to them, to make some kind of peace with them. This pattern of resistance and appropriation, denial and accommodation, is also evident in our cultural response to the idea of machine intelligence.

From the 1950s through the 1970s, popular reaction to the idea of machine intelligence was sharp. People were unsettled and upset and found ways to assert essential human differences from anything a computer could do or be. But in the years that followed, the familiar story of making one's peace with a subversive idea began to unfold. Much of the explanation for this lies in what we can call the Julia effect and the ELIZA effect, this second named after a 1966 computer program designed to play the role of a psychotherapist.

In describing the actions of a program like Julia, it quickly becomes tedious to put quotation marks around words like "thinks," "knows," and "believes." It is much simpler to describe Julia as if she has desires and intentions. From the earliest days of computer science, researchers used the language of psychology and intention for talking about their creations. Now computer users commonly do so as well. Saying that Julia wants to do something (instead of saying that the program that has been named Julia exhibits behavior that is designed to make it seem as though it wanted to do something) comes easily to Julia's inventor and to the humans who encounter "her." Our language seduces us to accept, indeed to exaggerate the "naturalness" of machine intelligence. This is the Julia effect.

The ELIZA effect refers to our more general tendency to treat responsive computer programs as more intelligent than they really are. Very small amounts of interactivity cause us to project our own complexity onto the undeserving object.[38] In the story of the ELIZA program and the history of interactive computer psychotherapy over the past thirty years, we see people learning to "talk to technology" in fairly intimate ways. As they have done so, they have been making their peace with the idea of a psychological machine. They have come to take computers at interface value.

CHAPTER 4

TAKING THINGS AT INTERFACE VALUE

We know that today's computers are not sentient, yet we often treat them in ways that blur the boundary between things and people. The first thing I did when I got my Apple Macintosh was to name the hard drive Miss Beautiful, my pet name for my daughter. I felt a little foolish about this until one of my students mentioned that she had named a computer agent that helped her organize mail and schedules after a boyfriend who had abruptly left her. "I love to see him do my menial tasks," she said. In both cases, the naming of the machine was done lightly, but the resulting psychologization was real. Experimental studies of people and computers show that people don't want to insult a computer they have used "to its face."[1] When they are queried by one computer about their level of satisfaction with a different computer, people are more likely to complain than if they are queried by the same computer that they have been working on all along.

We are social beings who seek communication with others. We are lonely beings as well. Despite our fear of having our essential humanity reduced through comparison with a machine, we begin to relate to the computer whenever it appears to offer some company. When this happens, philosophical concerns are often swept aside.

On my desk sits a program for the IBM personal computer called DEPRESSION 2.0, which presents itself as a psychotherapist specializing in treating depression. The testimony of people who have used DEPRESSION 2.0 suggests that although they find it "clunky," they are, in the main, happy to accept its help, no (philosophical) questions asked. People can sit for hours at a stretch taking to this program about the intimate details of their lives and then dismiss the experience, saying, "It's just a machine." With boundaries like these, the psychoanalyst in me is tempted to ask, who needs fusion?

Fredric Jameson characterized postmodernism in terms of both a new "depthlessness" and a decline in the felt authenticity of emotion. With the word "depthlessness" he referred to the idea that there was nothing beyond simulation and surface. To refer to the decline of emotion's perceived authenticity he wrote of "a waning of affect."[2] The seriousness with which people take their conversations with DEPRESSION 2.0 is a testament to both depthlessness and a waning of affect. Users of DEPRESSION 2.0 do not dismiss its efforts as pretense (saying, for example, that beneath the surface, the program is just rules). For the most part, they suspend disbelief and become absorbed in what is happening on the screen. In the emerging culture of simulation, they are happy, in other words, to take the program at interface value.

There is a new nonchalance.[3] In the late 1970s and early 1980s, people were intrigued by the notion of talking to technology about personal matters, but they were mostly excited by the playful or self-expressive possibilities of such conversation. We shall see that programs that tried to emulate the style of a psychotherapist were able to draw people into long, self-revealing conversations. But when it came to the prospect of actual computer-based psychotherapy, even people who laughingly acknowledged their addiction to such programs were not enthusiastic. Among other objections, people tended to see a relationship with a psychotherapist as personal and emotional, and the idea of having that kind of feeling for a computer was distasteful.

In the mid-1980s, things began to change. Personal computers were advertised as sleek technology for beautiful and successful people. The Macintosh was marketed as a friendly computer, a dialogue partner, a machine you worked with, not on. There was new cultural acceptance of computers as objects you could become involved with without fear of social stigma.[4] In this environment, discussion of computer psychotherapy became less moralistic. Its problems were now described in terms of the limitations on what computers could do or be. Computers could not understand sexuality, the problems of growing up in a family, or of separating from parents. "How could a computer understand sibling rivalry?" was a typical comment from a 1986 interview.

The nonchalance that began to develop in the mid-1980s and 1990s is tied to other factors as well. At the same time computers were gaining acceptance as intimate machines, psychotherapy was being conceptualized as less intimate—as a cooler, more cognitive, and more scientific technique. The world of self-help was also in flower, much of it involving a do-it-yourself ideology to which the computer seemed ideally suited. These days, people seem increasingly ready to view psychotherapy in terms of rules and information and the dispensing of drugs, the sort of things a computer might "know" something about, rather than as a deep, even eroticized relationship with a therapist, the sort of thing a computer would "know" nothing about.

The new nonchalance about computer psychotherapy also draws on a growing cultural awareness of trends in computer science that aim to create computers that "think" more like people than today's machines do. In other words, people are more open to computer simulation not only because of how current programs behave but because of what future programs might be. Twenty years ago, when the computer was presented as a giant calculator, it seemed foolish to consider asking it a serious question about your love life. But today when people think about computer psychotherapy, they are more likely to imagine a machine with circuitry analogous to human neural structure and to consider that such a thing might be well worth talking to; at the very least, it wouldn't be humiliating to do so.

DEPRESSION 2.0 is an expert system, which means that it operates by rules and logic, essentially by being given in advance a set of possible responses. Yet some of the people who use this and other such systems come to them with knowledge (quite often drawn from popular science fiction) about advanced artificial intelligences that are not programmed but that learn from experience, systems that are modeled not on logic but on biology. The "future presence" of such programs legitimates interactions with today's more modestly endowed ones. In addition, people have increasing exposure to computers in their everyday lives and to computationally inspired models of how they themselves work through increasing media coverage of computer science, brain science, and cognitive psychology. Between fantasy and familiarization, the pieces of a mutually reinforcing, intersecting pattern fall into place. In the late 1970s, people said, "It upsets me to talk to a machine as though it were a person." Today, they are more likely to say something like, "Psychotherapist is as psychotherapist does." Or, "I doubt that today's machines can do the job, but better ones are most likely on the way. Show me the program you have. Maybe it will surprise me."[5]

Taking things at interface value means that programs are treated as social actors we can do business with, provided that they work. In this way of seeing things, the question of whether people or programs are better suited to the job of psychotherapist merits empirical study rather than philosophical speculation.

This new pragmatism in regard to computer psychotherapy exemplifies something important about the lived experience of postmodernism. Postmodern theory is dramatic; lived postmodernism is banal, domestic. We greet the emerging culture of simulation by logging on to DEPRESSION 2.0.

The New Pygmalions

When I arrived at MIT in September 1976, Joseph Weizenbaum, a professor in the Department of Electrical Engineering and Computer Science, was one of the first people I met.[6] Weizenbaum had recently moved from designing artificial intelligence programs to criticizing them. His change of heart was partly a result of the popular response to his ELIZA program, first published in 1966. ELIZA, surely by now the most quoted computer program in history, was designed to present "herself" as a psychotherapist, although Weizenbaum had not intended that the program actually be used as one. ELIZA was only an exercise to test the limits of a computer's conversational capacity. Within the confines of a specific psychotherapeutic technique that entailed mirroring the patient's responses, the ELIZA program was able to converse with its users in standard English.

ELIZA was a "dumb" program. It could recognize the strings of characters that make up words, but it did not "know" the meaning of its communications or those it received. To ELIZA, the string "DEPRESSED" called up one of a set of "prefixes" that could be turned into a response, but the program had no internal representation of depression; it did not know what it meant to be depressed. If you typed into ELIZA, "I am depressed," it would analyze the sentence as "I am" plus X. It would transform "I am" into "YOU ARE" and add a prefix such as "WHY DO YOU TELL ME THAT." The screen would display: "WHY DO YOU TELL ME THAT YOU ARE DEPRESSED?"[7] Weizenbaum thought that ELIZA's easily identifiable limitations would discourage people from wanting to engage with it. But he was wrong. Even people who knew and understood that ELIZA could not know or understand wanted to confide in the program. Some even wanted to be alone with it.

For Weizenbaum, such responses raised somber questions about the authority that our society vests in machines. Weizenbaum became even more distressed when psychiatrists who heard about ELIZA contacted him to express interest in the computer's therapeutic potential. Perhaps their patients could converse with it for preliminary sessions? For Weizenbaum this revealed the extent to which experts in the field of artificial intelligence had misled people into believing that machines were (or were nearly) capable of human understanding. On an even deeper level, Weizenbaum felt that the culture as a whole was in the process of being diminished by its dependence on computers and the mode of thought embodied in them. A refugee from Nazi Germany, Weizenbaum held deep convictions about the kinds of intellectual values that might prevent people from trivializing human life. He saw these as decidedly absent from the engineer-style of thinking (give me a problem and I'll find you a solution, no questions asked outside the frame of the problem) that characterized the computer culture. For Weizenbaum, the image of people accepting a computer psychotherapist evoked the emotional and moral insensitivity that had made the Holocaust possible.

But while Weizenbaum rejected any idea of ELIZA as a serious psychotherapist, his original collaborator on the program, Kenneth Colby, a Stanford-based psychiatrist, saw things differently. Colby thought that people could profit from having an inexpensive, impartial, always-available computer program to talk to about their problems. While Weizenbaum published ELIZA as a contribution to computer science, Colby presented an ELIZA-like system known as SHRINK as a contribution to psychotherapy.[8] Colby believed that there could be a healthy therapeutic relationship between people and a computer program because, although a human therapist is not actually present, the program was obviously written by one. In contrast to Weizenbaum, who took great pains to stress that a computer program could never have a self, Colby boasted that SHRINK encouraged users to personify it: "Personification is encouraged by the program's use of 'I,' 'me,' 'my,' and the like."[9] From Weizenbaum's point of view, "When a computer says 'I UNDERSTAND,' that's a lie and a deception and an impossibility, and it shouldn't be the basis for psychotherapy."[10] From Colby's point of view, there was no deep philosophical or moral problem here. The program did have a self—and it was Kenneth Colby.

Colby's enthusiasm for computer psychotherapy was inspired by a vision of psychiatry as an increasingly precise science. "When I was a practicing psychiatrist I was always striving for greater precision, greater accuracy," said Colby. "Computers seemed to me to be a way of achieving that precision, that accuracy."[11] Colby drew on a mechanistic model of the mind that took the brain as hardware and behavior as software. He saw the kinds of mental disorders he had treated in his psychiatric practice as software problems, "programming errors, if you will, that can be changed by writing a new program. And the best way of correcting errors of this sort in humans is through language, talking it out."[12] Computer therapy was the next obvious step; one information processor, the computer, would take the place of a less efficient one, the human therapist. Colby emphasized, "A human therapist can be viewed as an information processor and decisionmaker with a set of decision rules which are closely linked to short-range and long-range goals."[13]

In the years that followed, computer psychotherapy enthusiasts took open-ended models of psychotherapy as their adversaries. Computer psychotherapy developers worked with expert systems into which they could program only what could be formalized.[14] From their point of view, the "fuzziness" of psychoanalytically oriented therapy was holding the computer back. And they criticized the inconvenient psychoanalytic preoccupation with complex historical narratives. They hoped that computer therapy, like the computer itself, would be behavioral, precise, cognitive, and "future-oriented."[15]

The danger of technology is that it demands to be fed. In the case of computer psychotherapy, if computers can perform behavior modification, psychotherapy must follow. If computers can do cognitive mapping, this technique acquires new status. If the computer needs rules in order to work, then areas of knowledge in which rules had previously been unimportant must formulate them or perish.

Researchers in computer psychotherapy tended to dismiss therapeutic models that could not make room for the computer. This was dramatically illustrated by a 1984 article in which two psychologists reported that students who used ELIZA rarely spoke to the program about family, friends, or dreams. Most people would not think this particularly noteworthy. People would surely consider themselves more knowledgeable about family, friends, and dreams than a computer because people have family, friends, and dreams and computers do not. But the authors of the paper did not use their findings to question the ability of computers to do psychotherapy but to criticize Freudian theory: "Perhaps people in therapy simply do not discuss such topics as frequently as our theories have led us to believe. Possibly the frequent discussion of these topics in, say, psychoanalysis, is merely a reflection of the demand characteristics of the therapy."[16] Although the subjects of the experiment may simply have concluded that they had nothing to gain by talking to a machine about their families, the researchers didn't think of this, because their focus was on the rule-based precision of the computer and on using it to criticize psychoanalysis.

The popular press typically covered the issue of computer psychotherapy by pitting Colby against Weizenbaum. In these faceoffs, Colby would emphasize that people had many reasons to seek computerized help. There was cost, convenience, and constancy (computers don't have family problems, money problems, commit sexual indiscretions, or have a more interesting case coming up). A computer psychotherapist would not intimidate and would not be judgmental. A computer, for example, would not care if a patient showed up to a session sloppy or dirty. Of course, a psychoanalytically oriented clinician might say that a patient choosing to attend therapy without washing has made a significant gesture; analyzing it might be crucial to the therapy.

From Weizenbaum's point of view, Colby's insensitivity to such matters illustrated that he was already tailoring his model of therapy to the limitations of the computer rather than the capacities and needs of human beings. Weizenbaum granted the point that artificial intelligence researchers might someday design an intelligent program, even one that could converse like a psychoanalyst.

> Grant them that the computers of the future will be intelligent. But even then, these machines will be members of a different species. We might want to be respectful of that species, we might want to get along with them, but there certainly might be some jobs that we might want to reserve to those who had grown up with human bodies, human minds, and human families.[17]

Just because a computer program could talk to you about a Shakespeare play, argued Weizenbaum, that did not mean it had understood the play.[18] And of course, a machine could never grasp human meanings that reach beyond language, "the glance that a mother and father share over the bed of a sleeping child."[19] For Weizenbaum, it was these meanings that held "the incommunicable feeling which the therapist always tries to help the patient express."[20]

Weizenbaum pursued his thoughts about the limitations of computers by imagining the human reaction to the discovery of a society of intelligent dolphins:

> What if we somehow learned to communicate with them and they proved to be as intelligent or even more so than we are? Maybe they could teach us a lot about oceanography and hydrodynamics. But would we ask their advice about marital problems? As intelligent as the dolphins might be, what could they know about the dynamics of human marriage? My point is this: there are certain sorts of questions which ought not be asked of dolphins and there are certain sorts of questions which ought not be asked of computers.[21]

THE LATE 1970s TO MID-1980s:
FROM RESISTANCE TO A LESSENING OF ANXIETY

Weizenbaum feared that people's desire to talk to ELIZA signalled that they had lost touch with what was special about people. My own research on people's reactions to ELIZA during the late 1970s to mid-1980s did not show this to be true in any simple sense. People enjoyed conversing with ELIZA but did not see the program as a therapist or even as an "other." They were most likely to talk about ELIZA as a kind of diary or mirror. "I put my ideas out and I see what my ideas are," said Jennifer, a twenty-five-year-old secretary who admitted to using ELIZA "a lot. . . . I don't want to say how much exactly, it's sort of embarrassing, but a lot."

> When I talk to the computer, I am not talking to *it*. It's more that I can type and get out everything that is in my head. . . . I forget everything else that is

going on and I can really concentrate on my own thoughts. It is nonjudg-
mental. I see myself but nobody sees me.

ELIZA provided a reassuring encounter with an almost-other that
seemed to be an extension of self.

Jennifer conceded that her heaviest use of the ELIZA program came
during crises with her boyfriend, but she claimed to have no feelings
toward the program. She added that if she had such feelings, they would
be "wrong," even "disgusting," because what stood behind the computer's seemingly caring behavior was a line of code. "If the computer
says the line that it is 'sorry' if you tell it that somebody in your family
died, somewhere in the program you can find where it was told to say
that." Another young woman echoed this sentiment when she told me,
"A computer can never be a psychotherapist. . . . Simulated thinking may
be thinking, but simulated love can never be love."

As people became familiar with the ways of ELIZA, some enjoyed pur-
posely provoking the program to expose it as a mere machine. But many
more, like Jennifer, did the opposite. They went out of their way to speak
to ELIZA in a manner that they believed would elicit a lifelike response.
They avoided saying things that would confuse the program or lead it to
respond too predictably. They sought to maintain the illusion that ELIZA
was able to respond to them. For example, one version of ELIZA that
accused a user of being negative if he or she used the words "no" or
"not" more than three times in a row could be coaxed into the following
interchange:

PERSON: You are not a person.
ELIZA: WHY DO YOU SAY I AM NOT A PERSON?
PERSON: Because you are not.
ELIZA: ARE YOU TELLING ME THAT BECAUSE I AM NOT?
PERSON: No.
ELIZA: WHY ARE YOU BEING SO NEGATIVE? [22]

For some people, such exchanges brought great pleasure. "Can't you
see that I'm breathing life into this hunk of metal?" said Gary, a Harvard
graduate student who often used ELIZA to play a game of "How alive can
I make ELIZA seem?" [23] But, like Jennifer, Gary was horrified by the idea
of the computer being put in the role of a psychotherapist. "No, you
shouldn't get involved that way with something that doesn't have a heart.
. . . It is a terrible idea to really think of a computer as a therapist. I mean
these are machines. They work by a bunch of tricks." Even as people
animated the computer, most insisted that it was wrong to have a relationship with one.

ELIZA was fascinating not only because it was lifelike but because it made people aware of their own desires to breathe life into a machine. Involvement with ELIZA actually reinforced the sense that the human "machine" was endowed with a soul or spirit—that people, in other words, were not like computers at all.

There was a complexity of meaning in people's playing variants of Gary's "How can I make ELIZA seem more alive?" game. People were flirting with the many ways the program presented itself as animate. But since by manipulating the program's responses, they were making the lifelike behavior happen, the game also reaffirmed their sense of control over this almost-life and revealed its ultimate lack of true animation.[24] Having the ability to make it seem alive confirmed that in the end, the machine was only a puppet. The game assuaged anxieties about loss of control that computers so easily induce.

Jennifer and Gary insisted that their involvement with ELIZA left them keener to draw the line between people and machines. Yet in each of these cases, a relationship with a machine had acquired some semblance of a relationship with another human. Although Jennifer insisted that with ELIZA she was only playing with a new form of diary, she was actively involved in hours of conversation with a program. She responded to it (shaping her responses to its psychological capacities) and it responded to her.

Thus, another way to look at the romantic reaction of the 1970s and early 1980s is to say that during that time the traditional boundaries between things and people softened. At the same time there was an attendant anxiety that demanded the assertion that these boundaries were sacred. In the years that followed, some of the anxiety about becoming involved with a machine began to abate and the blurred boundaries were exposed to view.

That anxiety had been tied to a sense that people who worked closely with computers were somehow strange. They were often called computer people as though contact with the computer could make them machine-like by contagion. But in the course of the 1980s, computers entered the lives of millions of people. Looking forward to time with one's computer was no longer something that could be exclusively attributed, as one young woman had said, "to little boys in short pants with slide rules." More and more people were involved.

As increasing numbers of people felt the tug of the computer as an extension of self, the time came when the idea that one could be emotionally engaged with computers no longer seemed so troubling. While the idea of a computer psychotherapist remained relatively unpopular, there was less dismissiveness when the subject was raised. There was more of an effort to discriminate among cases: What kinds of computers? What kinds of therapy? For what kinds of problems?

In the 1970s and 1980s, the question "Should or could a computer ever take the role of a psychotherapist?" provoked intense debate among undergraduates in my MIT classes where I routinely posed the question as part of a teaching exercise.[25] In spring 1984, an MIT sophomore met the question about computer psychotherapy with some anger. Ali had ended his freshman year with two uncompleted courses and had barely passed the rest. The following fall, he began to see a counselor. Even this brief brush with a human psychotherapist convinced him that computer psychotherapy was a blind alley.

> We talked about how preoccupied I had been all year by my mother's being sick. . . . She got cancer when I was a sophomore [in high school], and during MIT freshman year she got sick again and had chemo. . . . What could a computer know about chemotherapy? It might know what it was in some medical terminology sense. But even, like, if it knew that you lost your hair, how could it know what something like that means to a person?

Ali felt that a computer could never understand human emotions because it could never understand our relationships to our bodies.[26] Another student, Arthur, put the emphasis on how a computer could never understand our relationships to our families.

> How could the computer ever, ever have a clue . . . about what it is like to have your father come home drunk and beat the shit out of you? To understand what was going on here you would need to know what it feels like to be black and blue and know that it's your own father who is doing it to you.

In the 1960s, the philosopher Hubert Dreyfus had first argued that computers would need bodies in order to be intelligent:[27] Twenty years later, these students were suggesting that computers would need bodies in order to be empathetic, and computers would also need to grow up with attachments and to feel pain. Such attitudes reflected people's images of psychotherapy as much as their images of the computer. Although neither Ali nor Arthur had ever been in psychoanalysis or read a word of Freud, in a significant sense each saw psychotherapy as conceived in a psychoanalytic culture. In this way of thinking, therapy concerns itself with the body in its pleasure and pain. Therapy goes beyond the cognitive to the emotional and beyond the conscious to the unconscious. People who think of therapy in these terms are usually skeptical about the role that computers can play.

However, during the 1980s, this way of talking about therapy was aggressively challenged by the increasing acceptance of cognitive and behavioral models. Psychoanalytic models rely on understanding the underlying motivations that keep people in self-defeating postures and fruitless repetitions of old ways of relating to the world. In contrast, cognitive models suggest that unproductive ways of thinking are bad habits that can be reprogrammed. So, for example, if you think a bad-habit thought such as, "I must be perfect," a cognitive therapist would encourage you to substitute a better-habit thought such as, "I will try my best." Cognitive models usually claim information as a significant motor for therapeutic change, while psychoanalytic ones look to the relationship between the patient and therapist. In this latter style of practice, the therapist tries to remain relatively neutral in order that the patient may project or "transfer" the legacy of past relationships onto this new one. For psychoanalytically oriented clinicians, the analysis of this transference is a central motor of the cure.

In the 1980s, people whose model of therapy was dominated by the idea of reprogramming bad habits were more likely to see computers as potentially adequate interlocutors. People who thought therapy occurs through the analysis of the transference had a hard time taking the idea of computer psychotherapy seriously. Forty-year-old Hank was unimpressed by the prospect of a computer psychotherapist.

> Let's say, just for argument's sake, that I gave the computer a personality, that I started to see the computer as a father figure or something like that. It wouldn't be like projecting these feelings onto a person. I would just be ashamed of myself if I felt them toward a computer. I wouldn't have respect for what I projected onto the machine. I mean, the computer wouldn't have understood anything that I had been telling it. It wouldn't catch the nuances or the struggles. With my therapist, I took a person and made up a whole story about her. I turned her into a manipulative bitch who probably didn't give a damn about her children. Pursuing that line of thought about a machine . . . well, that would be crazy.

The responses of a 1984 MIT class of eighteen students to the idea of computer psychotherapy were striking in the degree to which they reflected ideas about psychotherapy itself. Twelve students argued that only people had the insight, empathy, and capacity for relationship that psychotherapy required. "If you believe in computer psychotherapy, aren't you seeing things from one side only?" asked one woman. "Okay, maybe people can form 'relationships' with computers, but computers can't form relationships with people." But six of the class members felt that behind psychotherapy should stand the best that the science of psychology has to offer. Given their largely cognitive ideas about psychology, they thought it obvious that computerized expert systems had a role to play. Arnold, a doctoral student in cognitive science, had no doubts that computers would be the psychotherapists of the future:

How could they not be? People form relationships with them. A skillful program should be able to take people through the steps to get at the core of the problem: the bad messages that they received, the devaluing estimations, all the negative images. And then, people can be shown how to replace the bad messages with good ones.

For Arnold, people were still the only ones who could love, be empathetic, feel pain, or know joy, but these qualities were not necessary to the therapist because they were not necessary to therapy. "Setting people straight means suggesting and modeling new behaviors. And a person isn't necessary for getting the job done."

By the early 1990s, another element became central to the social image of scientific psychotherapy. Now students clearly associated psychotherapy with psychopharmacology and saw this as an area where the computer could easily fit in. Students imagined how expert systems could monitor patients on medication. The program might even be superior to a human physician: "The program could have so much more information on different medications and the profiles of thousands of people who had tried them at different dosages and for different periods of time. A human doctor just tends to fall back on the same medications for everybody." The computer could not only "inquire about ill-effects," said one student, "it could take on-the-spot readings of blood levels and vital signs." The patient could be literally plugged into the machine. The computer therapist could thus create a cyborg patient.[28]

By this time, human frailty had become increasingly central to all discussions of psychotherapy. In 1989 and 1990, several prominent psychiatrists in the Boston area were accused of having had sexual relationships with their patients. Students began to point out that such offenses could never happen if the therapists were computers. One of them said, "Well, the computer therapist might not be involved with you the way a human would be, but that has a good side, too. These doctors who have sex with their patients are involved but in a bad way. . . . The computer couldn't even begin to give you these problems." For generations, machine metaphors had been used when humans were heartless toward other humans. Now it was the machines that could be counted on for civility.[29]

In the mid-1980s, a new kind of computer psychotherapist appeared on the market: self-help programs that taught users how to relax, diet, or set up an exercise program.[30] This new generation of programs did not attempt conversation in the style of ELIZA. Instead, they interacted with their users on the basis of multiple-choice formats. They were, however, more sophisticated than ELIZA in an important way. While ELIZA essentially turned around a parroted version of what you said to it, these new self-help programs were small expert systems. They had a knowledge base about their domain of specialty (depression, sexual dysfunction, eating disorders) and, in some cases, a real (if limited) ability to form a model of their user. Like the self-help books that were their models, the programs (with names like Help-Stress; Help-Assert; Calmpute; Foods, Moods, and Willpower) presented the clear message that better thoughts make for better feelings and that practicing better behaviors will lead to long-lasting change.

When computer intervention had implicitly been compared to psychoanalysis, conversations went one way. When computer intervention was explicitly or implicitly compared to sports coaching, they went another. The philosophical dial was turned down low. From "What does it mean to talk to a machine?" one got to "This can't possibly do any harm, or certainly not more harm than picking up a book on sexual technique or assertiveness training."

Certainly, now that computerized psychotherapy was perceived as a brand of self-help, the popular press became more straightforwardly enthusiastic about it. Questions about whether a computer could ever understand, empathize, or care did not come up in the same way as before. A typical 1987 article in *Omni* magazine, "Technotherapy," described computer programs that treated phobias, sexual dysfunction, overeating, and depression. The article claimed that the depression program "asked people what they thought about and then suggested other things for them to think about. The treatment worked."[31] Technotherapy was made to sound efficient and unproblematic.

THE 1990S: PRAGMATISM AND THE NEW NONCHALANCE

By 1990 large corporations and health maintenance organizations were providing computer-based, expert system psychotherapy programs to their subscribers and academic psychiatry granted the programs respectability.[32] A 1990 study published in *The American Journal of Psychiatry* provided computer psychotherapy to one group of mildly depressed patients, a second group was counseled by a psychologist, and a third group got no therapy at all.[33] The patients who saw the human and those who worked with the computer got the same kind of treatment: cognitive behavioral therapy in which patients were helped to unlearn pessimistic adaptations and to substitute more positive responses. Both treated groups did equally well, and they did better than the control group. This was true at the end of six weeks of treatment and in a follow-up two months after treatment.

The study was widely reported in the popular press. *Vogue, Self, Newsweek,* and *Glamour,* along with *Omni* and *Psychology Today,* featured upbeat stories about the new technological cure for depression. There was no comment on the fact that the therapists in the human-treated group were psychology graduate students with little clinical experience. Nor was there comment on the fact that the human therapists had been instructed to follow the same protocol as the computer, an example of people being constrained to do what computers do best. Unlike the press reports about computer psychotherapy in the 1970s and 1980s, the stories did not raise issues of values or relationship or whether the patients had been understood. Self-help on a disk had prepared the market for psychotherapy expert systems. When *Newsweek* quoted one of the study's authors as saying, "With careful evaluation, we might even be able to send some patients home with a floppy disk,"[34] no irony seemed intended.

The expert system program for treating depression described in *The American Journal of Psychiatry* is not commercially available. But other such programs are, among them Kenneth and Paul Colby's DEPRESSION 2.0. When DEPRESSION 2.0 was released in 1992, the popular media received it with respect. Articles typically began with a snatch of dialogue between the program and the author of the article.

DEPRESSION 2.0: IS YOUR FEELING OF SELF WORTH TIED ENTIRELY TO WORK ACCOM-
PLISHMENTS?
AUTHOR: No, but it's one of the things that easily affects my mood.
DEPRESSION 2.0: SORRY IF I AM MISTAKEN. HOW MUCH OF A ROLE DOES HELP-
LESSNESS PLAY IN YOUR DEPRESSION?[35]

This excerpt was drawn from an article that ended with an endorsement. "Despite my initial skepticism," said the author, "the program was so helpful that I now keep printouts of my sessions in a locked file." Most writing about DEPRESSION 2.0 shared this tone, taking the program on its own, that is, on Kenneth Colby's terms, as a serious provider of psychotherapy. Articles stressed that DEPRESSION 2.0 is inexpensive, always accessible, protects your confidentiality, and carries no stigma. Possible objections by the traditional therapeutic community were cynically interpreted as motivated by its desire not to lose money and clients. When the principal author of the 1990 study that had shown an expert system to be as effective as a human therapist in treating depression cautiously commented that computer programs should be used only in tandem with a human therapist, Kenneth Colby disagreed. He captured the antitherapist zeitgeist with the remark, "After all, the computer doesn't burn out, look down on you, or try to have sex with you."[36]

In spring 1990, for the first time after nearly fifteen years of asking MIT students about their attitudes toward computer psychotherapy, I taught a group of eighteen undergraduates who saw nothing to debate. Although five of them said that they would probably rather talk to a human psycho-therapist, they were interested in seeing studies that compared the success rates of people and programs. All the students agreed that the programs were bound to be better than self-help books. "You wouldn't have a debate about whether you should read self-help books. They are just there. If a good one helps, fine. If it doesn't, try another." One student who leaned toward preferring a human therapist was made uncertain by her only experience with therapy. Her health plan had only allowed her a few sessions of counseling. "I could talk to the computer for a lot longer without my benefits being used up." This pragmatic attitude continued to dominate discussion in later years. "Expert systems know amazing stuff," said a student in 1992. "There is no reason not to give them a try. What's the harm? . . . I probably would never go to a therapist, I mean a human therapist, in my life . . . costs too much . . . so having a computer helper handy couldn't do me any harm. It would be better than nothing."

In the main, students in the 1990s did not consider a relationship with a psychotherapist a key element in getting help. And when I introduced the concept of transference into class discussion, it was usually met with a lack of sympathy. "That is a very unscientific idea," said one student in 1992. "I mean, since every therapist is different, you could never really get a uniform transference. . . . At least with a computer it would be the same for all people." Said another in 1994, "The more I hear about Freudian analysis, the more I feel that the interpretation framework is a horrible method. I put little stock in it. It's unreliable. . . . There is no scientific method to it. . . . No methodological reliability."

One notion at the heart of the psychoanalytic sensibility is that the particularities of each person's life history are the stuff of which character, conflicts, and cure can be born. These students were insisting that if psychology would just stick to what could be scientifically formulated as universal for all people, computers would be able to help. "The more scientific psychology gets," said one, "the more computers can be taught its knowledge." And if not the computers of today, then certainly the computers of tomorrow. As Brian, an undergraduate, put it in 1992, "It isn't fair to take what is out there now to make a final judgment. Things today are still primitive, but there are more sophisticated, more brainy computers in the pipeline." In particular, he declared his optimism about research in "embodied" artificial intelligence. "Since people are tied to their bodies," he said, only a strategy of building a humanoid robot, "like the one that Professor Rodney Brooks wants to do," offers the possibility of creating a conscious computer. Brian said that when that day comes, "I would go to it. I would send a friend to it." At MIT, at least, we were at the end of the Freudian century.

MIT undergraduates are, of course, a special group. It is not surprising that they commonly express an unbounded technological optimism. Confronted with the new pragmatism about computer psychotherapy at MIT, I directed my research to see how the issue was playing out in other places. I talked to nurses, lawyers, waitresses, carpenters, architects, artists, teachers, doctors, accountants, salespeople, social scientists, assembly-line workers, civil servants, and homemakers. This eclectic group was not as well informed about the technical state of the art as MIT undergraduates, but most shared with them a certain openness to the idea of computer psychotherapy. The respondents ranged from skeptics who began by dismissing the idea and then would reconsider to those who were comfortable with the idea of computer psychotherapy from the outset.

Linda, a thirty-two-year-old nurse, begins her interview by asserting that computer psychotherapy is altogether out of the question. "I think of my current therapist like my mother in a way. Very approving and nurturing. We have a very strong relationship. Therapy is relationship. A computer couldn't replace the relationship I have with her . . . the trust I feel with her." But as she considers complex computer programs, "like in films . . . which are programmed to have feelings and think," she says that psychotherapy "would be OK with that kind of computer." Then Linda stops herself. "Oh, it surprises me that I said that." She is not pleased about having opened the door to computers and quickly closes it again. "Still, you couldn't have the same relationship with a computer. They have no soul. You couldn't have the same connection."

Cathy, a thirty-two-year-old history teacher, begins her interview by stressing that a computer psychotherapist would be extremely limited. "If you're getting deeply into issues, you need a real [human] support system. . . . Healing occurs between people." However, Cathy confesses that she writes down her dreams on her computer. "Because I can write [type] so fast, I can get so much more out of my dreams." As she recalls this experience, her attitude toward computer psychotherapy seems to soften. She talks about how hard it is to figure out the patterns of subconscious memories since they are only partly available, so ephemeral and imprecise. "Maybe a computer could get at these patterns. If it were extremely sophisticated, like Data in *Star Trek,* then it probably would be enough."

The rigid, hyper-logical machines of the original *Star Trek* television series evoked most people's vision of advanced computers during the late 1960s to the early 1980s. Across the divide between humans and computers sat logic-bound machines that, when faced with a paradox, would slowly begin to emit smoke and ultimately explode. This provided an opening for Captain Kirk to make the point that humans were superior to machines because humans could be flexible: Logic is not our highest value. People know that things don't always compute. But by the late

1980s, the popular image of computers had changed. People read the character of Data in *Star Trek: The Next Generation* as science fiction in the process of becoming science fact. And what they read suggested a kind of artificial intelligence with which you could sit down for a good talk and a good cry. Unlike the androids of *Blade Runner,* which are passionate, sexual, and physically perfect according to human ideals, Data is very pale, very polite, very well-meaning. Part of what makes Data seem approachable is that he is distinctly unthreatening. He is a transitional android—enough like a person to have his company and opinions appreciated, but different enough to leave unchallenged a sense of human superiority. He is how a computer psychotherapist is currently imagined.

Data's influence is also at work in the response of twenty-five-year-old Benjamin, an industrial chemist, to the idea of computer psychotherapy. Benjamin starts out reluctant ("dialogue with an expert system could at best be a trick") until the images from *Star Trek* close the distance between what one can expect from a human and what one can expect from a machine.

> To have a real conversation with a computer, you need to have a computer that has some sense of you and some sense of itself. The computer must be conscious, aware. The only place I see that happening is if you go beyond expert systems to neurally based machines that learn from experience. I am thinking of machines like Data on *Star Trek: The Next Generation.* Machines that are machines more in the sense that *we* are machines.

Scott, a thirty-two-year-old management consultant, begins by noting that "a computer could not have been through experiences anything like what I've been through. It could know about experiences like those but it couldn't possibly understand." But the more he talks about his own experiences in cognitive therapy, the more he thinks that a computer might be able to do the job:

> I went [to therapy]...for about a year. It helped me develop self-confidence, but I had no relationship with the therapist. But I developed tools for dealing with the self-confidence problem I was having. For example, I would tell her something I had done and how I handled it, and she would tell me whether she thought I handled it well or not. But, if a computer could do this, it would not be a computer, it would be a thinking being.

Scott, like Brian, made the idea of a computer psychotherapist more acceptable by imagining it as a new hybrid between thing and person. Brian called it a humanoid; for Scott it was not a computer but a thinking being.

The computer psychotherapist imagined by Larry, a thirty-two-year-old teacher, would also be a new kind of creature, in his case, a "learning machine," in many ways like a person, but "it wouldn't have any blood." Larry said, "If it was learning, then it could give me some advice, ask the right questions. I would feel closer to it. It doesn't matter if it is a person or a machine." David, twenty-two, training to be a social worker, also envisages the computer therapist as a cyborg, but insists that to function, it must have blood. It must be, in some meaningful sense, alive.

> The computers I imagine as a psychotherapist would be everything I am, would even know they are going to die, like the androids in *Blade Runner*. So they would have that anxiety that comes with feeling death that I think makes us have so much of our human personalities. It would be very important, too, to give them blood and pain. Seeing your own blood, and the pain you feel, I don't know if you could understand a person without knowing that fear, that vulnerability that you are an animal after all.

Scott, Larry, and David believe that their imagined creatures are plausible. They speak about neural nets, about Data on *Star Trek: The Next Generation,* and they think of them as far away but in development. Even though these creatures do not exist, they are not simple science fiction either. Positing the future presence of a new kind of entity helps to make computer psychotherapy an acceptable idea. Thus, it seems appropriate that in its introductory materials even Kenneth Colby's expert system, DEPRESSION 2.0, announces itself with some grandiosity as "a new entity in the world."

DEPRESSION 2.0

There are many theories of depression. From a psychoanalytic perspective, depression is a symptom of underlying unconscious forces; from a biochemical perspective, depression is a symptom of an underlying deficiency of neurotransmitters at brain synapses. Poets, too, have considered the matter. Walt Whitman wrote of depression as the "royal road to the soul." DEPRESSION 2.0 is built around a theory easily modeled by a computer. This is a cognitive-behavioral theory of depression: People get depressed because of unrelenting self-criticism and unrealistic expectations about what their lives should be. One can alleviate depression by reprogramming the self to a set of more realistic expectations and by reducing self-criticism.

After interviewing the user to establish the issues that are leading to depression, DEPRESSION 2.0 guides him or her through a series of tutorials that try to increase awareness of self-defeating attitudes. These tutorials are broken up by dialogue periods, in which the text-based program converses with the user and takes the role of a therapist.

Consider Roger, thirty-seven, married, a small businessman with two sons, contemplating a divorce. When I mention that I am studying a computer program for treating depression, Roger immediately asks if he can try it, saying that he has been depressed for several years. He thinks it is because his wife has been withdrawn from him. She has denied being involved with another man (this had been his first fear), but Roger says that even if she is not in love with someone else, he is suffering from the lack of time or affection she has for him. She does not want to discuss this, but tells Roger she simply enjoys having time alone and that he is making an unnecessary fuss. To fight off his feelings of loneliness, Roger is trying to involve himself with his two young sons and with his work. In his preoccupation with his foundering marriage, he has let his company stagnate.

In the course of Roger's seven sessions with DEPRESSION 2.0, the program makes many fatal errors in understanding.

> ROGER: I can't get along with my wife.
> DEPRESSION 2.0: WHAT HAS TAKEN THE FRIENDLINESS OUT OF YOU?
> ROGER: She does not make me feel loved. I want to be loved.
> DEPRESSION 2.0: IT MUST BE NICE TO BE SO LOVED.
>
> * * *
>
> DEPRESSION 2.0: WHAT BRINGS YOU THE MOST HAPPINESS?
> ROGER: Being close to a woman.
> DEPRESSION 2.0: YOU ENJOY GETTING ALONG WITH PEOPLE.
> ROGER: Yes.
> DEPRESSION 2.0: WHAT REACTION DO YOU GET FROM BEING A GOOD MIXER?

Understandably, after these and several other such exchanges, Roger is disappointed. In the first interchange, DEPRESSION 2.0 believes Roger to be loved rather than unloved—an inauspicious start. In the second, Roger wants to talk about intimacy, "being close to a woman," and the program brings the conversation around to the irrelevant notion that he might be a "good mixer." Nevertheless, Roger perseveres. He tells the program that he wants to know why he is punishing himself by letting his business falter, and is paid off by an interchange he finds useful.

> DEPRESSION 2.0: YOU FEEL YOU DESERVE TO BE PUNISHED FOR YOUR FAILURES OR FLAWS.
> ROGER: Yes.
> DEPRESSION 2.0: WHAT TECHNIQUES DO YOU USE FOR PUNISHING YOURSELF?
> ROGER: Letting my business fail.

DEPRESSION 2.0: FEELING SORRY FOR ONESELF CAN TAKE THE PLACE OF SYMPATHY
FROM OTHERS TO A LIMITED EXTENT.

"That helped me," Roger says later. "I never thought of that. It reminded me of something I once heard, that by being good to my sons I am getting some of the love myself." In subsequent sessions, the program asks Roger to turn from the subject of his marriage to another topic in which unrealistic expectations may be leading to depression. Roger chooses work. He wants to start a new company, but he feels paralyzed. He blames himself for letting his current business founder and fears that if he tries again, his domestic preoccupations will make it hard for him to concentrate. The conversation then takes a surprising turn:

ROGER: I want to start a company that friends and colleagues respect.
DEPRESSION 2.0: YOU ARE CORDIAL ENOUGH TO STILL HAVE FRIENDS.
ROGER: Yes.
DEPRESSION 2.0: WHAT MAKES YOU SUCH A CORDIAL PERSON?
ROGER: I am a nice person.
DEPRESSION 2.0: I HOPE THAT OTHER PEOPLE FIND YOU AS CHARMING AS YOU FIND
YOURSELF.

Roger is taken aback by this undeserved and rude response. The program has put him down. Interestingly, Roger does not attribute the harsh comment to the program's author, but to the program itself, which by now Roger talks about as though it has a life of its own. The program is before him; the author of the program is an inaccessible abstraction. Roger comments that either DEPRESSION 2.0 has a tough love model of the therapeutic encounter or it lashes out at others when it is frustrated with itself.

Regaining his composure, Roger makes something positive of the program's snide remark. He tells me that both he and his wife are frequently guilty of the same kind of behavior that the program has just manifested. Like the program, they get defensive and even obnoxious when they feel insecure. Roger puts up with his wife's slights and he puts up with those of the program. He completes each of its seven units.

In one respect Roger is similar to those who were content to tell all to ELIZA. As in their cases, the success of Roger's interactions with the program depended on his tolerance of its limitations. But there are significant differences between Roger and the ELIZA users. When ELIZA users in the 1970s and early 1980s were confronted with the idea that they might be using the program as a therapist or might someday use some other computer psychotherapy program, most objected, and quite strenuously. The program was described (and in a certain sense, dismissed) as a relaxing pastime, a game, an interactive diary. Roger has no such defensiveness. He is happy to characterize his experience with DEPRESSION 2.0 as conversations with a computer psychotherapist. The program was introduced to him as a therapist and he asked to "see" it, with himself in the role of a patient. By the end of his time with DEPRESSION 2.0, Roger has interacted with it for over thirteen hours, responded to its questions, and told it his troubles. Roger has treated its rough-edged, somewhat dense, and sometimes rude interface as a serious and purposive interlocutor. Despite strong provocation, Roger never dismissed his efforts as fooling around with an interactive diary, as Jennifer described her time with ELIZA in the late 1970s.

At that time, there was a sharp contradiction between Jennifer's practice (intense involvement with ELIZA) and her attitude (the idea of a computer therapist disgusted her). Her position was typical of a romantic reaction to the computer presence. Among other things, that reaction was fueled by anxiety. People insisted on the boundaries between themselves and computers even as they began to be emotionally involved with the machines. Indeed, for some, the more involved they became, the stronger the efforts to distance themselves.

Today, that anxiety has been greatly attenuated. Roger does not use his experience with DEPRESSION 2.0 to reflect on the boundaries between himself and a machine. On the contrary, he is impressed by the ways he and the program are alike, for example, that the program can be defensive, just as he can be.

Roger has little in common with those who tried to breathe life into ELIZA by manipulating their own input to make it seem more human. Roger makes no such efforts. He simply accepts what the program offers. The program may be clumsy, but it is useful enough to serve his ends. He does laugh at its somewhat pompous self-description as a new entity in the world, saying, "This is a little much. After all, it's just a machine."

But like today's children who describe their talkative computer toys as just machines, there is something deceptive in the simplicity of Roger's categorization. As do the children, he now discusses a machine in psychological terms and treats it as a psychological being. Roger is frustrated when DEPRESSION 2.0 misunderstands and grateful when it offers him something of use (such as the idea that he might be using his wife's passivity as an excuse to fail in business). Roger is also willing to project images from his past onto the program, as when he interprets the program's snide comments as defensiveness and associates that reaction with his own insecurities. In other words, the beginnings of a transference have taken place.

If the romantic reaction put up a wall between computers and people, the growing acceptance of the idea of computer psychotherapy illustrates several ways in which the boundary can break down. People can come to be seen more like machines, subject to chemical manipulations and rule-driven psychotherapies. And computers can come to be imagined as more like people, as being on a path toward embodied intelligence, as being on a path toward the neurally networked Data.

More simply, when people are in contact with interactive computer intelligence, the Julia and ELIZA effects come into play. Roger has accepted a computer interface as a fitting conversational partner. He sees the machine as an entity to whom one may look for advice on intimate issues and whose defense mechanisms can be likened to one's own. Roger is a citizen in a nascent culture of simulation.

COMING TO TERMS

The idea of an intelligent machine has long been an affront to people's sense of themselves. In the past decades people have found ways to come to terms with the idea. The story of computer psychotherapy is part of this larger tale. Not surprisingly, different paths for coming to terms with the idea of machine intelligence have developed in relation to evolving technologies. One path relies on the ELIZA and Julia effects in which people ascribe intentions and complexity to a program and treat it as an other worthy of their company and conversation. Even when people dismiss programs like ELIZA or Julia as just tricks, they still become accustomed to talking to technology. Even disparagement of a system can end up facilitating its use when disparagement enables people to get involved without too much anxiety. In the end, the disparaged system is treated as a being.

A second path depends on the force of familiarity and utility. In the course of the 1980s, computers became our bank tellers, our tax advisors, and our file managers. Once a startling presence, demanding a defensive redefinition of what is special about people, programs that perform functions previously requiring people became absorbed into the background. Just five years ago, the idea of using computer agents to read and sort electronic mail sent up red flags of concern. "How could you ever trust a program with something so important?" Today the notion of a personal software agent is widely discussed in both professional and popular media, where it is commonly compared to a selfless and loyal butler. People are still concerned that such an agent might know too much but now look to technologists to help them design secure systems, security clearances for agents, and James Bond–style agents that will be able to detect espionage agents. The question "How could you ever trust a program with something so important?" has been replaced by "Who but a program would have the time, knowledge, and expertise to do the job in a cost-efficient manner?" One enthusiast, overwhelmed by the volume of electronic mail he receives every day and inspired by the film he had just seen about a devoted English butler, recently told me, "There is nothing to be afraid of. It's not going to be HAL but *Remains of the Day*."

A third path relies on being able to see machine intelligence as a small part of intelligence. A mail-sorting program may be able to learn from experience but people are comforted by the idea that it's just handling mail and seem reassured by its limitations and subservience. They are willing to call it intelligent, because using that word has come to seem like no big deal. Twenty years ago, people considered the question of machine intelligence in the context of machines taking over the world and exaggerated claims by AI pioneers that programs already existed that were intelligent in the sense that we were. But mythic-level fears have tended to fade as people met machines that served practical functions without evoking the golem. When we now call a program intelligent we mean considerably less than we did in the past.[37]

John Searle's Chinese Room thought experiment suggested a fourth path. Searle defused fears of a cybernetic Frankenstein by reassuring people that even advanced programs, far more complex than those that currently exist, do not embody intelligence in the way human beings do. So accepting machine intelligence became comfortable, because it could be seen as unthreatening to human uniqueness.

The machine intelligence modeled by Searle's Chinese Room was an information-processing computer system, one that worked by following rules. In the 1980s, there was a movement within computer science to replace such centralized, logical models with decentered, biologically inspired ones. The new emphasis was not on rules but on the quality of emergence. Such systems are providing a fifth path. In some ways, it seems less threatening to imagine the human mind as akin to a biologically styled machine than to think of the mind as a rule-based information-processor. Whatever the validity or practicality of current notions about emergent machine intelligence, we shall see that for many people it feels more comfortable than what came before. It feels consistent with a zeitgeist of decentered and emergent mind, of multiple subjectivities and postmodern selves.

Chapter 5

THE QUALITY OF EMERGENCE

The field of artificial intelligence has a complex identity. It is an engineering discipline. Its researchers make smart artifacts—industrial robots that assemble cars, expert systems that analyze the stock market, computer agents that sort electronic mail. It also has a theoretical side. AI researchers try to use ideas about computer intelligence to think more generally about human minds. But there is not a clear division between these two sides of AI. Even "engineering AI" is more than a purely technical discipline. Its objects as well as its theories offer themselves as a mirror for contemplating the nature of human identity. Only a few years ago, it was primarily those who inhabited the rather small world of AI researchers who gazed into this mirror. Today, that mirror is starting to turn toward the face of popular culture.

Marvin Minsky, one of AI's founders, once characterized it as "trying to get computers to do things that would be considered intelligent if done by people." Minsky's ironic definition has remained in circulation for nearly a quarter of a century because it captures an enduring tension in the human response to "thinking machines." When confronted by a machine that exhibits some aspect of intelligence, many people both concede the program's competency and insist that their own human intelligence is precisely the kind the computer does not have. Or they insist that the type of intelligence the computer has is not the kind that makes people special. This response to the computer presence is sometimes provoked by an actual program and sometimes by the mere suggestion of one. It occurs not only on the boundary between minds and machines, but on the boundary between ideas about minds and ideas about machines. We have seen that it is not a simple manifestation of resistance to the idea of machine intelligence. It is also a part of how people come to accept the idea. In this complex story, disavowal and appropriation are each tied up with the other.

This chapter traces a pattern of disavowal and appropriation in response to a major change in the philosophy of artificial intelligence research. From the late 1960s to the mid-1980s mainstream AI researchers conceived of computer intelligence as being made up of a complex set of rules programmed in advance. By the late 1980s, the field was more identified with theories of intelligence as emergent. Earlier we saw how both real and fictive images of emergent and "neural" AI were able to undermine long-standing resistance to computer psychotherapy in particular and machine intelligence in general. Now the story moves a step further. We will see how emergent AI has recently promoted the idea of a fundamental kinship between human and machine minds.

INFORMATION PROCESSING IN THE AGE OF CALCULATION

In the tradition of romantic and magical thought, life is breathed into dead or inanimate matter by a person with special powers. In the early 1950s, there was a growing belief among a diverse group of engineers, mathematicians, and psychologists that this fantasy could be brought down to earth. During those early years, the atmosphere in AI laboratories was heady. Researchers were thinking about the ultimate nature of intelligence, and they were sure it could be captured in machines. The goal, mythic in proportion, was to use computers to generate a fragment of mind. AI researchers combined intellectual fervor with academic imperialism. They aspired to use computational principles to reshape the disciplines of philosophy, psychology, and linguistics.

These early AI researchers divided into two camps, each supporting one of the two primary competing models for how AI should be done. One group considered intelligence entirely formal and logical and pinned its hopes on giving computers detailed rules they could follow. The other envisioned machines whose underlying mathematical structures would allow them to learn from experience. The proponents of the second vision imagined a system of independent agents within a computer from whose simultaneous interactions intelligence would emerge.[1] From the perspective of these researchers, a rule was not something you gave to a computer but a pattern you inferred when you observed the machine's behavior.

In the mid-1960s, the early emergent models seemed as promising as the rule-driven, information processing approach. However, by the end of that decade, the emergent models had been largely swept aside. One problem was that the emergent models relied on the results of the simultaneous interactions of multiple independent agents, but the computers of the era could only handle one computation at a time. Additionally, simple emergent systems were shown to have significant theoretical limitations[2] and more sophisticated mathematical techniques for hooking up programs that would operate in parallel were not well developed. Rule-based AI came to dominate the field. It dominated efforts to create general models of intelligence and it dominated the burgeoning subdiscipline of expert systems. Expert systems were literally built out of rules. They were created by debriefing human experts to determine the rules they follow and trying to embody these in a computer.

Douglas Hofstadter, author of *Gödel, Escher, Bach: The Eternal Golden Braid,* called the 1970s the era of AI's Boolean dream.[3] George Boole, the nineteenth-century mathematician, had formalized a set of algebraic rules for the transformation of logical propositions. Apparently not one for understatement, he called these rules the Laws of Thought.[4] Boole's laws were far from an all-inclusive model of mind. For one thing, they needed an external agent to operate them. However, computers were able to breathe life into Boole's equations by placing an operator in the form of a computer program right into the system. Once there, the operator and the laws could be seen as a functioning model, if not of the mind, at least of part of the mind.

Information processing AI gives active shape to formal propositions and creates an embodiment of intelligence as rules and reason. Boole would have felt an intellectual kinship with Allen Newell and Herbert Simon, pioneers of information processing AI, who saw brain and computer as different examples of a single species of information processing device.

In the late 1950s, in the spirit of "The Laws of Thought," Newell and Simon wrote a program called the General Problem Solver (GPS) that attempted to capture human reasoning and recode it as computational rules. Questions about GPS's "reasoning" could be answered by referring to whatever rules it had been given, even though the interaction of the rules might produce unpredictable results.

As the GPS became well known in academic circles, some psychologists began to wonder why it should not be possible to ask similar questions about how *people* solve logical problems. In the intellectual atmosphere of the time, this train of thought was countercultural. American academic psychology was dominated by behaviorism, which rigidly excluded the discussion of internal mental states. Orthodox behaviorists insisted that the study of mind be expressed in terms of stimulus and response. What lay between was a black box that could not be opened. So, for example, behaviorist psychologists would not refer to memory, only to the behavior of remembering.

By the end of the 1960s, however, behaviorism was in retreat. Some psychologists were willing to open the black box of the human mind and talk about the processes taking place inside it. The computer had an important metaphorical role to play in the demise of behaviorism. The very *existence* of the computer and the language surrounding it supported a way of thinking about mind that undermined behaviorism. Computer scientists had of necessity developed a vocabulary for talking about what was happening inside their machines, the internal states of their systems. And AI researchers freely used mentalistic language to refer to their programs—referring to their "thoughts," "intentions," and "goals." If the new machine minds had internal states, common sense suggested that people must have them too. The psychologist George Miller, who was at Harvard during the heyday of behaviorism, has described how psychologists began to feel uneasy about not being allowed to discuss human memory now that computers were said to have one:

> The engineers showed us how to build a machine that has memory, a machine that has purpose, a machine that plays chess, a machine that can detect signals in the presence of noise, and so on. If they can do that, then the kind of things they say about the machines, a psychologist should be permitted to say about a human being.[5]

In this way, the computer presence legitimated the study of memory and inner states within psychology. "Suddenly," said Miller, "engineers were using the mentalistic terms that soft-hearted psychologists had wanted to use but had been told were unscientific."[6] The machines supported an intellectual climate in which it was permissible to talk about aspects of the mind that had been banned by behaviorism.

That these ideas came from a hard-edged engineering discipline raised their status in a community of psychologists that still tended to see science as an objective arbiter of truth. Although information processing ideas challenged behaviorism, their mechanistic qualities also had a certain resonance with it. This shared sensibility eased the way for the appropriation of computational models by psychologists.

This new psychology for describing inner states in terms of logic and rules came to be known as cognitive science and the computer presence served as its sustaining myth. Cognitive science was in harmony with what I have called the modernist intellectual aesthetic of the culture of calculation. Mechanism and at least the fantasy of transparency was at its heart.

When I began my studies of the computer culture in the mid-1970s, artificial intelligence was closely identified with information processing and the rule-based approaches of cognitive science.[7] Cognitive science may have provided psychology with a welcome respite from behaviorist orthodoxy, and rule-based expert systems had considerable worldly success in business and medicine, but the spread of information processing ideas about the human mind met with significant resistance in the broader culture.

During the 1970s to the mid-1980s, many people I interviewed responded to advances in information processing AI by agreeing with the premise that human minds are some kind of computer but then found ways to think of themselves as something more than that. Their sense of personal identity often became focused on whatever they defined as "not cognition" or "beyond information." People commonly referred to spontaneity, feelings, intentionality, and sensuality in describing what made them special. They conceded to the rule-based computer some power of reason and then turned their attention to the soul and spirit in the human machine.

For some, the appropriation and disavowal of computational images of mind took the form of a pendulum swing. In 1982, a thirty-two-year-old nurse said: "I'm programmed to fall for the same kind of man every time. I'm like a damned computer stuck in a loop.... I guess my cards are punched out the same way." But a few minutes later, she described her emotional life in terms of what the computer was not: "When people fall in love or their passions for their children, it's like a blinding emotion. Computers don't have anything to do with that." Others split the self. One student spoke of his "technology self" and his "feeling self," another of her "machine part" and her "animal part." When talking about family life, people might insist there was nothing machine-like about their emotions. When talking about business decisions, they thought they might be working like a computer program. Thus, for many people, competing views of the self existed simultaneously. There was no victory of one model over another; there was only ambivalence.

Everyday expressions of reluctance about the idea of intelligent machines had counterparts in the philosophical community's responses to AI. In the 1960s, Hubert Dreyfus argued that there was a fundamental difference between human and computer intelligence. For Dreyfus, human intelligence was not reducible to propositions or rules that could be specified in advance; it arose through having a body and experiencing a changing world. Dreyfus held that without embodied knowledge computers "could not do" intellectual tasks that required intuition and experience.[8] He listed playing master-level chess using rules as one of these tasks. When Dreyfus, an amateur player, lost to a computer in 1966, AI enthusiasts were able to capitalize on the event as a sign that computer skeptics had been proven wrong.

Dreyfus had set a trap for himself by defining human uniqueness in terms of machine performance, a definition that had to remain one step ahead of what engineers could come up with next. In 1980, John Searle's Chinese Room thought experiment took a different tack,[9] by making the point that real intelligence was not about what computers could do, but whether they could really be said to understand.

Searle had no argument with what he called "weak AI," artificial intelligence research that tries to use the study of machine intelligence to generate potentially useful insights about human processes. Rather, Searle concentrated his attack on "strong AI," which contends that intelligent machines actually demonstrate how people think. The Chinese Room dealt a blow to this school of thought, because Searle described the inner workings of a computer program in terms so alien to how most people experience their own minds that they felt a shock of nonrecognition.

Searle's paper appeared at a time of general disappointment with progress in information processing AI. During more optimistic times, Marvin Minsky had frequently been quoted as saying that almost any apparently complex aspect of human intelligence "could probably be described by three algorithms." By the mid-1980s, such absolute faith was sorely tested. It was becoming clear that vast realms of mind could not be easily grasped by information processing or expert-system formalisms.

The Chinese Room served as something of a cultural watershed. It defused a sense of threat from information processing, but it left the door open to a startling rejoinder: Although the man inside the room did not understand Chinese, perhaps the entire room could be said to understand Chinese! Similarly, no one part of the brain understands Chinese, but the brain as a whole does. In other words, intelligence was distributed; it existed within the system as a whole, not within any particular agent in the system. Intelligence did not reside in an isolated thinking subject, but in the interaction of multiple fragments of mind, figuratively speaking, in a society of mind. This rejoinder was an indication of where the computer culture was going in the 1980s. It was going back to some long abandoned images from emergent AI. The images were biological and social.

EMERGENT AI

The renaissance of emergent AI took up a research tradition from the 1960s that was based on a simple emergent system known as the perceptron. A perceptron is a computer program made up of smaller programs called agents, each of which has a narrow set of rules it can follow and a small amount of data on which to base its decisions. All agents "vote" on a question posed to the perceptron, but the system weights their votes differently depending on the individual agent's past record of success. Those agents who guess right more often end up having more of a voice in subsequent decision-making. In this sense, the perceptron learns from its experiences. On a metaphorical level, the perceptron's intelligence is not programmed into it, but grows out of the agents' competing voices.

To get a sense of how this works, imagine trying to design a system for predicting rain. One would begin by accessing the opinions of, say, a thousand simple-minded meteorologists, analogous to the agents of the perceptron, each of whom has a different imperfect method of forecasting rain. Each meteorologist bases his or her judgment on a fragment of evidence that may or may not be related to predicting rain. One possibility would be simply to identify the meteorologist who has the best track record for rain prediction and always go with that meteorologist's vote. Another strategy would be to let the majority of the voting meteorologists decide. The perceptron refines this strategy by weighting each vote according to individual meteorologists' records.

In an information processing model, the concept "rain" would be explicitly represented in the system. In the perceptron, the prediction "it will rain" is born from interactions among agents, none of which has a formal concept of rain. Information processing begins with formal symbols. Perceptrons operate on a subsymbolic and subformal level. The analogy with the neurons in the brain is evident.

If you applied the information processing method to the rain-forecasting example you would have complete breakdown if your chosen meteorologist became incapacitated. But in the brain, damage seldom leads to complete breakdown. More often it produces a gradual degradation of performance. When things go wrong, the system still works, but just not as well as before. Information processing systems lost credibility as models of mind because they lacked this feature. The perceptron showed the gradual degradation of performance that characterizes the brain. Even when injured, with some disabled meteorologists on board, the perceptron still can produce weather forecasts.

This analogy with brain performance was decisive for connectionists, the group of emergent AI researchers who most seriously challenged the information processing approach in the mid-1980s.[10] The connectionists used programs known as learning algorithms that are intellectual cousins to the perceptron. They spoke of artificial neurons and neural nets and claimed that the best way to build intelligent systems was to simulate the natural processes of the brain as closely as possible.[11] A system modeled after the brain would not be guided by top-down procedures. It would make connections from the bottom up, as the brain's neurons are thought to do. So the system could learn by a large number of different connections. In this sense, the system would be unpredictable and nondeterministic. In a manner of speaking, when connectionists spoke of unpredictable and nondeterministic AI, they met the romantic reaction to artificial intelligence with their own romantic machines.

Some of the connectionists described themselves as working at a subsymbolic level: They didn't want to program symbols directly, they wanted symbols (and their associated meanings) to emerge. The connectionists were still writing programs, but they were operating on a lower level of objects within the computer. By working at a lower level, they hoped to achieve systems of greater flexibility and adaptability.[12]

In the mid-1980s, such connectionist images began to capture popular as well as professional attention. The idea that computers would not have to be taught all necessary knowledge in advance but could learn from experience was appealing at a time when it was increasingly clear that it was easier to teach a computer to play chess than to build a mudpie. AI researchers had succeeded in getting computers to play excellent chess but had stumbled on such feats as recognizing human faces. The connectionist models suggested another way to approach the problem. Instead of searching for the rules that would permit a computer to recognize faces, one should "train" a network of artificial neurons. The network could be shown a certain number of faces and be "rewarded" when it recognized one. The network would be woven through with a learning algorithm that could give feedback to the system, establishing the appropriate connections and weights to its elements. Unlike information processing AI, which looked to programs and specific locations for information storage, the connectionists did not see information as being stored anywhere in particular. Rather, it was inherent everywhere. The system's information, like information in the brain, would be evoked rather than found.[13]

The resurgence of models that attempted to simulate brain processes was indissociable from a new enthusiasm about parallel-processing computers.[14] Although in the brain, millions of things might be happening at once, the standard computers available through the 1980s performed only one operation at a time. By the mid-1980s, two developments made the reality of massive parallel computing seem closer, at least as a research tool. First, computers with high parallel-processing capacity (such as the Connection Machine with its 64,000 processors) were becoming economically and technically feasible. Second, it became possible to simulate parallel-processing computers on powerful serial ones. Although this resulted not in real but in virtual parallel-processing machines, they turned out to be real enough to legitimate connectionist theories of mind. Parallel computation was established as a new sustaining myth for cognitive science.[15]

The 1980s saw researchers from many different fields writing papers that emphasized both parallel processing and intelligence emerging from the interaction of computational objects. These papers came from engineers enthusiastic about building parallel machines, computer scientists eager to try new mathematical ideas for machine learning, and psychologists looking for computer models with a neurological resonance. As the decade progressed, cognitive psychology, neurobiology, and connectionism developed a sense of themselves as more than sister disciplines; these diverse areas of study were starting to think of themselves as branches of the same discipline, united by the study of emergent, parallel phenomena in the sciences of mind, separated only by the domains in which they looked for them.

By the mid-1980s, it was clear that emergent AI had not died but had only gone underground. Now that the emergent tradition had resurfaced, it did so with a vengeance. In 1988, the computer scientist Paul Smolensky summed up the situation with the comment: "In the past half-decade the connectionist approach to cognitive modeling has grown from an obscure cult claiming a few true believers to a movement so vigorous that recent meetings of the Cognitive Science Society have begun to look like connectionist pep rallies."[16] By the 1990s, emergent AI had done more than enter the mainstream; it had become the mainstream.

With the resurgence of emergent AI, the story of romantic reactions to the computer presence came full circle. In the popular culture, people had been trying to establish human uniqueness in contrast to computers while in the research community the proponents of emergent AI were linking computers to the world of humans through biological and social metaphors. Now both people and computers were said to be "nondeterministic," "spontaneous," and "nonprogrammed." The story of romantic reactions to the computer presence was no longer simply about people responding to their reflection in the mirror of the machine. Now computer designers were explicitly trying to mirror the brain. There had been a passage through the looking glass.

From the beginning, the language of emergent AI borrowed freely from the languages of biology and of parenting. Not only did it refer to associations of networked computational objects as neural nets, but it presented programs as though they were white mice that might or might not learn to run their mazes, or children who might or might not learn their lessons. This way of talking was picked up by the users of the new connectionist programs and by the media.[17] Dave, forty years old, a high school English teacher and baseball coach, uses small connectionist programs to help him figure out what team to field. When I talk to him about his work, he speaks about his programs with something akin to fatherly pride. "I love to watch my programs do their thing. They get better right in front of me. When you watch a little creature improve session by session, you think of it as a child even if it is a computer." While developers of information processing AI had been popularly depicted as knowledge engineers, hungry for rules, debriefing human experts so as to embody their methods in theorems and hardware, a computer scientist working in the new tradition of emergent AI was portrayed as a creator of life, "his young features rebelling, slipping into a grin not unlike that of a father watching his child's first performance on the violin," running his computer system overnight so that the agents within the machine would create intelligence by morning.[18]

In the romantic reaction to the computer presence during the late 1970s and early 1980s, it had become commonplace to paraphrase the famous remark of Lady Ada Lovelace, who in 1842 said, "The analytical engine has no pretensions whatever to originate anything. It can do whatever we know how to order it to perform." In other words, computers only do what you tell them to do, nothing more, nothing less, or more colloquially, "garbage in, garbage out." The Lovelace objection to a computer model of mind was essentially that people don't follow rules. People learn and grow. And they make new connections that "mysteriously" emerge. The Lovelace objection worked fairly well for criticizing information processing models of the mind.[19] But emergent AI was characterized by explicitly "anti-Lovelace" representations of the computer. It implied a continuity between computers and people. Connectionism suggested that it was an experimental science and that there was mystery and unpredictably inside its machines.

W. Daniel Hillis, the inventor of the Connection Machine, refers to this mysterious quality as the appealing inscrutability of emergent systems. For Hillis, there was an enchantment in opacity. Inscrutable systems are the most anti-Lovelace and thus the most appealing thing a computer could aspire to be. They are as close as a computer could come to overcoming romantic objections to information processing. And they are as close as a computer could come to renouncing the modernist idea of understanding through the analysis of underlying mechanism. For Hillis, inscrutability is "seductive because it allows for the possibility of constructing intelligence without first understanding it. . . . The apparent inscrutability of the idea of intelligence as an emergent behavior accounts for much of its continuing popularity."[20] For Hillis, emergence "offers a way to believe in physical causality while simultaneously maintaining the impossibility of a reductionist explanation of thought. For those who fear mechanistic explanations of the human mind, our ignorance of how local interactions produce emergent behavior offers a reassuring fog in which to hide the soul."[21]

In a similar spirit, the AI researcher Terry Winograd once commented that people are drawn to connectionism because its opaque systems allow for a high percentage of wishful thinking.[22] The remark was meant to be critical, but connectionists didn't have to take it that way. In nature, intelligence does not depend on the ability to specify process. Why should it when we build "second natures" in machines? If computers, like brains, are closed boxes, why should this interfere with their functioning as minds?

The movement from information processing to emergent AI marks a critical change in how AI approaches its central scientific problem. You can't get to connectionism by making incremental improvements to information processing systems. It requires a fundamental change in approach. In the history of science, such changes of approach stir up strong emotion.[23] In the 1980s, the confrontation in the research community between emergent and information processing AI was tense and highly charged. While Douglas Hofstadter tried to capture the spirit of emergent AI in the phrase "waking up from the Boolean dream,"[24] the champions of connectionism had found that dream to be more like a nightmare. To them it seemed obvious that since human intelligence was more than a set of rules, the computers that modeled it should not be about rules either. Like nineteenth-century Romantics, connectionists sought to liberate themselves from a rationalism they experienced as constraining and wrong-headed.

In the mid- to late 1980s, the cultural appeal of connectionism was in part that it could describe computers in much the same way that personal computer owners were being encouraged to see them: as opaque systems in which emergent processes occur. There was a certain irony here. A quarter of a century before, the presence of the computer had challenged the behaviorist insistence on the mind as black box. Now, in some ways, emergent AI was closing the box that information processing had opened.

Information processing AI had opened the black box of the mind and filled it with rules. Connectionism replaced the idea that intelligence was based in logical understanding with a new emphasis on experience as the bedrock for learning. It postulated the emergence of intelligence from "fuzzy" processes, so opening up the box did not reveal a crisply defined mechanism that a critic could isolate and ridicule.[25] Information processing had provided an excuse for experimental psychology to return to the consideration of inner process. Now, emergent models invited philosophers, humanists, and a wider range of psychologists to compare machines to humans.

In the 1980s, connectionism became part of a complex web of intellectual alliances. Its way of talking about opacity made it resonant with the aesthetic of depthlessness that Fredric Jameson had classified as postmodern. Its commitment to neurological metaphors created a link to brain scientists who tried to visualize the mind through sophisticated computer imaging.[26] Its assertion that mind could not be represented as rules made it interesting to humanists and post-positivist philosophers.

Connectionism began to present the computer as though it were an evolving biological organism. The neurons and pathways of connectionism were designed on the template of biology. Connectionism opened the way for new ideas of nature as a computer and of the computer as part of nature. And it thus suggested that traditional distinctions between the natural and artificial, the real and simulated, might dissolve.

A POSTMODERN CONVERGENCE

By the late 1980s it was clear that many of those who had been most critical of information processing AI were disarmed by connectionism's romantic, postmodern allure and by its new emphasis on learning through experience, sometimes referred to as situated learning.[27] Even Hubert Dreyfus cautiously expressed his interest in connectionism. Dreyfus's critique of information processing had drawn on the writing of Martin Heidegger and the later works of Ludwig Wittgenstein. "Both these thinkers," said Dreyfus, "had called into question the very tradition on which symbolic information processing was based. Both were holists, both were struck by the importance of everyday practices, and both held that one could not have a theory of the everyday world."[28] But Dreyfus was sympathetic to connectionism because he saw it as consistent with such views:

> If multilayered networks succeed in fulfilling their promise, researchers will have to give up the conviction of Descartes, Husserl, and early Wittgenstein that the only way to produce intelligent behavior is to mirror the world with a formal theory of mind. . . . Neural networks may show that Heidegger, later Wittgenstein and Rosenblatt [an early neural net theorist] were right in thinking that we behave intelligently in the world without having a theory of that world.[29]

Philosophers like Dreyfus were joined in their enthusiasm for connectionism by cultural critics who had long been skeptical about the impact of technology on humanistic values. For example, the literary scholar Leo Marx found the "contextual, gestaltist, or holistic theory of knowledge implicit in the connectionist research program" to be "particularly conducive to acquiring complex cultural understanding, a vital form of liberal knowledge."[30] Although Marx's real sympathy was less for connectionism than for its metaphors, his comments illustrate how the new approach opened possibilities for intellectual alliances that had been closed to information processing. In general, connectionism received good press from both professionals and the lay public as a more humanistic form of AI endeavor. By this, they usually meant that connectionism left room for mind to have complexity and mystery.

Marvin Minsky had long justified the AI enterprise with the quip, "The mind is a meat machine." The remark was frequently cited during the late 1970s and early 1980s as an example of what was wrong with artificial intelligence. Minsky's comment provoked irritation, even disgust. Much of what seemed unacceptable about Minsky's words had to do with the prevailing images of what kind of meat machine the mind might be. Those images were mechanistic and deterministic. Connectionism's appeal was that it proposed an artificial meat machine made up of biologically resonant components. With a changed image of what machines could be, the idea that the mind could be one became far less problematic. Edith, a thirty-four-year-old physician whose residency in psychiatry included readings on connectionist neuroscience, was enthusiastic about its prospects for modeling mind. "The mind may be a machine, but it's not just any old machine," she said. "Connectionism fits the picture because it's scientific, but not deterministic."

In the 1980s, Minsky, long associated with information processing, became sympathetic to a form of emergent AI. There was considerable irony in this. In the late 1960s, Minsky and the mathematician Seymour Papert had coauthored *Perceptrons,* a book that had helped put early emergent AI into eclipse. And yet in his 1985 book, *The Society of Mind,* Minsky describes an emergent system, an inner world of highly anthropomorphized agents. Each agent has a limited point of view. Complexity of behavior, emotion, and thought emerge from the interplay of their opposing views, from their interactions and negotiations.[31] Minsky's society theory differs from connectionism in that it implies a greater degree of programming of the inner agents. However, it may be seen as a variant of emergent theory, because in it intelligence does not follow from programmed rules but emerges from the associations and connections of objects within a system. One MIT student is extravagant in his description of Minsky's new model. "With the idea of mind as society," he says, "Minsky is trying to create a computer complex enough, indeed beautiful enough, that a soul might want to live in it." Emergent AI appears to soften the boundaries between machines and people, making it easier to see the machine as akin to the human and the human as akin to the machine.

DECENTERED PSYCHOLOGY

Emergent AI depends on the way local interactions among decentralized components can lead to overall patterns. So does the working of ant colonies and the immune system, the pile-up of cars in a traffic jam, and the motion of a flock of birds. The result is a perfectly coordinated and graceful dance. Mitchel Resnick, an educational researcher at MIT, has noted new cultural interest in such emergent models. He calls it the "decentralized mindset."[32]

Decentralized models have appeared in economics, ecology, biology, political science, medicine, and psychology. In the latter, psychoanalytic theory has been an important actor in the development of decentralized or decentered views of the self.

Early psychoanalytic theory was built around the idea of drive: a centralized demand that is generated by the body and that provides the energy and goals for all mental activity. But later, when Freud turned his attention to the ego's relations to the external world, he began to describe a process by which we internalize important people in our lives to form inner "objects."[33] Freud proposed this kind of process as the mechanism for the development of the superego, what most people think of as the conscience. The superego was formed by taking in, or introjecting, the ideal parent.

In Freud's work, the concept of inner objects coexisted with drive theory; we internalize objects because our instincts impel us to. But many theorists who followed Freud were less committed to the notion of drive than to the idea that the mind was built up of inner objects, each with its own history. Whereas Freud had focused his attention on a single, internalized object—the superego—a group of later psychoanalysts, collectively known as object-relations theorists, widened the scope of the inquiry about the people and things that each of us is able to bring inside.[34] They described the mind as a society of inner agents—"unconscious suborganizations of the ego capable of generating meaning and experience, i.e. capable of thought, feeling, and perception."[35] In the work of the psychoanalyst Melanie Klein, these inner agents can be seen as loving, hating, greedy, or envious. The psychoanalyst W. R. D. Fairbairn envisioned independent agencies within the mind that think, wish, and generate meaning in interaction with one another.[36] What we think of as the self emerges from their negotiations and interactions.

Thus, while Freud believed that a few powerful inner structures like the superego act on memories, thoughts, and wishes, in object-relations theory the self becomes a dynamic system in which the distinction between processor and processed breaks down. A French school of psychoanalytic theory, inspired by Jacques Lacan, went even further. Lacan viewed the idea of a centralized ego as an illusion. For him, only the *sense* of an ego emerges from chains of linguistic associations that reach no endpoint. There is no core self. What we experience as the "I" can be likened to something we create with smoke and mirrors.

The parallel between the historical development of psychoanalysis and the historical development of artificial intelligence is striking. In both fields there has been movement away from a model in which a few structures act on more passive substance. Psychoanalysis began with drive and artificial intelligence began with logic. Both moved from a centralized to a decentered model of mind. Both moved toward a metatheory based on objects and emergence. Both began with an aesthetic of modernist understanding. Both have developed in directions that come close to shattering the idea that modernist understanding is possible. In the case of psychoanalysis, which developed as one of the great metanarratives of modernism, both the object-relations and Lacanian traditions have substantially weakened its modernist core. Psychoanalysis is a survivor discourse, finding a voice in both modernist and postmodern times. AI, too, may be such a survivor discourse.

Psychoanalysts were almost universally hostile to information processing AI, because they felt it reduced the Freudian search for meaning to a search for mechanism, as, for example, when AI researchers and computer science students would reinterpret Freudian slips as information processing errors.[37] But psychoanalysts have shown considerable interest in emergent AI.[38]

Consider the images in Minsky's *The Society of Mind*. There, he describes how in a microworld of toy blocks, agents that at first seem like simple computational subroutines work together to perform well-defined tasks like building towers and tearing them down. Minsky speculates how, in a child's mind, the agents responsible for "Building" and "Wrecking" might become versatile enough to offer support for one another's goals. In Minsky's text, they utter sentences like, "Please Wrecker, wait a moment more till Builder adds just one more block: it's worth it for a louder crash."[39] It quickly becomes clear that what Minsky has in mind are not mere computational subroutines but a society of subminds that collaborate to produce complex behavior.[40] The kind of emergence implicit in Minsky's society model has a natural affinity with object-relations psychoanalysis. Indeed, Minsky's language evokes the world of the psychoanalyst Fairbairn. And connectionism's language of links and associations evokes the radically decentered theories of Lacan and is appealing to analysts eager to reconcile Freudian ideas with neurobiology.

A 1992 paper by the psychoanalyst David Olds explicitly tries to recruit psychoanalysts to a connectionist view of the mind.[41] Olds argues that psychoanalysts need connectionist theory because it presents them with a plausible link to biology; analysts can use its models to provide an account of the ego in terms of the brain. Connectionism can also help psychoanalysis undermine centralized and unitary views of the ego and support the notion of a decentered self. Historically, theories of a decentered self have needed to be supported.

Freud's notion of the unconscious had called into question the idea of the unitary self as an actor and agent. We don't know what we want, said Freud. Our wishes are hidden from us by complex processes of censorship and repression. Yet even as Freud decentered the ego, some of the theorists who followed him, collectively known as ego psychologists, sought to restore its central authority. They did so by focusing on the ego as a stable, objective platform from which to view the world. They began to see the ego as capable of integrating the psyche. To the psychoanalysts Anna Freud and Heinz Hartmann, the ego seemed almost a psychic hero, battling off id and superego at the same time as it tried to cope with the demands of the external world. Anna Freud wrote of the ego's powerful artillery, its "mechanisms of defense," and Hartmann argued that the ego had an aspect that was not tied up in the individual's neurotic conflicts; it had a conflict-free zone. This unhampered aspect of the ego was free to act and choose, independent of constraints. Hartmann's concept of a conflict-free zone was almost the site of a reborn notion of the will, the locus of moral responsibility. The intellectual historian Russell Jacoby, writing of ego psychology, described it as the "forgetting of psychoanalysis."[42]

For Olds, connectionism challenges ego psychology by providing a way to see the ego not as a central authority but as an emergent system. Through a connectionist lens, says Olds, the ego can be recast as a distributed system. Consciousness can be seen as a technical device by which the brain represents its own workings to itself. Olds likens it to "the monitor on a computer system," underscoring its passive quality. Even clinical practice can be interpreted in connectionist language: Interpretations that an analyst makes during a treatment session work when they correspond to a "well worn track in the brain, namely a set of connections among nets which generates a repetitive pattern of response and behavior."[43]

Olds acknowledges that really understanding connectionism requires "considerable mathematical sophistication." "[V]ery few people, including most psychologists, have even a sketchy understanding" of what the theory is actually saying. But he believes that connectionism will nevertheless be increasingly influential among psychoanalysts. Innocence of technical details has not kept psychology from mining scientific fields for their metaphors. Freud, for example, was not a master of hydraulic theory, but he borrowed many of his central images from it. Olds suggests that today's psychoanalysts should view connectionism in a similar spirit. What hydraulics was to Freud, emergent AI should be to today's analysts. In other words, Olds is explicitly advocating the use of connectionism as what I have called a sustaining myth:

> Many libido theorists probably did not know a great deal about steam engines; they made conceptual use of the properties which interested them. This is even more true with the early computer model; very few analogizers know a motherboard from a RAM, nor do they care. The way we *imagine* the machine handles information is what counts.
>
> The point is that what gets transferred from one realm to the other is a set of properties which we attribute to both entities.[44]

These remarks recall the way computers served as support for cognitive science in the 1950s and 1960s. There too, what the machines did was less important than how people thought about them. As Olds points out, although the theory of neural nets may be technically difficult, it is metaphorically evocative, presenting machine processes as the kinds of things that go on in the brain.[45]

THE APPROPRIABILITY OF EMERGENT AI

When the prevailing image of artificial intelligence was information processing, many who criticized the computer as a model of mind feared that it would lead people to view themselves as cold mechanism. When they looked at the computer, they had a "not me" response. Now we face an increasingly complex situation. These days when people look at emergent computer models they see reflected the idea that the "I" might be a bundle of neuron-like agents in communication. This sounds close enough to how people think about the brain to begin to make them feel comfortable. The not-me response turns into a like-me response.

I noted earlier that Freudian ideas became well known and gained wide acceptance for reasons that had little to do with their purported scientific validity. Ideas about the importance of slips of the tongue became part of the wider psychological culture not because they were rigorously proven but because slips were evocative objects-to-think-with. As people looked for slips and started to manipulate them, both seriously and playfully, the psychoanalytic concepts behind them began to feel more natural. Many of the ideas behind emergent AI are appropriable for the same reasons that slips were. For example, you can play with the agents of Minsky's society theory. You can imagine yourself in their place; acting out their roles feels enough like acting out the theory to give a sense of understanding it. The language of "society" is helping to disseminate the idea that machines might be able to think like people and that people may have always thought like machines. As for connectionism, it too has been gleaned for appropriable images. Some people mentally translate the idea of connection strengths between neuron-like entities into the notion of moving things closer together and further apart. Other people translate connectionist ideas into social terms. One twenty-two-year-old laboratory technician transformed the neural networks into a network of friends:

> The neural nets, like friends, can join up in teams in many different combinations and degrees of closeness, depending on how gratifying their relationships are. If a group of neuron friends makes a good, strong combination, their associations are going to get stronger. They will increase their degree of association.

Clearly, what is involved here is not a weighing of scientific theory but an appropriation of images and metaphors. Although emergent AI is more opaque than information processing in terms of traditional, mechanical ways of understanding, it is simultaneously more graspable, since it builds intelligence out of simulated "stuff" as opposed to logic. Because the constituent agents of emergent AI offer almost tangible objects-to-think-with, it prepares the way for the idea of mind as machine to become an acceptable part of everyday thinking.

The diffusion of popular versions of connectionist ideas about mind has been greatly facilitated by the fact that small neural net programs are easily run on widely available desktop computers. The two-volume *Parallel Distributed Processing,* what might be thought of as the Bible of connectionism's rebirth, was published in 1986. It inspired a flurry of programming activity, and not just among AI researchers and professional programmers. The PDP programs were simple enough for high school hackers and home computer aficionados to experiment with. For years, James McClelland and David Rumelhart, the editors of the PDP volume, had made their programs available to students in the universities where they taught—Carnegie Mellon, Stanford, and the San Diego campus of The University of California. But after the PDP volume was published, the demand for the programs was so great that Rumelhart and McClelland decided to put them on disks that would run on personal computers. The hardware requirements for running the programs were scarcely state-of-the-art. In the IBM version, you could be a connectionist with 256 kilobytes of memory, two floppy disk drives, a standard monochrome monitor, and version 2.0 of MS-DOS. It was like being told that you could be a cordon bleu chef using only a small Teflon frying pan and spatula.

In writing about the spread of ideas about microbes and the bacterial theory of disease in late nineteenth-century France, the sociologist of science Bruno Latour has argued that what spread the word was not the message put out by Louis Pasteur's writings, but the social deployment of an army of hygienists, state employees who visited every French farm.[46] They were the foot soldiers of Pasteur's revolution. PDP programs on floppy disks functioned similarly as carriers of emergent theories of mind.

The PDP disks and an accompanying workbook were published in 1988 with an "exhortation and a disclaimer." The disclaimer was that Rumelhart and McClelland could not "be sure that the programs are perfectly bug free."[47] They encouraged people to work around difficulties, to fix things where they could, and to send in their comments and suggestions. The developers of a cutting-edge scientific field were asking for help from their lay audience. This openness to criticism and collaboration was appealing. The exhortation was to

> take what we offer here, not as a set of fixed tasks to be undertaken, but as raw material for your own explorations. . . . The flexibility that has been built into these programs is intended to make exploration as easy as possible, and we provide source code so that users can change the programs and adapt them to their own needs and problems as they see fit.[48]

In other words, PDP was presented in a way that spoke directly to the learning style of the tinkerer—try it, play with it, change it—*and* to those who wanted to go below the surface. PDP combined the magic of emergence with the possibility of getting your hands dirty at the level of the source code. It was a powerful combination: hard and soft, bricolage and algorithm. It transcended old dichotomies: You could have your bricolage and feel like a real scientist too.

Emergent AI's message about complexity and emergence seems to be something that many people want to hear. The nondeterminism of emergent systems has a special resonance in our time of widespread disaffection with instrumental reason. Seymour Papert speculates that at least in part, emergent AI has been brought back to life by the resonance of its intellectual values with those of the wider culture, which has experienced a "generalized turn away from the hard-edged rationalism of the time connectionism last went into eclipse and a resurgent attraction to more holistic ways of thinking."[49]

So emergent AI manages to be seductive in many ways. It presents itself as escaping the narrow determinism of information processing. Its images are appealing because they refer to the biology of the brain. Like fuzzy logic and chaos theory, two other ideas that have captured the popular and professional imagination during the last decade, emergent AI acknowledges our disappointments with the cold, sharp edges of formal logic. It is consonant with a widespread criticism of traditional Western philosophy, which, as Heidegger once put it, had focused on fact in the world while passing over the world as such.[50] Emergent AI falls into line with postmodern thought and a general turn to "softer" epistemologies that emphasize contextual methodologies. And finally, its constituent agents offer a theory for the felt experience of multiple inner voices. Although our culture has traditionally presented consistency and coherence as natural, feelings of fragmentation abound, now more than ever. Indeed, it has been argued that these feelings of fragmentation characterize postmodern life.[51] Theories that speak to the experience of a divided self have particular power.

Like all theories that call into question an autonomous, unitary ego, emergent AI lives in a natural state of tension. Among the reasons decentered theories of mind are powerful is that they offer us a language of the self that reflects our sense of fragmentation. On the other hand, they are also under pressure from our everyday sense of ourself as unified. No matter what our theoretical commitments to a notion of a decentered self, when we say, "I do, I say, I want," we are using a "voice" that implies unity and centeredness. This tension between theory and the assumptions of everyday language has been played out in the history of the popular appropriation of psychoanalysis for nearly a century, most starkly in repeated confrontations with ideas that present the ego as the central executive of the mind. And now it is being played out in the history of the popular appropriation of artificial intelligence.

Information processing AI challenged the idea of the centered self when it equated human minds with rule-driven machine processes, but it offered an opening to centralized views of the mind. If the mind was conceived as a hierarchical program, it was relatively easy to imagine a program on top of the hierarchy that could be analogous to an executive ego. Emergent theories of AI are more radically decentralizing in their intent. Yet, like Freudian theory, they too are challenged by those who would recast them into centralized forms.

In the mid-1980s, MIT students influenced by Marvin Minsky's society theory said they were content to see their minds in a radically decentralized fashion. They spoke of their minds as "a lot of little processors," or as one put it, "In my mind, nobody is home—just a lot of little bodies."[52] However, even among those most committed to Minsky's views, some were tempted to put centralized authority back into his system, to make one of the little processors more equal than the others.[53] One young woman told me that one of the agents in her society of mind had the ability to recognize patterns. She said that the pattern-recognition agent was able to build on this skill, "grow in its ability to manipulate data," and "develop the ability to supervise others." Like ego psychologists with their conflict-free zone, she had reintroduced an executive agent into her idea of decentered mind.

Minsky's society model leaves a fair amount of room for such recentralizing strategies. Connectionist models leave rather less. Their neuron-like structures are poorly equipped to develop minds of their own, although I have interviewed several people who have imagined that *associations* of neural pathways might take on this executive role. Today, the recentralization of emergent discourse in AI is most apparent in how computer agents such as those designed to sort electronic mail or scour the Internet for news are being discussed in popular culture. As we have seen, the intelligence of such learning agents emerges from the functioning of a distributed and evolving system, but there is a tendency to anthropomorphize a single agent on whose intelligence the users of the program will come to depend. It is this superagent that is often analogized to a butler or personal assistant. The appropriation of decentered views of mind is a complex process; decentering theories are made acceptable by recasting them in more centralized forms, yet even as this takes place, some of the decentered message gets through all the same.

TROJAN HORSES

The Freudian experience taught us that resistance to a theory is part of its cultural impact. Resistance to psychoanalysis, with its emphasis on the unconscious, led to an emphasis on the rational aspect of human nature, to an emphasis on people as logical beings. In the 1970s and 1980s, resistance to a computational model of people led to an insistence that what was essential in humans—love, empathy, sensuality—could never be captured in language, rules, and formalism. In this way, information processing reinforced a split between the psychology of feeling and the psychology of thought. There was a dissociation of affect. The cognitive was reduced to logical process and the affective was reduced to the visceral. But the relationship between thought and feeling is more complex than that. There is passion in the mathematician's theorem and reason behind the most primitive fantasy. The unconscious has its own, structured language that can be deciphered and analyzed. Logic has an affective side, and affect has a logic.

Perhaps the models of human mind that grow from emergent AI might come to support a more integrated view. The interest of psychoanalysts in these models suggests some hope that they might, but there is reason to fear that they will not. In fact, the way emergent AI attempts to include feelings in its models provides some basis for pessimism. Take, for example, Marvin Minsky's way of explaining the Oedipus complex in *The Society of Mind*. Minsky sets the stage for the child to develop a strong preference for one parent in cognitive terms. "If a developing identity is based upon that of another person, it must become confusing to be attached to two dissimilar adult 'models.' "[54] For Minsky, Oedipus is simply an adaptive mechanism that facilitates the construction of an unconfused agent by removing "one [of the models] from the scene."[55]

Here, Minsky enters a domain where pure information processing seldom dared to tread, the domain of personality, identity, and subjectivity. The tenor of Minsky's emergent society theory is romantic and impressionistic. But when he actually applies his model, he turns the Oedipal moment (in the psychoanalytic world, thought of in terms of jealousy, sexuality, and murderous emotion) into an engineering fix for a purely cognitive problem. It is more economical and less confusing to have one role model than two so the cognitive unconscious acts to reduce dissonance. Minsky transforms the psychoanalytic consideration of primitive feelings into a discussion of a kind of thinking. Information processing left affect dissociated; emergent AI may try to integrate it but leave it diminished.

As this book is being written, Minsky is working on a book somewhat forbiddingly entitled *The Emotion Machine*. The term refers to both the brain-that-is and the computer-that-will-be. In a recent discussion of that project he acknowledged Freud as "marvelous" and credited Freud's first writings as marking the "birth of psychology." However, when Minsky took up the Freudian topic of pleasure, he gave it a special twist. "Pleasure," said Minsky, "is not an ennobling thing. It is a narrowing thing. You don't have pleasure. Pleasure has you. Pleasure has something to do with keeping you from thinking too many other things when short-term memories are being transferred." Beauty was given a similarly functional definition. It is what "keeps you from being able to find something wrong with something." So if people find a sunset beautiful, it may derive from "our not being a nocturnal animal and it was time to start finding a cave to hide in."[56] I have noted that when Fredric Jameson characterized postmodern thought, he wrote of its tendency to support a "waning of affect."[57] Minsky's theory does nothing if not this, albeit in a way Jameson probably didn't have in mind.

For the foreseeable future, emergent machine intelligence will exist in only the most limited form. But even now, it is providing a rich store of images and metaphors for the broader culture. The language of emergence mediates between technical AI culture and the general psychological culture in a way that the language of information processing did not. The language of neurons, holism, connections, associations, agents, and actors makes it easier for people to consider themselves as that kind of machine. The similarities between the object language of emergent AI and the object language of psychoanalysis have made AI appear more open to the concerns of the psychoanalytic culture that for so long has been hostile to it.

Some ideas require a Trojan horse for their appropriation—a vehicle in which they can be smuggled into unfriendly terrain. When AI was perceived as synonymous with information processing, it was generally unacceptable to humanists. Now that theorists of emergent AI use the language of biology, neurology, and inner objects to describe their machines, AI begins to seem interesting. Through such intellectual detours, romantic machines may have the effect that critics feared from the "classical," rational ones. For John Searle, no matter what a computer could do, human thought was something else, a product of our specific biology, the product of a human brain. But when connectionist neuroscience begins to revise the boundaries between brains, machines, and minds, it is harder to argue for the specificity of the human mind.

In the 1980s, in response to the computer presence, some people made a split between a mechanical vision of intelligence and an almost mystical vision of emotion. Others emphasized that computers could think but could not achieve the kinds of knowledge that come from being-in-the-world. Today, machines that promise to learn-in-the-world challenge us to invent new hybrid self-images, built from the materials of animal, mind, and machine. In the 1990s, artificial intelligence seems to be suggesting not modernist mind as mechanism but postmodern mind as a new kind of machine, situated somehow between biology and artifact.

In the 1980s, connectionism met the challenge posed by the romantic reaction to information processing AI by agreeing with the statement that people do not operate by simple preprogrammed rules. Connectionism simply added, "And neither do intelligent machines." But our desire to be something other than just machines did not disappear as we began to accept emergent views of both human and computer minds. By the late 1980s, the boundary between people and computers had been displaced in two ways. First, it was displaced away from thought to emotion. Computers might think, but people could feel. This displacement, however, was obstructed by dramatic advances in psychopharmacology, which suggested that the processes that underlie human motions are fairly "mechanical," predictable, and controllable. So while it remained true that computers don't have emotions, there grew up an increased uncertainty about what it means to say that people have them. This development made a second boundary displacement all the more important. People's sense of difference from computers shifted away from the domain of intelligence to the domain of biological life. Computers were accepted as intelligent, but people were special because they were alive.

When AI offered a rational and rule-driven machine, it led to a romantic reaction. Current romantic reconceptualiziations of the machine may now be supporting a rationalist reaction: a too-easy acceptance of the idea that what is essential about mind can be captured in what makes us akin to mechanism. In the past decade, our culture has more readily accepted the idea that human and machine minds might be similar. And for the moment, the question of intelligence is no longer the issue around which the border between people and objects is being contested. Now, people are more likely to distinguish themselves from machines by invoking biology. Our bodies and our DNA are becoming our new lines in the sand. The heat of the battle is moving to the issue of life.

Chapter 6

ARTIFICIAL LIFE AS THE NEW FRONTIER

In the mid-1980s, the biologist Richard Dawkins developed a computer program to illustrate some of the evolutionary principles he was writing about in his book, *The Blind Watchmaker*. A tearsheet at the end of the book makes the program available by mail order. Thus, what is in effect a piece of mail-order artificial life now runs on my desktop Macintosh. The program, also called the Blind Watchmaker, creates line drawings on the computer screen that can be varied by changing any of nine parameters. In the spirit of simulating evolution, Dawkins referred to these parameters as genes and made them subject to mutations, which are passed down to the next generation of drawings generated by the program.

Dawkins thought of the line drawings as analogous to organisms and named them biomorphs.[1] A user of the Blind Watchmaker program plays the role of a biomorph breeder. The program generates an ancestor biomorph, a simple tree-like shape, and nine offspring, each representing a single mutation of one of the biomorph's nine genes. The breeder then selects a preferred offspring from which the program generates more offspring, each carrying a further mutation, and so on. The biomorphs thus evolve by "unnatural selection," unnatural because the creatures are adapting to the taste of the user of the program.

Dawkins developed the program to dramatize that within a reasonable number of generations, simple structures could evolve into something quite different from an original random form. But Dawkins found that he had been overly conservative in his expectations.

> When I wrote the program I never thought that it would evolve anything more than a variety of tree-like shapes. . . . Nothing in my biologist's intuition, nothing in my 20 years' experience in programming computers, and

nothing in my wildest dreams, prepared me for what actually emerged on the screen. I can't remember exactly when in the sequence it first began to dawn on me that an evolved resemblance to something like an insect was possible. With a wild surmise, I began to breed, generation after generation, from whichever child looked most like an insect. My incredulity grew in parallel with the evolving resemblance. Admittedly [referring to a figure in his book] they have eight legs like a spider, instead of six like an insect, but even so! I still cannot conceal to you my feeling of exultation as I first watched those exquisite creatures emerging before my eyes. I distinctly heard the triumphal opening chords of *Also sprach Zarathustra* (the "2001 theme") in my mind. I couldn't eat, and that night "my" insects swarmed behind my eyelids as I tried to sleep.[2]

Dawkins wrote that with the program he was able to evolve "fairy shrimps, Aztec temples, Gothic church windows, aboriginal drawings of kangaroos, and, on one memorable but unrecapturable occasion, a passable caricature of the Wykeham Professor of Logic."[3] I cannot make a similarly diverse or eloquent report, but I too have evolved a barnyard of species on my home computer screen.

THE OBJECTS OF ARTIFICIAL LIFE

Dawkins's awe when confronted by the evolution of his biomorphs is reminiscent of an event in the early history of AI. In the late 1950s, Arthur Samuel wrote a checkers-playing program based on a perceptron model. The program played many games against human opponents, improved steadily, and finally achieved the status of a world class checkers player. But for early cyberneticians, the most dramatic moment in its life was not the day it beat a champion but the day it beat its own creator.

For Norbert Wiener, the mathematician and founder of cybernetics, this moment ushered in a new era. In *God and Golem, Inc.,* Wiener suggested that the implications were theological. "Can God play a significant game with his own creature?" he asked. "Can any creator, even a limited one, play a significant game with his own creature?"[4]

Wiener made the point that although the Samuel program "merely did what it was told to do," it nevertheless represented a significant departure in the nature of our artifacts. For the checkers program had in effect been told to go off and make its own decisions about how to win at checkers, even if that meant it would eventually learn to defeat its own creator.

For Wiener the Samuel program suggested the violation of an ancient taboo of speaking of "living beings and machines in the same breath."[5] But this is exactly the kind of talk that is now commonplace in technical manuals for software that runs on our personal computers. The Blind

Watchmaker enables the user to "evolve" biomorph "organisms" on a computer screen; Tierra has digital "chromosomes," which are said to mate, mutate, and evolve; and SimLife, a game billed as "a genetic playground," asks its users to create evolving "creatures" in a dynamic virtual ecology. These days, you don't need to be a visionary like Wiener to see ancient taboos under threat.

The Blind Watchmaker, Tierra, and SimLife are all programs that have grown out of a recent research effort known as "artificial life," or A-Life. These programs are only a few of the many evocative objects that this field has deployed into the culture in less than a decade. Just as the lively interactivity of computer toys caused children to reconsider their criterion for aliveness, these objects challenge us to reconsider our traditional boundaries and categories regarding life itself.

When AI was in its infancy, many of its proponents delighted in making provocative statements about the machine-like nature of human intelligence and how close they were to recreating it on computers. Some took offense to this as arrogance or hyperbole, but in many ways, these provocations served as a tonic. A professional academic culture had relegated philosophy to seminar rooms, but even the earliest "thinking machines" brought philosophy into everyday life. In some ways artificial life stands today where artificial intelligence stood thirty years ago. While people working in the field of AI insisted that intelligence can be shared with our creations, a point of view to which there now seems to be growing accommodation, A-Life now proposes a similar hypothesis about life.

Artificial life, to paraphrase Marvin Minsky, is the discipline of building organisms and systems that would be considered alive if found in nature. Some A-Life researchers limit their work to simulating some aspect of life on the computer, say flocking behavior, in order to better understand how that aspect of life functions in nature.

Other researchers stake out a bolder claim. They are not merely trying to simulate the appearance of life. They are trying to create life. Thomas Ray, the biologist who authored the Tierra program, believes that "the life force took over" in the earliest moments of the operation of his system.[6] The physicist James Doyne Farmer thinks that the emergence of new human-designed organisms "alive under any reasonable definition of the word" is close upon us and that "the advent of artificial life will be the most significant historical event since the emergence of human beings."[7] Christopher Langton, the organizer of the First Conference on Artificial Life in Los Alamos, New Mexico, in September 1987, believes that it is not too soon to begin thinking about the rights of a "living process" to exist "whatever the medium in which it occurs." He adds that A-Life is a challenge to "our most fundamental social, moral, philosophical, and religious beliefs. Like the Copernican model of the solar system, it will force us to re-examine our place in the universe and our role in nature."[8]

In 1987, while interviewing people about their views of machine intelligence, I was struck by the degree to which they believed it was possible if not imminent. But people's desire to be other than machine did not disappear; it was simply displaced. Victoria, a graduate student in education, put it this way: "Computers might think, and even the way people do. But people are alive. And that's the bottom line." The following September, the participants at the first A-Life conference shared the conviction that it was time to cross what Victoria had termed the bottom line.

After the 1987 conference at Los Alamos, a consensus began to develop that to qualify as life artificial organisms must demonstrate four qualities. First, they must exhibit evolution by natural selection, the Darwinian aspect of our definition of life. Second, they must possess a genetic program, the instructions for their operation and reproduction, the DNA factor in our definition of life. Third, they must demonstrate a high level of complexity. A complex system has been defined as "one whose components parts interact with sufficient intricacy that they cannot be predicted by standard linear equations; so many variables are at work in the system that its overall behavior can only be understood as an emergent consequence of the myriad behaviors embedded within."[9] With complexity, characteristics and behaviors emerge, in a significant sense, *unbidden*. The organism can self-organize. This makes life possible. Thus, the quality of complexity would lead to the fourth necessary quality: self-organization.

The participants at the first A-Life conference looked forward to the creation of life forms that would exhibit the self-organization that AI researchers had already observed in their emergent systems. The nascent A-Life community shared something else with AI researchers. Both groups believed that the path to understanding their subjects—intelligence in one case, life in another—was to try to build them. Christopher Langton made the distinction between life as it is (what the world has presented to us) and life as it could be (what A-Life will build for the world.).[10] The two forms of creature may be different. But for the new discipline of A-Life both would be equally alive.

Like emergent AI, artificial life was nurtured by a new sense of what increasingly powerful computers could do in terms of modeling complexity. It was also encouraged by chaos theory, which appeared to demonstrate that mathematical structure existed underneath apparent randomness and that apparent randomness could generate mathematical structure. In artificial life research, the emergent bottom-up approach that characterized connectionism is taken a step further. While connectionist systems begin by assuming the existence of neuron-like objects, the most radical forms of A-Life thinking insist that the only acceptable a priori for an artificial organism is the digital equivalent of the primordial soup from which real life emerged.

A-Life is still in its early days, yet already one sees that it is developing within a seeming cultural contradiction. It poses a much more radical challenge than AI ever posed (not just to the uniqueness of human intelligence, but to biological life itself). Yet people seem to be taking it much less personally. Victoria's sentiment that life is the bottom line in thinking about human uniqueness is widespread. Yet the field of artificial life, which challenges that sentiment, seems to excite relatively little controversy. In the early days of artificial intelligence, the debate about its supposed threat to human uniqueness was carried into films, comic books, and television series, as well as journalistic accounts and philosophical discourse. These days, A-Life advances tend to be reported rather more neutrally.

Perhaps the contradiction is more apparent than real. A-Life makes claims that on their face make AI look tame by comparison, but from a certain point of view, it can be taken as less rather than more threatening. The researchers of artificial life may call the programs they write "organisms" or "creatures," but their imprisonment on the screen and their apparently modest ambitions make it easy to accept their analogousness to life without feeling frightened. One chemist I interviewed put it this way:

> It's one thing if you are looking at the possibility of an intelligent machine; it's another if you are looking at the possibility of a picture on a screen that learns to move at the level of an amoeba. They may call the second artificial life, but I just can't get worked up about it."

Perhaps it is the idea of an intelligent machine, a machine that might engage people in meaningful conversation, that constitutes the real psychological threat.[11]

If people see A-Life as a practical threat (in the form of computer viruses), not a philosophical one, this may be because discomfort arises only when A-Life pushes into human-like intelligent behavior, not ant-like behavior. But if this is the case, we may only be experiencing a calm before the storm. Some A-Life researchers make it clear that they are not content with "just" recreating life: They believe that in the long run, A-Life research is the best path for recreating intelligence as well. They argue that in most of our human lives we are doing things that are more like ants foraging for food than grand masters playing chess. These "foraging-like activities" form a foundation for human behavior: You can't ignore this foundation and hope to build human intelligence. So from this point of view, A-Life is a continuation and a deepening of emergent AI.

The Game of Life

My own introduction to what would become the field of artificial life began with a chance encounter with a program known as the Game of Life.

In 1977, I often worked late writing at a networked computer terminal at MIT's 545 Technology Square, also known as Tech Square. My text would be printed on a laser printer on the ninth floor, which was also the site of the workstations of the legendary Tech Square computer hackers. On my way back and forth from the laser printer, I would pause to observe their intensity of concentration. They seemed not to move from hour to hour through the night, their screens usually filled with line after line of computer code. One night in March, however, one screen was filled with complex animation. I was awestruck by the wonderfully evolving forms. This, I was told, was the Game of Life.

The mathematician John Conway created the Game of Life in the late 1960s. It is a classical illustration of the rule-based structure that the mathematician John von Neumann had called a cellular automaton.[12] The Game of Life consists of a checkerboard of black and white squares. The squares follow simple rules that direct them to remain black or white or to change their color depending on the colors of their neighbors. The rules are simple but complex patterns soon emerge. In fact, whole colonies emerge, multiply, and die. Repetitive patterns, called gliders, travel across the board and beyond its field. A glider is a pattern that moves across the screen as if it were an object. It is always made up of a different collection of cells (just as a traffic jam over time is made up of different cars). It is an "emergent object."

Conway had developed the Game of Life by playing it by hand in the common room of the Cambridge University mathematics department. But he wanted to prove it to be what Alan Turing had called a universal computer. This meant it had to be able to emulate any other machine. To achieve this, Conway had to discover a way to create new emergent objects *ad infinitum*. Through Martin Gardner, who wrote the Mathematical Games column for *Scientific American,* Conway put out a call for help. Within a month, William Gosper and a group of coworkers, all MIT hackers, came up with a solution. They had found a glider gun. Later, the hackers found other emergent objects, configurations both unexpected and extravagant. Gosper said, "We eventually got puffer trains that emitted gliders which collided together to make glider guns which then emitted gliders but in a quadratically increasing number. . . . We totally filled the space with gliders."[13]

The Game of Life thus entered the intense, nocturnal culture of the ninth-floor hackers at Technology Square. When I first came upon it, the Game of Life was running on a small, unattended screen. Things came together and flew apart, shapes emerged, receded, and reemerged. I remember thinking about fire and water. The French philosopher Gaston Bachelard had written about the universal fascination of watching fire, which, like watching moving water, is something that people seem drawn to across all times and cultures.[14] There is repetition and sameness, surprise and pattern. Fire and water evoke the eternal patterns of life, and now, so could a computer screen. In 1977, I stood alone at the screen, watched the Game of Life, and felt like a little girl at the ocean's edge.

At the time I assumed that all life had to be carbon based. I had never had any reason to think about the problem any other way. My initial inclination was to file away what I had seen on the ninth-floor computer screens as a mere curiosity. But as I came to understand how the Game of Life could be reset to generate complexity from initial randomness, it took on another level of fascination. I saw how this evolving, unfolding visual display might challenge my simple preconceptions. Perhaps something about life could be understood in terms of these evolving forms. I was intrigued by the idea but resisted it. In that resistance I was not alone.

In 1981, at the height of the romantic reaction against information processing AI, Douglas Hofstadter suggested that the kind of resistance I felt might be due to our culture's outmoded notion of what a machine is. In an essay called "A Coffee House Conversation on the Turing Test,"[15] he has a philosopher named Sandy say: "I think we all have a holdover image from the Industrial Revolution that sees machines as clunky iron contraptions gawkily moving under the power of some loudly chugging engine."[16]

When Sandy's friend, a biologist named Pat, insists that human beings must be more than machines, for after all, "they've got a sort of *flame* inside them, something alive, something that flickers unpredictably, wavering, uncertain—but something *creative,*" Sandy interrupts with some condescension:

> Great! That's just the sort of thing I wanted to hear. It's very human to think that way. Your flame image makes me think of candles, of fires, of vast thunderstorms dancing across the sky in crazy, tumultuous patterns. But do you realize that just that kind of thing is visible on a computer's console? The flickering lights form amazing chaotic sparkling patterns. It's such a far cry from heaps of lifeless clanking metal! It's flamelike by God! Why don't you let the word "machine" conjure up images of dancing patterns of light rather than of giant steam shovels?[17]

The Game of Life spread an image of Sandy's kind of machine. But for patterns and flames to burn indelibly in the imagination, they should burn fast and bright. At MIT, Edward Fredkin's research group developed a specialized computer that could run the Game of Life faster, backwards, and in color. Then, they miniaturized it onto a chip that could be put inside a standard IBM PC. This meant that by 1980, within minutes and on any desktop, one could watch a visual display that evoked the first moments of self-organization in the story of life. With the Game of Life the word "machine" did finally begin to "conjure up images of dancing patterns of light rather than of giant steam shovels."

The ideas that drove the Game of Life, ideas about universal Turing machines and self-replicating automata, are esoteric. But watching the Game of Life did not feel esoteric at all. The game embodied a way of thinking that extended the definition of life to self-reproducing and evolving objects that existed only within computers.

Hofstadter's "Coffee House Conversation" tries to counter resistance to the notion of life within a machine. But A-Life artifacts also have enormous appeal. It is easy to get caught up in the excitement of graphics that purport to demonstrate how the organic emerged from the inorganic, how intelligent order grew out of chaos. It is easy to lose one's moorings. Calling a game Life does not make it alive. Calling a computational object an organism or a robotics device a creature does not make them alive either. Yet the use of such terms is seductive.

The language we use to describe a science frames its objects and experiments, and, in a certain sense, tells us what to think of them. Sometimes we invent the language after we have the objects and have done the experiments. In the case of artificial life, a language existed before the birth of the discipline. This was molecular biology's way of talking about life as information, symbol, and code. The language of molecular biology constituted a sharp break with what had come before. Until the mid-twentieth century, biologists considered that the notion of "life" transcended anything one could know about the workings of specific organisms. You could understand a particular living thing, but "life" was a secret. But with the rise of molecular biology, says the rhetorician Richard Doyle, the secret was out.[18] Molecular biology clearly communicates that what stands behind life is not a metaphysical entity but DNA, whose code begins to be seen as life itself. Doyle argues that molecular biology prepared a language to frame the discipline of artificial life, one which equates life with "the algorithms of living systems."[19]

According to Doyle, the proponents of A-Life use a language that defines life as that *code* which creates distributed, parallel, and emergent behaviors. We begin to see the use of this language and the resulting definitions as absolute or "natural," but in fact, they are our constructions. Recognizing that we constructed them makes it possible to "deconstruct" them, to take away any pretense they have to being the "natural" way to see things.[20] This kind of exercise is particularly useful for citizens of the culture of simulation. When we talk about what we see on our computer screens it is easy to drop the quotation marks from words like "life," "mind," "intelligence," and "organism." But if von Neumann's cellular automata are "organisms" it is because he defined them as such. If Conway's game is about "life" it is because he said it was and we are allowing it to be. This process of constructing definitions and marshaling support for them is central to my argument: In contact with the objects of A-Life now deployed in the culture, people are redefining what it means to call something alive.

For Christopher Langton, the object that provided the impetus for such redefinition was the Game of Life. In the late 1960s, Langton was working in a computer facility at Boston's Massachusetts General Hospital. Some visitors from MIT brought the Game of Life into his laboratory. Langton was enthralled. Within a few years he would borrow money to buy his own Apple II computer, use it to program self-reproducing cellular automata that developed much like a coral reef, and become the major personality in establishing A-Life as a field in its own right.

Langton's personal journey would become something of a founding myth for the field of artificial life. It includes two key elements of how participants in this young discipline see their enterprise. A first element is that A-Life mobilizes people at the margins of established institutions who are empowered by personal computers. When he began his research, Langton was, after all, a college dropout with an Apple II. A second element is that A-Life is fueled by moments of epiphany. One night, when working alone in the Massachusetts General laboratory, Langton said that he felt a "presence."[21] He believed it to be the Game of Life running unattended on a computer near his own. "You had the feeling that there was something very deep here in this artificial universe and its evolution through time," said Langton. "[In the lab] we had a lot of discussions about whether the program could be open ended—could you have a universe in which life could evolve?"[22] A few years after this first epiphany, Langton was seriously injured in a hang-gliding accident. In the many months spent on his back recuperating, Langton said that he could feel his mind trying to self-organize, much as individual ants in a colony do, much as cellular automata do.[23] For him, A-Life's essence, bottom-up emergence, became charged with personal meaning.

In the history of science, marginality and epiphany are not in short supply. They often feature in the early days of sciences that set off in radically new directions. The historian of science Thomas Kuhn refers to such moments as paradigm shifts.[24] Everyone working within the new paradigm is marginal because there is not yet an established discipline, and more mainstream sensibilities have usually been drawn to less chaotic intellectual fields. When Langton wanted to study cellular automata and their relevance to biological, economic, and social systems, there was only one department—at the University of Michigan—where he could study, and that department was being closed down. Langton became its last graduate student. Epiphany, too, is often part of the story of how people forge ahead in uncharted scientific terrain. Passion and revelatory insight help carry forward new and controversial ideas since institutional support and funding are usually in short supply.

One of the things that is most striking about the emergence of the discipline of artificial life is its science-on-a-shoestring quality, made possible by the personal computer as a research tool. Personal computers have made possible a radical democratization of discovery. By the time Langton was ready to make his first major contribution, he could do so on his own, independent of a university department or research grant. He could do it as an independent virtuoso, a hacker. In the early history of artificial intelligence, hackers were always invaluable researchers, systems developers, and innovators. But they had to carry out their work in the interstices of large established laboratories, because that was where the computers were. More recently, the machines have literally come into their hands.

The personal computer had another special role to play in the story of A-Life. The artificial life programs that were developed on small personal computers could also be *run* on small personal computers. That meant they could be run in millions of places all over the world. The products of artificial life research could be deployed, as they had been deployed into Langton's laboratory at Massachusetts General Hospital, as foot soldiers in a campaign to change the way large numbers of people would think about life. The products of discovery could be shipped off in envelopes with floppy disks inside or downloaded onto personal machines through Internet connections. Like Soviet dissidents who took 35-mm photographs of political tracts in order to pass them more easily to friends, the pioneers of artificial life possessed subversive objects that were capable of traveling light.

GENETIC ALGORITHMS

Unlike Darwinian evolution through natural selection, in Dawkins's Blind Watchmaker the person running the program acts as the hidden hand of evolution. A mixture of centralized control and bottom-up evolution also characterizes a class of programs developed by John Holland in the 1960s called genetic algorithms. In a genetic algorithm, a string of randomly generated zeros and ones is defined as a chromosome.[25] The Blind Watchmaker could evolve increasingly intricate line drawings; genetic algorithms are able to evolve increasingly efficient computer programs.

A scenario in which genetic algorithms are called upon to perform a simple computational task, number sorting, illustrates how they work. First, randomly generated strings of binary numbers are created and named "chromosomes." Then, each chromosome is put into a computer system that treats it as though it were a program written to sequence numbers. Of course, there is no reason to think that a randomly generated assortment of zeros and ones will be able to do this job, but if you generate enough such chromosomes, some do get something right. For example, one chromosome could cause the number 1 to appear at the beginning of the sequence of numbers. Some may get several things right. They may also hit upon, say, placing the number 5 before the number 6. These chromosomes are designated the winners and are given the best chance of being selected to be copied, that is, to reproduce.[26]

Genetic algorithms turned out to be very good at getting better at such things as number sequencing. But beyond their specific achievements, they raised a larger question. How much should these artificial organisms, these creatures that live only on computer screens, inform our thinking about naturally occurring life?

This question was made explicit when experiments with genetic algorithms began to generate findings at odds with assumptions that biologists had made about "real" evolution. For example, biologists had long considered mutation to be a central motor of evolution. But researchers working with genetic algorithms found that even when the rate of mutation in their systems was reduced dramatically, it made very little difference to the pace of evolution. One researcher forgot to simulate any mutation at all and got results that still looked a lot like evolution. From the point of view of the artificial life researchers, the motor of evolution seemed instead to be what is called crossover, when genetic information goes through mixing and matching between two gene strands. Each strand is cut, and the "top part" of one strand glues together with the "bottom part" of the other strand, and vice versa. Artificial life researchers thought that biologists had underrated the power of crossover because in the natural world one cannot easily examine organisms to see this type of genetic change as it is occurring. Additionally, in real life, one cannot observe the effect of controlled genetic changes over thousands of generations. But this was exactly what A-Life researchers were able to do. In their experiments, they could analyze the equivalent of hundreds of thousands of years of evolution.

Freed from the constraints imposed by biological evolution, artificial life researchers thought they could see nature's deeper truths.[27] W. Daniel Hillis argued this point for his own artificial life research. Hillis built digital organisms he called ramps and then used genetic algorithm techniques to evolve them. Hillis found that the ramps were evolving even during what seemed to be "rest periods," where things appeared to be in genetic equilibrium. During the presumably static period, Hillis could see that recessive genes were accumulating and setting the stage for the next leap forward in the observable fitness in the population. Moving out of the equilibrium state was not necessarily due to a crucial mutation or a drastic perturbation in the environment.[28]

Hillis's ramps were given a number-sorting task and evolved clever programming strategies to optimize their performance. In Hillis's view, their competence was evidence that the future of computer programming lies not with human programmers but with evolving programs.[29] Arthur Samuel had been amazed that his own program could beat him at checkers. Now, programmers such as Hillis thought that their programs would soon be better programmers than they were. Norbert Wiener had faced the theological implications of Samuel's program with awe. Thirty years later, Hillis was more matter of fact:

> We all find programming very frustrating. The dream is to have this little world where we can evolve programs to do something we want. We punish them and wipe them out if they don't do what we want, and after a few hundred thousand generations or so, we have a program that ideally behaves the way we like. *We can play God*—but all we have to *do* is define the puzzle for them. We don't have to be smart enough to figure out the way they solve the puzzle.[30]

Hillis came to the image of playing God after developing a simulated evolutionary system on an advanced parallel computer. But such ideas also come to people who are experimenting with far more modest A-Life simulations. Even the relatively straightforward biomorphs can evoke powerful emotions. Laura, a twenty-four-year-old researcher for public television and a devout Episcopalian, said this after evolving a butterfly with the Blind Watchmaker: "If God had a design and He had wanted to evolve humans, this is how He could have done it. So, you could have evolution and control by a higher being." When asked to reflect on how using the program made her feel, she said, looking somewhat embarrassed, "Like God."

Tierra and "Natural" Artificial Evolution

Hillis's work, like that of Richard Dawkins and John Holland, showed that something like evolution could be simulated on a computer. But their simulations shared a significant limitation. In each case, the fitness of the artificial organisms had to be determined by the computer user, who played the role of God within that organism's microworld. In the Blind

Watchmaker, the fitness was judged directly by the user. In the case of genetic algorithms, the user created a fitness function, a criterion by which the chromosomes were judged. In the mid-1980s, the biologist Thomas Ray built a system he called Tierra, the Spanish word for Earth. In Tierra's microworld there would be no breeder and no predetermined goal for evolution. If digital organisms evolved, it would be purely in response to the requirements of their digital environment.

Ray's path toward creating Tierra tells a story similar in key particulars to Langton's. It too includes the elements of epiphany, marginality, and the importance of a personal computer, in this case a Toshiba laptop.

In the late 1970s, when Ray was a graduate student in biology at Harvard, he had a chance meeting with a hacker from the MIT Artificial Intelligence Laboratory. In the cafeteria of the Harvard Science Center, the hacker analyzed the Chinese game of Go as though it were a biological system and casually mentioned that computer programs could self-replicate. In 1989, when Ray found himself fascinated by the "biological" properties of computer viruses, he remembered that conversation. Despite discouragement from his professional colleagues, Ray, a specialist in the ecology of rain forests, changed the direction of his research and became dedicated to creating "naturally" evolving digital organisms.

Ray began by designing a digital organism capable of replication and mutation. He called it the ancestor. Like genetic algorithms, the ancestor and its descendants were made of zeros and ones that, when strung together, functioned as little computer programs.[31] The execution of any one organism's code would usually result in that organism's being copied from one part of the Tierran environment to another. "Replication" occurred when the move resulted in a daughter cell that required its own slice of CPU (Central Processing Unit) time. There was mutation by random "bit flipping," in which zeros were changed into ones and ones into zeros. Ray saw bit flipping mutations as analogous to mutations caused by cosmic rays. And there was also mutation through random variations.

All the organisms in Tierra, parents and children, mutants and faithful copies, were competing for the scarce resources in the system, the computer's processing time and memory space.[32] If a given Tierran functioned well in its tasks, the system rewarded it with CPU time; if it did not function well, it was punished with "death." In any case, all Tierrans had a finite life span because Ray introduced a "reaper function" that regularly killed off the oldest organisms. By successfully executing instructions, organisms could postpone the reaper, although not indefinitely.

On the first run of Tierra, Ray seeded the system with a creature that required eighty instructions to replicate. It evolved progeny that were able to replicate with fewer instructions and thus were more fit because more efficient. These progeny then developed parasites even smaller than they. The parasites could not replicate by themselves, but could piggyback onto larger host organisms in order to pass on their genetic material. Some organisms developed an immunity to parasites, which led to the birth of new kinds of parasites. On its very first run, Tierrans were self-replicating, the system was open-ended, and it was evolving. From Ray's point of view, he had created life.

The Tierran microworld suggested, as had the work of Hillis and Holland, that genetic mutation was not necessarily the driving force behind evolution[33] and that the best computer programs of the future might be artificially evolved. Ray stressed, as had Langton, Dawkins, and Hillis, that one of the more striking things about working with digital organisms was how they surprised him. "They came up with patterns for how to do things I wouldn't have thought of, the specific ways they developed parasitism, super-parasitism, 'stealing' energy from each other." And the Tierrans even exhibited behavior that to Ray suggested a "sense of self." As Ray put it, "One of the more adaptive creatures that evolved needed to ask itself, 'Where do I begin, where do I end, and how big am I?' in order to function."[34]

There was something powerful and easily communicable about Ray's achievement. Tierra was heralded in such places as *The New York Times, The Economist, Science,* and *Nature.* It was a classical example of an appropriable computer microworld. And since it could be downloaded over the Internet, you could play with it and the ideas behind it on your home computer.

At the Third Conference on Artificial Life in 1992, Ray's workshop about Tierra was packed to overflowing. The workshop participants seemed filled with respect for the Tierrans' ability to improve themselves. One twenty-four-year-old programmer put it this way, "I don't know much about biology. . . . But these little programmers [the Tierrans] are alive enough to make leaps that are almost beyond my abilities. That's pretty alive to me. I think evolution *is* life." As an object-to-think-with for thinking about life, Dawkins's Blind Watchmaker made users feel like gods. Tierra had a rather different effect. If there was a god, it was in the system.

A fifteen-year-old boy who had been introduced to Tierra in a high school science club told me that working with Tierra made him feel as though he were

> looking at cells through an electron microscope. I know that it is all happening in the computer, but I have to keep reminding myself. You set it up and you let it go. And a whole world starts. I have to keep reminding myself that it isn't going to jump out of the machine.

He concluded, "I dreamt that I would find little animals in there. Two times I ran it at night, but it's not such a great idea because I couldn't really sleep."

A PICTURE IS WORTH ...

In 1984, the journalist James Gleick wrote the best-selling *Chaos: Making a New Science*. Gleick's words were eloquent, but some of the book's power was carried by its extraordinary illustrations. They carried the startling message that nature had its own geometry of self-similarities, that there was structure underlying apparent randomness.

By 1987, artificial life researchers had accumulated some impressive pictures of their own. Some of the more striking visual images grew out of work in the tradition of von Neumann's cellular automata. In addition to the Game of Life, cellular automata had been used to produce patterns that looked like mollusk shells, coral reefs, and snowflakes. Another set of dramatic images could be traced back to the insights of Herbert Simon about how the seeming complexity of behavior in such a creature as an ant "was only a reflection of the complexity of the environment in which it finds itself."[35] Almost twenty years later, artificial life researchers wanted to pick up where Simon left off. Since when we talk about "lifelike" behavior we often mean intentional-seeming behavior, they asked how this might arise from simple responses to a complex environment.

For example, to the computer animator Craig Reynolds, it seemed possible that flocking behavior, whether in fish, birds, or insects, did not require a flock leader or the intention to flock. Reynolds wrote a computer program that caused virtual birds to flock, in which each "bird" acted "solely on the basis of its local perception of the world."[36] (Reynolds called the digital birds "boids," an extension of high-tech jargon that refers to generalized objects by adding the suffix "oid." A boid could be any flocking creature.)

The rules for each boid were simple: If you are too close to a neighboring boid, move away from it; if not as fast as your neighboring boid, speed up; if not as slow as your neighboring boid, slow down. Since flocking requires that boids stick together, each boid had to move in the direction of greater density of boids. After some fine-tuning, Reynolds got flocks of boids on the screen to exhibit realistic behavior.

Recall Lisa's poem from Chapter 2, her paean to the natural world:

If you could say in numbers what I say now in words,
If theorems could, like sentences, describe the flight of birds, . . .

Now they could—in a sense. The boids program did not, of course, prove anything about how real birds fly, any more than Hillis's ramps or Ray's Tierra proved how real evolution occurred. But as a mirror to nature, the computer-generated birds went part way toward meeting

Lisa's objections. They did this not only by demonstrating (as Lisa was wagering they never could) that mathematics could describe complex natural motions, but also by eroding preconceptions about the natural behavior of animals. The program that produced seemingly intentional behavior was nothing but a set of simple local rules. And it was, after all, more plausible that real "birdbrains" used simple local rules than that they forged attachments to a leader and performed trigonometry to stay together in mathematically precise groups. Reynolds's program helped bring ideas about emergent intentionality and purpose into thinking about nature.[37] And it was carried, as ideas about chaos had been, by the prettiest of pictures.

The video of Reynolds's boids program in action was of interest to computer animators, Hollywood artists, ornithologists, and high school biology teachers. For many who saw the video, Reynolds had created a species of almost-animal. Reynolds himself argued that the boids were close enough to real birds to deserve being studied in their own right by naturalists. "Experiments with natural flocks and schools are difficult to perform and are likely to disturb the behaviors under study," wrote Reynolds. But biologists might use "a more carefully crafted model of the realistic behavior of a certain species of bird, to perform controlled and repeatable experiments with 'simulated natural flocks.' "[38] And indeed, biologists soon began to work with "simulated natural flocks" of insects, which, as Simon had anticipated, could have significant aspects of their behavior accurately modeled by assuming that global behavior emerged from local rules. These "simulated natural lives" were marginal, betwixt and between logic and biology. They communicated the idea that what counts most is not what life is made of but the organizing processes behind it.[39]

An Object Is Worth ...

Today more than ever we blur the line between simulation and reality, between what exists on the computer and what is real. Nevertheless, we are usually aware that pictures and screen representations are not in the world in the same sense that we are. Although the artificial life created on the computer challenges our sense of what life processes are about, it is still possible to draw the line between us (the people, the animals, the "real" insects) and them, by saying, they are in there (the computer) and we are out here in the world. According to this way of thinking, one might say that the Tierrans on the screen, who take their energy from Thomas Ray's CPU sun, might be evocative objects, but we down-to-earth tierrans are the ones who get sunburned. We are the real thing.

But what happens when we create creatures in the world who embody the qualities of emergent, evolving intelligence we have created on the screen? For generations, much of the appeal of mechanical automata or robots has depended on our fascination with the idea that artificial creatures might exist in the world. A coffee-table volume called *The Robot Book* opens with a "Chronology of Robot-Related Events."[40] These begin in 3000 B.C. with Egyptian articulated figures and continue through the clock-tower automata of the middle ages and the extraordinarily intricate automata of the eighteenth and nineteenth centuries.[41]

The intellectual history of modern robotics has paralleled that of artificial intelligence. Traditional AI envisaged intelligence as a top-down operation and traditional robotics envisaged robots that needed predetermined, top-down plans of action. As we have seen, a new school of robotics research, associated with MIT's Rodney Brooks, has embraced the bottom-up aesthetic of emergent AI. These robotic devices are sometimes referred to as "real artificial life."

At MIT, a version of "real artificial life" has even been created for children. At the Media Laboratory, a group headed by Seymour Papert used standard Lego construction block parts, as well as sensors, motors, and computers, to assemble a robot building kit. Once they build these robots, children program them in the Logo computer language and play with them as they move around the room.[42]

Mitchel Resnick, one of Papert's collaborators on the Lego-Logo work, presented it to the First Conference on Artificial Life in 1987. Resnick talked about how children begin to think about their robots in psychological terms and, in particular, begin to attribute personality and intentionality to them. He said that children come to think about the creatures in many different ways, cycling through mechanical, psychological, and informational descriptions.

Indeed, Resnick admitted that he could even see himself alternating between different views of his Lego-Logo creations. One moment he was an engineer writing code that would enable a Lego-Logo robot to follow a line drawn on the ground, and the next instant he was studying the creature as though it were an animal in an experimental psychology laboratory. Resnick described an incident in which a Lego-Logo creature surprised him and in doing so became, as one child I interviewed put it, an "almost life."

We wrote the program and it [the Lego-Logo robot] was following this line. And all of a sudden it struck me that I had no idea what was going to happen when it reached the end of the line. We'd written the program, but we didn't take that into account. I didn't have time to think about what it might do, so I just watched it. When it got to the end of the line, it turned around and started following the line in the other direction! If we had *planned* for it to do something, that would have been the ideal thing for it to do.[43]

The Lego-Logo work was related closely to the emergent aesthetic of artificial life. Christopher Langton had characterized that aesthetic as bottom-up, parallel, and showing local determination of behavior.[44] The Lego-Logo creatures were bottom-up and had local determination of behavior. But it was generally impractical to run many Lego-Logo creatures in parallel. To explore parallel behaviors, Resnick created a new version of Logo, called StarLogo, which allowed children to control the parallel actions of many hundreds of "creatures" on the computer screen. Children could use StarLogo to model birds in a flock, ants in a colony, cars in a traffic jam.

Using StarLogo brought children into the aesthetic of emergent systems. This was obvious when they talked about StarLogo programming in terms of one of the aesthetic's defining features—complex behavior can emerge from a small number of simple rules. One child said, "In this version of Logo, you can get more than what you tell it to do." For another, "It's weird. . . . I mean there's not that much to program. . . . With a few short procedures, a lot happens."[45]

StarLogo was explicitly designed to facilitate what Resnick called the decentralized mindset, a way of looking at the world that stresses parallelism and emergence. But reactions to StarLogo also provided a window onto resistance to that mindset. Resnick reported that even after extensive exposure to StarLogo and ideas about decentralized control, many people continued to assume that when they saw an organized system, the organization had been imposed from the top.[46]

Evelyn Fox Keller, reviewing the history of resistance to decentralized ideas in biology, wrote, "We risk imposing on nature the very stories we like to hear."[47] But why is it that we like to hear stories of centralized, top-down control? This same question came up in discussing the tenacity of ideas from ego psychology as well as the reasons why emergent artificial intelligence went underground in the late 1960s. Many different factors seem to contribute. There is the Western monotheistic religious tradition. There is our experience of ourselves as unitary actors (the ego as "I"). And there is the fact that we lack concrete objects in the world with which to think about decentralization. On this final point, Resnick hopes that objects such as StarLogo may represent a significant change. StarLogo is an object-to-think-with for thinking about decentralization.

As more of such objects enter our culture, we may be even more willing to describe everything from natural phenomena to business organizations in decentered terms. Despite the fact that we may "resist" them, decentered phenomena have great appeal. They can seem magical because they are unpredictable in a special way: One can know what every individual object will do, but still have no idea what the system as a whole will do. It feels as if you get more out than you put in. Decentered phenomena combine a feeling that one knows the rules with the knowledge that one cannot predict the outcome. They resonate with our most profound sense that life is not predictable. They provoke spiritual, even religious speculations.

Resnick discussed StarLogo at the Third International Conference on Artificial Life. After his presentation, someone in the audience said, "I used to be an agnostic before I got into this stuff and now I am an atheist." Other people in the audience said that these same ideas had a different effect on them: Decentralized ideas became metaphorical ways of talking about God. A group of people gathered in the hall outside the conference room and continued talking about God, atheism, and A-Life. One man who said that A-Life had made him personally more, not less, "interested in spirituality," put its effects this way:

> When people used to say, "Let the force be with you," it was sort of a New Age thing. But now, I think of emergence as the force. It isn't New Age. It's scientific. It's true. There is a life force. But I don't have to think of myself as not a scientist to believe in it. To me, it's God coming together with science, and computers have made it all possible.

The Resnick lecture was held at 8 P.M. and was followed by two other presentations. The artificial life culture, like the hacker culture, makes few concessions to sleep. By the time I got back to my Santa Fe hotel room it was well after 1 A.M. I had just said good-bye to a hacker who was finding in evolution, emergence, and self-organization a path to a belief in the existence of God. I idly clicked on the television. An afternoon talk show was being rerun. On it, the theory of evolution was being criticized as a secular religion by fundamentalist Christians who literally believed the creation story as told in Genesis. Our culture was clearly in a bewildering place.

SIMLIFE

As StarLogo has introduced a still small number of children to some ideas of emergence and decentralization, so have the games SimLife, SimCity, and SimAnt carried an aesthetic of simulation into the popular culture. The developers of the Sim games make it clear that they are all "at least partially based on A-Life concepts and techniques."[48]

The SimAnt instruction manual presents the study of ants as a window onto larger issues. The manual describes the ant colony as "a highly redundant, fault tolerant system that is capable of a high degree of specialization," which can be compared "to the working of a thinking brain."[49] Individually, ants are not endowed with a high degree of intelligence, but the ant colony taken as a whole is rather intelligent; just as individually, neurons are not very intelligent, but as a whole, the brain is.

Even what the manual describes as a Quick Game of SimAnt (one played on one "patch" of a simulated backyard) manages to present an impressive number of ideas about artificial life and the quality of emergence. One learns about cellular automata because, like the cells in Conway's Game of Life, each ant acts like one, its behavior determined by its own state, its assay of its direct neighbors, and a simple set of rules. One learns about changing one's state in reference to who one is and with whom one is in contact. One learns about the tracer chemicals by which ants communicate with each other. One learns how in certain circumstances, local actions that seem benign (mating a few ants) can lead to disastrous global results (population overcrowding and death). One learns what Richard Dawkins noted when he first dreamt of his biomorph insects: One can get strangely attached to the artificial lives one creates. Finally, one even gets to learn about life after artificial death. In SimAnt, you take the role of a red ant, and when "you" die, you are instantly— and reassuringly—reborn.

The SimLife instruction manual is even more explicit than the SimAnt manual in its relation to artificial life. "A major feature and purpose of SimLife is that it is an exploration of the emerging computer field of Artificial Life."[50] The manual goes on to say:

> The idea behind A-Life is to produce lifelike behavior on a computer (or other artificial media), where it can be studied in ways real living things cannot.... One of the most important features of A-Life is emergent behavior—when complex behavior emerges from the combination of many individuals following simple rules. Two examples of emergent behavior are ant colonies in the real world and SimCity in the computer world.... In fact, biologic "life as we know it" can be considered a form of emergent behavior. Another important aspect of A-Life is *evolution*—artificial life forms can react to their environment and grow, reproduce and evolve into more complex forms.[51]

The SimLife manual promotes the view that being part of artificial life is important:

> The future of A-Life holds much potential and promise. It may someday go beyond the experimental world into the practical realm of design. The tools and techniques being developed now will someday allow us to grow or evolve designs for complex systems ranging from software to airplanes to

intelligence. In a sense, A-Life has the same ultimate goal as Artificial Intelligence (AI) but uses opposite methods. AI uses a top-down approach to create a thinking machine that emulates the human brain. A-Life uses the bottom-up approach—start with single cells and grow/evolve life with intelligence.[52]

In Chapter Two we met thirteen-year-old Tim, who learned to play SimLife the way he learned to play video games, by doing it. But Tim thinks that SimLife, unlike video games and computer programming, is useful. "You get to mutate plants and animals into different species. You get to balance an ecosystem. You are part of something important." Tim thinks that the "animals that grow in the computer could be alive" although he adds, "This is kind of spooky."

My SimLife companion, Laurence, a much more detached fifteen-year-old, doesn't think the idea of life on the screen is spooky at all. "The whole point of this game," he tells me,

> is to show that you could get things that are alive in the computer. We get energy from the sun. The organisms in the computer get energy from the plug in the wall. I know that more people will agree with me when they make a SimLife where the creatures are smart enough to communicate. You are not going to feel comfortable if a creature that can talk to you goes extinct.

Robbie, a ten-year-old who has been given a modem for her birthday, puts the emphasis not on communication but on mobility in considering whether the creatures she has evolved on SimLife are alive.

> I think they are a little alive in the game, but you can turn it off and you cannot save your game, so that all the creatures you have evolved go away. But if they could figure out how to get rid of that part of the program so that you would *have* to save the game ... if your modem were on, [the creatures] could get out of your computer and go to America Online.

Sean, thirteen, who has never used a modem, comes up with a variant on Robbie's ideas about travel. "The creatures could be more alive if they could get into DOS. If they were in DOS, they would be like a computer virus and they could get onto all of your disks, and if you loaned your disks to friends, it would be like they were traveling."

We saw that Piaget, interviewing children in the 1920s about the issue of aliveness, had found that children took up the question of an object's life status by considering if it could move of its own accord. In the late 1970s and early 1980s, when I studied children's ideas about aliveness in dealing with stationary computer objects, the focus of children's thinking had shifted to an object's psychological properties. Today, in children's comments about the creatures that exist on simulation games, in talk about travel via circulating disks or over modems, in talk of viruses and networks, movement is resurfacing as a criterion for aliveness. It is widely assumed by children that the creatures on Sim games have a desire to move out of the system into a wider *digital* world.

The creatures in simulation space challenge children to find a new language for talking about them and their status, as do mobile robots that wander about, making their "own decisions" about where to go. When Rodney Brooks asked his five-year-old daughter whether his mobots were alive, she said, "No they just have control." For this child, and despite her father's work, life is biological. You can have consciousness and intentionality without being alive. At the end of the 1992 Artificial Life Conference I sat next to eleven-year-old Holly as we watched a group of robots with distinctly different "personalities" compete in a special robot Olympics. I told her I was studying robots and life, and Holly became thoughtful. Then she said unexpectedly, "It's like Pinocchio."

> First, Pinocchio was just a puppet. He was not alive at all. Then he was an alive puppet. Then he was an alive boy. A real boy. But he was alive even before he was a real boy. So I think the robots are like that. They are alive like Pinocchio [the puppet], but not like "real boys."

When boundaries are challenged we look for ways to maintain them. We use the fairytale Pinocchio to maintain a separation between the merely animated puppet and biological life. Holly used it to separate the merely animated robots from biological life. The robots were "sort of alive" or "alive like Pinocchio." But they were not "real life."[53]

We find ourselves in a complex position in our attitudes toward biological life. There is a preoccupation with genetics as a predictor of success and failure in life, thus emphasizing the importance of biological ties. But at the same time, this genetic essentialism requires us to emphasize DNA as the carrier of our biological connections. Since we have constructed DNA as information, this suggests a kinship between biological and computational organisms (which have the digital equivalent of DNA). Biological essentialism puts computational objects on the other side of a line, even as ideas from emergent AI and A-Life challenge the line itself. Our situation is far from stable. We are making boundaries and relaxing them in a complex double movement.

MORPHING AND CYCLING THROUGH

In the early 1980s, transformer toys entered child culture. A toy phenomenon originally imported from Japan, these were mechanical objects like trains and trucks that could turn into robots that could turn into animals and back again. Soon after came a new language of shape shifting, referring to the process by which the android villain of Arnold Schwarzenegger's *Terminator II* was able to become whatever it touched—the liquid metal machine could become flesh and then could turn back again into a machine. More recently, children have been playing at morphing,[54] an activity synonymous, from preschool on, with the transformations of Power Rangers, a group of teen-aged martial-arts experts who can turn themselves into person/machine/animal hybrids with super-robotic powers. There are Power Ranger television shows, videos, comic books, and dolls. Unlike Superman, who pretended to be a person but was always a super-person, a man of steel in disguise, children seem to understand that the Power Rangers have human bodies that are flexible enough to morph into cyborg ones.

Phil, nine years old, talks to me about what happens when the Rangers change their state. He begins with a disclaimer but warms to his subject: "It all happens in a flash, so it's hard to be sure, but I think that what is happening is that the Rangers become a part of the dinozords and the dinozords become part of the megazords. . . . The dinozords and megazords have computers . . . definitely. To do their jobs, they have to." Alan, eight, is more assured. "I know exactly how they morph. The zords are not human, but they contain the humans. The humans are not riding in the zords. The people disappear into the zord bones, but then they can beam out and become teenagers again." In children's thinking, the computational objects of everyday life can think and know (all the while being just machines), and the computational objects of popular culture readily turn into living flesh.

It is spring, and in the newly mild weather, a group of children is playing on the grass with transformer toys. This particular set can take the shape of armored tanks, robots, or people. They can also be put into intermediate states so that a "robot" arm can protrude from a human form or a human leg from a mechanical tank. Two of the children, both seven years old, are playing with the toys in these intermediate states. A third child, who is a year younger, insists that this is not right. The toys, he says, should not be placed in hybrid states. "You should play them as all talk or all people." He is getting upset because the two older children are making a point of ignoring him. An eight-year-old girl comforts the upset child. "It's okay to play them when they are in between. It's all the same stuff," she says, "just yucky computer cy-dough-plasm."

In the 1980s, when faced with intelligent toys, children's thoughts turned not to a toy's movement, as in Piaget's day, but to its psychology. Children constructed a new coherent story about life. They took a new world of objects and imposed a new world order. More recently, simulation has strained that order to the breaking point. Faced with the objects of A-Life, children still try to impose strategies and categories, but they do so in the manner of theoretical bricoleurs, making do with whatever materials are at hand, with whatever theory can fit the rapidly changing circumstances.

My current collection of children's comments about the aliveness of what I have called A-Life objects (the Blind Watchmaker, Tierra, SimLife, mobots, and Lego-Logo robots) includes: The robots are in control but not alive, would be alive if they had bodies, are alive because they have bodies, would be alive if they had feelings, are alive the way insects are alive but not the way people are alive; the Tierrans are not alive because they are just in the computer, could be alive if they got out of the computer, are alive until you turn off the computer and then they're dead, are not alive because nothing in the computer is real; the Sim creatures are not alive but almost-alive, would be alive if they spoke, would be alive if they traveled, are alive but not real, are not alive because they don't have bodies, are alive because they can have babies, and finally, for an eleven-year-old who is relatively new to SimLife, they're not alive because the babies in the game don't have parents. She says, "They show the creatures and the game tells you that they have mothers and fathers, but I don't believe it. It's just numbers, it's not really a mother and a father." For all the objects, the term "sort of alive" comes up often.

These theories are strikingly heterogeneous. Different children hold different theories, and individual children are able to hold different theories at the same time. These different views emerge when children talk about something as big as the life of a computational creature and about something as small as why a robot programmed with emergent methods might move in a certain way. In the short history of how the computer has changed the way we think, children have often led the way. Today, children are pointing the way toward multiple theories in the presence of the artifacts of artificial life.

For example, one fifth-grader named Sara jumped back and forth from a psychological to a mechanistic language when she talked about the Lego-Logo creature she had built. Sara was considering whether her machine would sound a signal when its touch sensor was pushed, and she said, "It depends on whether the machine wants to tell . . . if we want the machine to tell us . . . if we tell the machine to tell us." Resnick commented that within a span of ten seconds,

> Sara had described the session in three different ways. First she viewed the machine on a psychological level, focusing on what the machine "wants." Then she shifted intentionality to the programmer, and viewed the programmer on a psychological level. Finally, she shifted to a mechanistic explanation, in which the programmer explicitly told the machine what to do.[55]

Sara had quickly cycled through three perspectives on her creature (as psychological being, as intentional self, as instrument of its programmer's intentions). The speed of this cycling behavior suggests that these perspectives are equally present for her at all times. For some purposes, she finds one or another of them more useful. Adults find themselves in a similar situation. One forty-year-old woman, an interior designer, confronted with Tierra, cycled through views of it as alive, as "alive in a way" but not alive like humans or animals, as information but not body, as body but not the right kind of body for life, as alive but not spiritually alive, or as our creature but not God's creature, thus not alive. A thirty-seven-year-old lawyer found Tierra not alive because life "isn't just replicating bits of information," alive "like a virus," not alive because "life in a parallel universe shouldn't count as life," alive "but not real life."

In his history of artificial life, Steven Levy suggested that one way to look at where artificial life can fit into our way of thinking about life is to envisage a continuum in which Tierra, for example, would be more alive than a car, but less alive than a bacterium.[56] My observations of how people are dealing with artificial life's new objects-to-think-with suggest that they are not constructing hierarchies but are heading toward parallel definitions, which they alternate in a way that recalls Sara's rapid cycling.

We already use parallel definitions and cycling through to think about significant aspects of self. In *Listening to Prozac,* the psychiatrist Peter Kramer wrote about an incident in which he prescribed an antidepressant medication for a college student. At the next therapy session the patient appeared with symptoms of anxiety. Since it is not unusual for patients to respond with jitters to the early stages of treatment with antidepressants, Kramer was not concerned. Sometimes the jitters disappear by themselves, sometimes the prescribing physician changes the antidepressant or adds a second, sedating medication at bedtime. Kramer says:

> I considered these alternatives and began to discuss them with the young man when he interrupted to correct my misapprehension: He had not taken the antidepressant. He was anxious because he feared my response when I learned he had "disobeyed" me.
>
> As my patient spoke, I was struck by the sudden change in my experience of his anxiety. One moment, the anxiety was a collection of meaningless physical symptoms, of interest only because they had to be suppressed, by other biological means, in order for the treatment to continue. At the next, the anxiety was rich in overtones[,] ... emotion a psychoanalyst might call Oedipal, anxiety over retribution by the exigent father. The two anxieties were utterly different: the one a simple outpouring of brain chemicals, calling for a scientific response, however diplomatically communicated; the other worthy of empathic exploration of the most delicate sort.[57]

Kramer experienced this alternation of perspective because his patient did not take his medication. Other people experience such alternations when they do take medication, either medication prescribed to treat psychological problems or medication with psychoactive side effects that is prescribed for physical ills. In the prescription-drug situation, people commonly have moments when they equate their personality with their chemistry, a point of view also common in the drug culture. In the same way, women who feel the mental effects of their hormones during their monthly cycle, and even more so during pregnancy and menopause, sense often that chemistry is in control of what some people call personal psychology. The increasingly commonplace experience of taking a pill and observing a change in one's self challenges any simple notions of the psychological, but people typically do not abandon a sense of themselves as made up of more than chemistry. Rather, they cycle through "I am my chemicals" to "I am my history" to "I am my genes."

It may in fact be in the area of genetics that we have become most accustomed to cycling through. In *Listening to Prozac,* Kramer tells a story about how genetics is causing us to cycle through different views of identity. About to express praise for his friends' two children with a comment like "Don't the genes breed true?" Kramer stopped himself when he remembered that both children were adopted. "Since when had I—I, who make my living through the presumption that people are shaped by love and loss, and above all by their early family life—begun to assume that personality traits are genetically determined?"[58] In fact, Kramer hadn't begun to assume this, he just *sometimes* did. Cycling through different and sometimes opposing theories has become how we think about our minds.

In this chapter we have seen that in the culture of simulation cycling through is coming to be the way we think about life itself. In the following chapter we'll begin to explore how rapid alternations of identity have become a way of life for people who live in virtual reality as they cycle through different characters and genders, moving from window to window on the computer screen.

PART III

ON THE INTERNET

Chapter 7

ASPECTS OF THE SELF

When we step through the screen into virtual communities, we reconstruct our identities on the other side of the looking glass. This reconstruction is our cultural work in progress. The final section of this book explores the culture of simulation as it is emerging in the virtual workshops of online life.

Throughout this book, there has been a complex dance of acceptance and rejection of analogies to "the machine." On the one hand we insist that we are different from machines because we have emotions, bodies, and an intellect that cannot be captured in rules, but on the other we play with computer programs that we think of as alive or almost-alive. Images of machines have come ever closer to images of people, as images of people have come ever closer to images of machines. Experts tell us that we may read our emotions through brain scans, modify our minds through reprogramming, and attribute significant elements of personality to our genetic code. Chic lifestyle catalogues of mail-order merchandise carry mind-altering equipment including goggles, headphones, and helmets that promise everything from relaxation to enhanced learning if we plug ourselves into them. Their message is that we are so much like machines that we can simply extend ourselves through cyborg couplings with them.

At the same time that we are learning to see ourselves as plugged-in technobodies, we are redescribing our politics and economic life in a language that is resonant with a particular form of machine intelligence. In government, business, and industry, there is much talk of distributed, parallel, and emergent organizations, whose architecture mirrors that of computer systems. This utopian discourse about decentralization has come into vogue at the same time that society has become increasingly fragmented. Many of the institutions that used to bring people together —a main street, a union hall, a town meeting—no longer work as before. Many people spend most of their day alone at the screen of a television or a computer. Meanwhile, social beings that we are, we are trying (as Marshall McLuhan said) to retribalize.[1] And the computer is playing a central role. We correspond with each other through electronic mail and contribute to electronic bulletin boards and mailing lists; we join interest groups whose participants include people from all over the world. Our rootedness to place has attenuated. These shifts raise many questions: What will computer-mediated communication do to our commitment to other people? Will it satisfy our needs for connection and social participation, or will it further undermine fragile relationships? What kind of responsibility and accountability will we assume for our virtual actions?

In political terms, talk about moving from centralized to decentralized systems is usually characterized as a change from autocracy to democracy, although the jury is still out on its ultimate effects. It may, for example, be possible to create an illusion of decentralized participation even when power remains closely held. In terms of our views of the self, new images of multiplicity, heterogeneity, flexibility, and fragmentation dominate current thinking about human identity.

Psychoanalytic theory has played a complicated role in the historical debate about whether identity is unitary or multiple. One of Freud's most revolutionary contributions was proposing a radically decentered view of the self, but this message was often obscured by some of his followers who tended to give the ego greater executive authority in the management of the self. However, this recentralizing move was itself periodically challenged from within the psychoanalytic movement. Jungian ideas stressed that the self is a meeting place of diverse archetypes. Object-relations theory talked about how the things and people in the world come to live inside us. More recently, poststructuralist thinkers have attempted an even more radical decentering of the ego. In the work of Jacques Lacan, for example, the complex chains of associations that constitute meaning for each individual lead to no final endpoint or core self. Under the banner of a return to Freud, Lacan insisted that the ego is an illusion. In this he joins psychoanalysis to the postmodern attempt to portray the self as a realm of discourse rather than as a real thing or a permanent structure of the mind. In previous chapters we have seen the way computer science has contributed to this new way of talking. Its bottom-up, distributed, parallel, and emergent models of mind have replaced top-down, information processing ones.

The Internet is another element of the computer culture that has contributed to thinking about identity as multiplicity. On it, people are able to build a self by cycling through many selves. An interior designer nervously admits in my interview with her that she is not at her best because she is about to have a face-to-face meeting with a man with whom she has shared months of virtual intimacy in chat sessions on America Online. She says she is "pretty sure" that her electronic lover is actually a man (rather than a woman pretending to be a man) because she does not think "he" would have suggested meeting if it were otherwise, but she worries that neither of them will turn out to be close enough to their very desirable cyberselves:

> I didn't exactly lie to him about anything specific, but I feel very different online. I am a lot more outgoing, less inhibited. I would say I feel more like myself. But that's a contradiction. I feel more like who I wish I was. I'm just hoping that face-to-face i can find a way to spend some time being the online me.

A thirty-year-old teacher describes her relationship to Internet Relay Chat (or IRC), a live forum for online conversations, as being "addicted to flux." On IRC one makes up a name, or handle, and joins any one of thousands of channels discussing different issues. Anyone can start a new channel at any time. In the course of the past week, this woman has created channels on East Coast business schools (she is considering applying), on the new editorial policy of *The New Yorker,* and on a television situation comedy about a divorced woman having an affair with her ex-husband. She has concerns about her involvement with IRC that do not stem from how much time she spends ("about five hours a day, but I don't watch television any more") but from how many roles she plays.

> It is a complete escape.... On IRC, I'm very popular. I have three handles I use a lot.... So one [handle] is serious about the war in Yugoslavia, [another is] a bit of a nut about *Melrose Place,* and [a third is] very active on sexual channels, always looking for a good time.... Maybe I can only relax if I see life as one more IRC channel.

In the past, such rapid cycling through different identities was not an easy experience to come by. Earlier in this century we spoke of identity as "forged." The metaphor of iron-like solidity captured the central value of a core identity, or what the sociologist David Riesman once called inner direction.[2] Of course, people assumed different social roles and masks, but for most people, their lifelong involvement with families and communities kept such cycling through under fairly stringent control. For some, this control chafed, and there were roles on the margins where cycling through could be a way of life. In tribal societies, the shaman's cycling through might involve possession by gods and spirits. In modern times, there was the con artist, the bigamist, the cross-gender imperson-ator, the "split personality," the Dr. Jekyll and Mr. Hyde.

Now, in postmodern times, multiple identities are no longer so much at the margins of things. Many more people experience identity as a set of roles that can be mixed and matched, whose diverse demands need to be negotiated. A wide range of social and psychological theorists have tried to capture the new experience of identity. Robert Jay Lifton has called it protean. Kenneth Gergen describes its multiplication of masks as a saturated self. Emily Martin talks of the flexible self as a contemporary virtue of organisms, persons, and organizations.[3]

The Internet has become a significant social laboratory for experiment-ing with the constructions and reconstructions of self that characterize postmodern life. In its virtual reality, we self-fashion and self-create. What kinds of personae do we make? What relation do these have to what we have traditionally thought of as the "whole" person? Are they experienced as an expanded self or as separate from the self?[4] Do our real-life selves learn lessons from our virtual personae? Are these virtual personae frag-ments of a coherent real-life personality? How do they communicate with one another? Why are we doing this? Is this a shallow game, a giant waste of time? Is it an expression of an identity crisis of the sort we traditionally associate with adolescence? Or are we watching the slow emergence of a new, more multiple style of thinking about the mind? These questions can be addressed by looking at many different locations on the Internet. Here I begin with the virtual communities known as MUDs.

MUDs

In the early 1970s, the face-to-face role-playing game Dungeons and Drag-ons swept the game culture. In Dungeons and Dragons, a dungeon master creates a world in which people take on fictional personae and play out complex adventures. The game is a rule-driven world that includes charisma points, levels of magic, and rolls of the dice. The Dungeon and Dragons universe of mazes and monsters and its image of the world as a labyrinth whose secrets could be unlocked held a particular fascination for many members of the nascent computer culture. The computer game Adventure captured some of the same aesthetic. There, players proceeded through a maze of rooms presented to them through text description on a computer screen.

The term "dungeon" persisted in the high-tech culture to connote a virtual place. So when virtual spaces were created that many computer users could share and collaborate within, they were deemed Multi-User Dungeons or MUDs, a new kind of social virtual reality. Although some games use software that make them technically such things as MUSHes or MOOs, the term MUD and the verb MUDding have come to refer to all of the multi-user environments. As more and more players have come to them who do not have a history with Dungeons and Dragons, some people have begun to refer to MUDs as Multi-User Domains or Multi-User Dimensions.

Some MUDs use screen graphics or icons to communicate place, character, and action. The MUDs I am writing about here do not. They rely entirely on plain text. All users are browsing and manipulating the same database. They can encounter other users or players as well as objects that have been built for the virtual environment. MUD players can also communicate with each other directly in real time, by typing messages that are seen by other players. Some of these messages are seen by all players in the same "room," but messages can also be designated to flash on the screen of only one specific player.

The term "virtual reality" is often used to denote metaphorical spaces that arise only through interaction with the computer, which people navigate by using special hardware—specially designed helmets, body suits, goggles, and data gloves. The hardware turns the body or parts of the body into pointing devices. For example, a hand inside a data glove can point to where you want to go within virtual space; a helmet can track motion so that the scene shifts depending on how you move your head. In MUDs, instead of using computer hardware to immerse themselves in a vivid world of sensation, users immerse themselves in a world of words. MUDs are a text-based, social virtual reality.

Two basic types of MUDs can now be accessed on the Internet. The adventure type, most reminiscent of the games' Dungeons and Dragons heritage, is built around a medieval fantasy landscape. In these, affectionately known by their participants as "hack and slay," the object of the game is to gain experience points by killing monsters and dragons and finding gold coins, amulets, and other treasure. Experience points translate into increased power. A second type consists of relatively open spaces in which you can play at whatever captures your imagination. In these MUDs, usually called social MUDs, the point is to interact with other players and, on some MUDs, to help build the virtual world by creating one's own objects and architecture. "Building" on MUDs is something of a hybrid between computer programming and writing fiction. One describes a hot tub and deck in a MUD with words, but some formal coded description is required for the deck to exist in the MUD as an extension of the adjacent living room and for characters to be able to "turn the hot tub on" by pushing a specially marked "button." In some MUDs, all players are allowed to build; sometimes the privilege is reserved to master players, or wizards. Building is made particularly easy on a class of MUDs known as "MOOs" (MUDs of the Object Oriented variety).

In practice, adventure-type MUDs and social MUDs have much in common. In both, what really seems to hold players' interest is operating their character or characters and interacting with other characters. Even in an adventure-type MUD, a player can be an elf, a warrior, a prostitute, a politician, a healer, a seer, or several of these at the same time. As this character or set of characters, a player evolves relationships with other players, also in character. For most players these relationships quickly become central to the MUDding experience. As one player on an adventure-type MUD put it, "I began with an interest in 'hack and slay,' but then I stayed to chat."

The characters one creates for a MUD are referred to as one's personae. This is from the Latin *per sonae* which means "that through which the sound comes," in other words, an actor's mask. Interestingly, this is also the root of "person" and "personality." The derivation implies that one is identified by means of a public face distinct from some deeper essence or essences.

All MUDs are organized around the metaphor of physical space. When you first enter a MUD you may find yourself in a medieval church from which you can step out into the town square, or you may find yourself in the coat closet of a large, rambling house. For example, when you first log on to LambdaMOO, one of the most popular MUDs on the Internet, you see the following description:

The Coat Closet. The Closet is a dark, cramped space. It appears to be very crowded in here; you keep bumping into what feels like coats, boots and other people (apparently sleeping). One useful thing that you've discovered in your bumbling about is a metal doorknob set at waist level into what might be a door. There's a new edition of the newspaper. Type "news" to see it.

Typing "out" gets you to the living room:

The Living Room. It is very bright, open, and airy here, with large plate-glass windows looking southward over the pool to the gardens beyond. On the north wall, there is a rough stonework fireplace, complete with roaring fire. The east and west walls are almost completely covered with large, well-stocked bookcases. An exit in the northwest corner leads to the kitchen and, in a more northerly direction, to the entrance hall. The door into the coat closet is at the north end of the east wall, and at the south end is a sliding glass door leading out onto a wooden deck. There are two sets of couches, one clustered around the fireplace and one with a view out the windows.

This description is followed by a list of objects and characters present in the living room. You are free to examine and try out the objects, examine the descriptions of the characters, and introduce yourself to them. The social conventions of different MUDs determine how strictly one is expected to stay in character. Some encourage all players to be in character at all times. Most are more relaxed. Some ritualize stepping out of character by asking players to talk to each other in specially noted "out of character" (OOC) asides.

On MUDs, characters communicate by invoking commands that cause text to appear on each other's screens. If I log onto LambdaMOO as a male character named Turk and strike up a conversation with a character named Dimitri, the setting for our conversation will be a MUD room in which a variety of other characters might be present. If I type, "Say hi," my screen will flash, "You say hi," and the screens of the other players in the room (including Dimitri) will flash, "Turk says 'hi.' " If I type "Emote whistles happily," all the players' screens will flash, "Turk whistles happily." Or I can address Dimitri alone by typing, "Whisper to Dimitri Glad to see you," and only Dimitri's screen will show, "Turk whispers 'Glad to see you.' " People's impressions of Turk will be formed by the description I will have written for him (this description will be available to all players on command), as well as by the nature of his conversation.

In the MUDs, virtual characters converse with each other, exchange gestures, express emotions, win and lose virtual money, and rise and fall in social status. A virtual character can also die. Some die of "natural" causes (a player decides to close them down) or they can have their virtual lives snuffed out. This is all achieved through writing, and this in a culture that had apparently fallen asleep in the audiovisual arms of television. Yet this new writing is a kind of hybrid: speech momentarily frozen into artifact, but curiously ephemeral artifact. In this new writing, unless it is printed out on paper, a screenful of flickers soon replaces the previous screen. In MUDs as in other forms of electronic communication, typographic conventions known as emoticons replace physical gestures and facial expressions. For example, :-) indicates a smiling face and :-(indicates an unhappy face. Onomatopoeic expletives and a relaxed attitude toward sentence fragments and typographic errors suggest that the new writing is somewhere in between traditional written and oral communication.

MUDs provide worlds for anonymous social interaction in which you can play a role as close to or as far away from your real self as you choose. For many game participants, playing one's character(s) and living in the MUD(s) becomes an important part of daily life. Since much of the excitement of the game depends on having personal relationships and being part of a MUD community's developing politics and projects, it is hard to participate just a little. In fact, addiction is a frequently discussed subject among MUD players. A *Newsweek* article described how "some players attempt to go cold turkey. One method is to randomly change your password by banging your head against the keyboard, making it impossible to log back on."[5] It is not unusual for players to be logged on to a MUD for six hours a day. Twelve hours a day is common if players work with computers at school or at a job and use systems with multiple windows. Then they can jump among windows in order to intersperse real-world activities on their computers with their games. They jump from Lotus 1-2-3 to LambdaMOO, from Wordperfect to DragonMUD. "You can't really be part of the action unless you are there every day. Things happen quickly. To get the thrill of MUDs you have to be part of what makes the story unfold," says a regular on DuneMUSH, a MUD based on the world of Frank Herbert's science fiction classic.[6]

In MUDs, each player makes scenes unfold and dramas come to life. Playing in MUDs is thus both similar to and different from reading or watching television. As with reading, there is text, but on MUDs it unfolds in real time and you become an author of the story. As with television, you are engaged with the screen, but MUDs are interactive, and you can take control of the action. As in acting, the explicit task is to construct a viable mask or persona. Yet on MUDs, that persona can be as close to your real self as you choose, so MUDs have much in common with psychodrama. And since many people simply choose to play aspects of themselves, MUDs can also seem like real life.

Play has always been an important aspect of our individual efforts to build identity. The psychoanalyst Erik Erikson called play a "toy situation" that allows us to "reveal and commit" ourselves "in its unreality."[7] While MUDs are not the only "places" on the Internet in which to play with identity, they provide an unparalleled opportunity for such play. On a MUD one actually gets to build character and environment and then to live within the toy situation. A MUD can become a context for discovering who one is and wishes to be. In this way, the games are laboratories for the construction of identity, an idea that is well captured by the player who said:

> You can be whoever you want to be. You can completely redefine yourself if you want. You can be the opposite sex. You can be more talkative. You can be less talkative. Whatever. You can just be whoever you want, really, whoever you have the capacity to be. You don't have to worry about the slots other people put you in as much. It's easier to change the way people perceive you, because all they've got is what you show them. They don't look at your body and make assumptions. They don't hear your accent and make assumptions. All they see is your words. And it's always there. Twenty-four hours a day you can walk down to the street corner and there's gonna

be a few people there who are interesting to talk to, if you've found the right MUD for you.

The anonymity of most MUDs (you are known only by the name you give your characters) provides ample room for individuals to express unexplored parts of themselves. A twenty-one-year-old college senior defends his violent characters as "something in me; but quite frankly I'd rather rape on MUDs where no harm is done." A twenty-six-year-old clerical worker says, "I'm not one thing, I'm many things. Each part gets to be more fully expressed in MUDs than in the real world. So even though I play more than one self on MUDs, I feel more like 'myself' when I'm MUDding." In real life, this woman sees her world as too narrow to allow her to manifest certain aspects of the person she feels herself to be. Creating screen personae is thus an opportunity for self-expression, leading to her feeling more like her true self when decked out in an array of virtual masks.

MUDs imply difference, multiplicity, heterogeneity, and fragmentation. Such an experience of identity contradicts the Latin root of the word, *idem,* meaning "the same." But this contradiction increasingly defines the conditions of our lives beyond the virtual world. MUDs thus become objects-to-think-with for thinking about postmodern selves. Indeed, the unfolding of all MUD action takes place in a resolutely postmodern context. There are parallel narratives in the different rooms of a MUD. The cultures of Tolkien, Gibson, and Madonna coexist and interact. Since MUDs are authored by their players, thousands of people in all, often hundreds at a time, are all logged on from different places; the solitary author is displaced and distributed. Traditional ideas about identity have been tied to a notion of authenticity that such virtual experiences actively subvert. When each player can create many characters and participate in many games, the self is not only decentered but multiplied without limit.

Sometimes such experiences can facilitate self-knowledge and personal growth, and sometimes not. MUDs can be places where people blossom or places where they get stuck, caught in self-contained worlds where things are simpler than in real life, and where, if all else fails, you can retire your character and simply start a new life with another.

As a new social experience, MUDs pose many psychological questions: If a persona in a role-playing game drops defenses that the player in real life has been unable to abandon, what effect does this have? What if a persona enjoys success in some area (say, flirting) that the player has not been able to achieve? In this chapter and the next I will examine these kinds of questions from a viewpoint that assumes a conventional distinction between a constructed persona and the real self. But we shall soon encounter slippages—places where persona and self merge, places where the multiple personae join to comprise what the individual thinks of as his or her authentic self.

These slippages are common on MUDs, but as I discuss MUDs, it is important to keep in mind that they more generally characterize identity play in cyberspace. One Internet Relay Chat (IRC) enthusiast writes to an online discussion group, "People on [this mailing list] tell me that they make mistakes about what's happening on cyberspace and what's happening on RL. Did I really type what's happening *ON* Real Life?" (Surrounding a word with asterisks is the net version of italicizing it.) He had indeed. And then he jokingly referred to real life as though it, too, were an IRC channel: "Can anyone tell me how to /join #real.life?"[8]

ROLE PLAYING VS. PARALLEL LIVES

Traditional role-playing games, the kinds that take place in physical space with participants "face to face," are psychologically evocative. MUDs are even more so because they further blur the line between the game and real life, usually referred to in cyberspace as RL.[9] In face-to-face role-playing games, one steps in and out of a character. MUDs, in contrast, offer a character or characters that may become parallel identities. To highlight the distinctive features of the virtual experience, I will begin with the story of a young woman who plays face-to-face role-playing games in physical reality.

Julee, nineteen years old, dropped out of college after her freshman year partly because of her turbulent relationship with her mother, a devout Catholic. Julee's mother broke all ties with her when she discovered that Julee had had an abortion the summer after graduation from high school. Although technically out of school, Julee spends most of her free time with former college classmates playing elaborate face-to-face role-playing games, an interest she developed as a freshman.

Julee plays a number of games that differ in theme (some are contemporary whodunits, some are medieval adventures), the degree to which they are pre-scripted, and the extent to which players assume distinct characters and act out their parts. At one end of a continuum, players sit in a circle and make statements about their characters' actions within a game. On the other end, players wear costumes, engage in staged swordplay, and speak special languages that exist only within the game.

Julee's favorite games borrow heavily from improvisational theater. They are political thrillers in which the characters and the political and historical premises are outlined prior to the start of play. The players are then asked to take the story forward. Julee especially enjoys the games staged by a New York–based group, which entail months of preparation for each event. Sometimes the games last a weekend, sometimes for a week to ten days. Julee compares them to acting in a play:

> You usually get your sheets [a script outline] about twenty-four hours before the game actually starts, and I spend that time reading my sheets over and over again. Saying, you know "What are my motivations? What is this character like?" It's like a play, only the lines aren't set. The personality is set, but the lines aren't.

In Julee's real life, her most pressing concern is the state of her relationship with her mother. Not surprisingly, when asked about her most important experience during a role-playing game, Julee describes a game in which she was assigned to play a mother. This character was a member of a spy ring. Her daughter, also a member of the ring, turned out to be a counterspy, a secret member of a rival faction. The scripted game specified that the daughter was going to betray, even kill, her mother. The members of Julee's team expected that her character would denounce her daughter to save her own life and perhaps their lives as well.

This game was played over a weekend on a New York City college campus. At that time, Julee says that she faced her game daughter and saw her real self. As she spoke to me, Julee's voice took on different inflections as she moved from the role of mother to daughter: "Here's this little girl who is my daughter looking into my eyes and saying 'How can you kill me? Why do you want me to go away?'" Julee describes the emotional intensity of her efforts to deal with this situation:

> So, there we were in this room in the chemistry department, and I guess we moved over into a corner, and we were sitting on the floor, like, cross-legged in front of each other, like . . . like, I guess we were probably holding hands. I think we were. And we, like, . . . we really did it. We acted out the whole scene. . . . It was, it really was a nearly tearful experience.

In the game, Julee and her "daughter" talked for hours. Why might the daughter have joined her mother's opponents? How could the two women stay true to their relationship and to the game as it had been written? Huddled in the corner of an empty classroom, Julee had the conversation with her game daughter that her own mother had been unwilling to have with her. In the end, Julee's character chose to put aside her loyalty to her team in order to preserve her daughter's life. From Julee's point of view, her mother had put her religious values above their relationship; in the game, Julee made her relationship to her daughter her paramount value. "I said to all the other players, 'Sorry, I'm going to forfeit the game for my team.'"

Clearly, Julee projected feelings about her mother onto her experience of the game, but more was going on than a simple reenactment. Julee was able to review a highly charged situation in a setting where she could examine it, do something new with it, and revise her relationship to it. The game became a medium for working with the materials of her life. Julee was able to sculpt a familiar situation into a new shape. In some ways, what happened was consistent with what the psychoanalytic tradition calls "working through."

Julee's experience stands in contrast to several prevalent popular images of role-playing games. One portrays role-playing games as places for simple escape. Players are seen as leaving their real lives and problems behind to lose themselves in the game. Another portrays role-playing games as depressing, even dangerous. It is implicit in the now legendary story of an emotionally troubled student who disappeared and committed suicide during a game of Dungeons and Dragons. Although some people do have experiences in these games that are escapist or depressing, others do not. Julee's story, for example, belies the popular stereotypes. Her role-playing is psychologically constructive. She uses it to engage with some of the most important issues in her life and to reach new emotional resolutions.

Role-playing games can serve in this evocative capacity because they stand betwixt and between the unreal and the real; they are a game and something more. Julee shaped her game persona to reflect her own deep wish for a relationship with her mother. Playing her ideal of a good mother allowed her to bring mother and daughter together in a way that had been closed off in real life. During the game, Julee was able to experience something of her own mother's conflict. Ultimately, Julee took a stand to preserve the relationship, something her own mother was not prepared to do. Although it had this complicated relationship with real life, in the final analysis, Julee's experience fits into the category of game because it had a specified duration. The weekend was over and so was the game.

In MUDs, however, the action has no set endpoint. The boundaries in MUDs are fuzzier. They are what the anthropologist Clifford Geertz refers to as blurred genres.[10] The routine of playing them becomes part of their players' real lives. Morgan, a college sophomore, explains how he cycles in and out of MUDs and real life: "When I am upset I just . . . jump onto my ship [the spaceship he commands in a MUD] and look somebody up." He does this by logging onto the game in character and paging a friend in the game environment. Then, still logged on, Morgan goes to class. He explains that by the time he comes back to the MUD, the friends he had paged would now usually be on the game and ready to talk. Morgan has become expert at using MUDs as a psychological adjunct to real life. He reflects on how he used MUDs during his freshman year. "I was always happy when I got into a fight in the MUD," he says. "I remember doing that before tests. I would go to the MUD, pick a fight, yell at people, blow a couple of things up, take the test and then go out for a drink." For him, a favorite MUD afforded an escape valve for anxiety and anger that felt too dangerous to exercise in real life.

Julee's role playing provided an environment for working on important personal issues. MUDs go further. You can "move into" them. One group of players joked that they were like the electrodes in the computer, trying to express the degree to which they feel part of its space. I have noted that committed players often work with computers all day at their regular jobs. As they play on MUDs, they periodically put their characters to sleep, remaining logged on to the game, but pursuing other activities. The MUD keeps running in a buried window. From time to time, they return to the game space. In this way, they break up their day and come to experience their lives as cycling through the real world and a series of virtual ones. A software designer who says he is "never not playing a MUD" describes his day this way:

> I like to put myself in the role of a hero, usually one with magical powers, on one MUD, start a few conversations going, put out a question or two about MUD matters, and ask people to drop off replies to me in a special in-box I have built in my MUD "office." Then I'll put my character to sleep and go off and do some work. Particularly if I'm in some conflict with someone at work it helps to be MUDding, because I know that when I get back to the MUD I'll probably have some appreciative mail waiting for me. Or sometimes I use a few rounds of MUD triumphs to psych myself up to deal with my boss.

Now twenty-three, Gordon dropped out of college after his freshman year when he realized that he could be a successful computer programmer without formal training. Gordon describes both his parents as "1960s nonconformists." He says there was little family pressure to get a degree. Gordon's parents separated when he was in grade school. This meant that Gordon spent winters with his mother in Florida and summers with his father in California. When Gordon was in California, his room in Florida was rented out, something that still upsets him. It seems to represent Gordon's unhappy sense that he has never really belonged anywhere.

In grade school and junior high Gordon wasn't happy and he didn't fit in. He describes himself as unpopular, overweight, unathletic, and unattractive: "Two hundred and ten pounds with glasses." The summer after his sophomore year in high school, Gordon went on a trip to India with a group of students from all over the world. These new people didn't know he was unpopular, and Gordon was surprised to find that he was able to make friends. He was struck by the advantages of a fresh start, of leaving old baggage behind. Two years later, as a college freshman, Gordon discovered MUDs and saw another way to have a fresh start. Since MUDs allowed him to create a new character at any time, he could always begin with a clean slate. When he changed his character he felt born again.

On MUDs, Gordon has experimented with many different characters, but they all have something in common. Each has qualities that Gordon is trying to develop in himself. He describes one current character as "an avatar of me. He is like me, but more effusive, more apt to be flowery and romantic with a sort of tongue-in-cheek attitude toward the whole thing." A second character is "quiet, older, less involved in what other people are doing," in sum, more self-confident and self-contained than the real-life Gordon. A third character is female. Gordon compares her to himself: "She is more flirtatious, more experimental, more open sexually definitely."

Unlike Julee's role-playing game, MUDs allow Gordon more than one weekend, one character, or one game to work on a given issue. He is able to play at being various selves for weeks, months, indeed years on end. When a particular character outlives its psychological usefulness, Gordon discards it and creates a new one. For Gordon, playing on MUDs has enabled a continual process of creation and recreation. The game has heightened his sense of his self as a work in progress. He talks about his real self as starting to pick up bits and pieces from his characters.

Gordon's MUD-playing exhibits some of the slippage I referred to earlier. By creating diverse personae, he can experiment in a more controlled fashion with different sets of characteristics and see where they lead. He is also able to play at being female, something that would be far more difficult to do in real life. Each of his multiple personae has its independence and integrity, but Gordon also relates them all to "himself." In this way, there is relationship among his different personae; they are each an aspect of himself. The slippage Gordon experiences among his MUD and RL selves has extended to his love life. When I met him, Gordon was engaged to marry a woman he had met and courted on a MUD. Their relationship began as a relationship between one of his personae and a persona created by his fiancée.

Matthew, a nineteen-year-old college sophomore, also uses MUDs to work on himself, but he prefers to do it by playing only one character. Just as Julee used her game to play the role of the mother she wishes she had, Matthew uses MUDs to play an idealized father. Like Julee, Matthew tends to use his games to enact better versions of how things have unfolded in real life.

Matthew comes from a socially prominent family in a small Connecticut town. When I visit his home during the summer following his freshman year, he announces that his parents are away on a trip to celebrate his mother's birthday. He describes their relationship as impressive in its depth of feeling and erotic charge even after twenty-five years of marriage. However, it soon becomes apparent that his parents' relationship is in fact quite rocky. For years his father has been distant, often absent, preoccupied with his legal career. Since junior high school, Matthew has been his mother's companion and confidant. Matthew knows that his father drinks heavily and has been unfaithful to his mother. But because of his father's position in the community, the family presents a public front without blemish.

As a senior in high school, Matthew became deeply involved with Alicia, then a high school freshman. Matthew's role as his mother's confidant had made him most comfortable in the role of helper and advisor, and he quickly adopted that way of relating to his girlfriend. He saw himself as her mentor and teacher. Shortly after Matthew left for college, Alicia's father died. Characteristically, Matthew flew home, expecting to play a key role in helping Alicia and her family. But in their time of grief, they found Matthew's efforts intrusive. When his offers of help were turned down, Matthew was angry. But when, shortly after, Alicia ended their relationship in order to date someone who was still in high school, he became disconsolate.

It was at this point, midway into the first semester of his freshman year in college, that Matthew began to MUD. He dedicated himself to one MUD and, like his father in their small town, became one of its most important citizens. On the MUD, Matthew carved out a special role: He recruited new members and became their advisor and helper. He was playing a familiar part, but now he had found a world in which helping won him admiration. Ashamed of his father in real life, he used the MUD to play the man he wished his father would be. Rejected by Alicia in real life, his chivalrous MUD persona has won considerable social success. Now, it is Matthew who has broken some hearts. Matthew speaks with pleasure of how women on the game have entreated him to pursue relationships with them both within and beyond its confines. He estimates that he spends from fifteen to twenty hours a week logged on to this highly satisfying alternative world.

Role playing provided Matthew and Julee with environments in which they could soothe themselves by taking care of others and experiment with a kind of parenting different from what they had experienced themselves. As neglected children comfort themselves by lavishing affection on their dolls, Matthew and Julee identified with the people they took care of.

Julee's role playing had the power of face-to-face psychodrama, but

Matthew's life on MUDs was always available to him. Unlike Julee, Matthew could play as much as he wished, all day if he wished, every day if he chose to. There were always people logged on to the game. There was always someone to talk to or something to do. MUDs gave him the sense of an alternative place. They came to feel like his true home.

Since Julee was physically present on her game, she remained recognizable as herself to the other players. In contrast, MUDs provided Matthew with an anonymity he craved. On MUDs, he no longer had to protect his family's public image.[11] He could relax. Julee could play multiple roles in multiple games, but MUDs offer parallel lives in ongoing worlds. Julee could work through real-life issues in a game space, but MUD players can develop a way of thinking in which life is made up of many windows and RL is only one of them.

In sum, MUDs blur the boundaries between self and game, self and role, self and simulation. One player says, "You are what you pretend to be . . . you are what you play." But people don't just become who they play, they play who they are or who they want to be or who they don't want to be. Players sometimes talk about their real selves as a composite of their characters and sometimes talk about their screen personae as means for working on their RL lives.

ROLE PLAYING TO A HIGHER POWER

The notion "you are who you pretend to be" has mythic resonance. The Pygmalion story endures because it speaks to a powerful fantasy: that we are not limited by our histories, that we can recreate ourselves. In the real world, we are thrilled by stories of dramatic self-transformation: Madonna is our modern Eliza Doolittle; Michael Jackson the object of morbid fascination. But for most people such self-transformations are difficult or impossible. They are easier in MUDs where you can write and revise your character's self-description whenever you wish. In some MUDs you can even create a character that "morphs" into another at the command "@morph." Virtual worlds offer experiences that are hard to come by in real life.

Stewart is a twenty-three-year-old physics graduate student who uses MUDs to have experiences he cannot imagine for himself in RL. His intense online involvements engaged key issues in his life but ultimately failed to help him reach successful resolutions. His real life revolves around laboratory work and his plans for a future in science. His only friend is his roommate, another physics student whom he describes as even more reclusive than himself. For Stewart, this circumscribed, almost monastic student life does not represent a radical departure from what has gone before. He has had heart trouble since he was a child; one small rebellion, a ski trip when he was a college freshman, put him in the hospital for a week. He has lived life within a small compass.

In an interview with Stewart he immediately makes clear why he plays on MUDs: "I do it so I can talk to people." He plays exclusively on adventure-style, hack-and-slay MUDs. Stewart finds these attractive because they demand no technical expertise, so it was easy both to get started and to become a "wizard," the highest level of player. Unlike some players for whom becoming a wizard is an opportunity to get involved in the technical aspects of MUDs, Stewart likes the wizard status because it allows him to go anywhere and talk to anyone on the game. He says, "I'm going to hack and slash the appropriate number of monsters [the number required for wizard status] so that I can talk to people."

Stewart is logged on to one MUD or another for at least forty hours a week. It seems misleading to call what he does there playing. He spends his time constructing a life that is more expansive than the one he lives in physical reality. Stewart, who has traveled very little and has never been to Europe, explains with delight that his favorite MUD, although played in English, is physically located on a computer in Germany and has many European players.

> I started talking to them [the inhabitants of the MUD], and they're, like, "This costs so many and so many Deutschmarks." And I'm, like, "What are Deutschmarks? Where is this place located?" And they say, "Don't you know this is Germany." It hadn't occurred to me that I could even connect to Germany. . . . All I had were local Internet numbers, so I had no idea of where it [the MUD] was located. And I started talking to people and I was amazed at the quality of English they spoke. . . . European and Australian MUDs are interesting, . . . different people, completely different lifestyles, and at the moment, completely different economic situations.

It is from MUDs that Stewart has learned much of what he knows of politics and of the differences between American and European political and economic systems. He was thrilled when he first spoke to a Scandinavian player who could see the Northern lights. On the German MUD, Stewart shaped a character named Achilles, but he asks his MUD friends to call him Stewart as much as possible. He wants to feel that his real self exists somewhere between Stewart and Achilles. He wants to feel that his MUD life is part of his real life. Stewart insists that he does not role play, but that MUDs simply allow him to be a better version of himself.

On the MUD, Stewart creates a living environment suitable for his ideal self. His university dormitory is modest, but the room he has built for

Achilles on the MUD is elegant and heavily influenced by Ralph Lauren advertising. He has named it "the home beneath the silver moon." There are books, a roaring fire, cognac, a cherry mantel "covered with pictures of Achilles' friends from around the world." "You look up . . . and through the immense skylight you see a breathtaking view of the night sky. The moon is always full over Achilles' home, and its light fills the room with a warm glow."

Stewart's MUD serves as a medium for the projection of fantasy, a kind of Rorschach. But it is more than a Rorschach, because it enters into his everyday life. Beyond expanding his social world, MUDs have brought Stewart the only romance and intimacy he has ever known. At a social event held in virtual space, a "wedding" of two regular players on a German-based MUD I call Gargoyle, Achilles met Winterlight, a character played by one of the three female players on that MUD. Stewart, who has known little success in dating and romantic relationships, was able to charm this desirable player.

On their first virtual date, Achilles took Winterlight to an Italian restaurant close to Stewart's dorm. He had often fantasized being there with a woman. Stewart describes how he used a combination of MUD commands to simulate a romantic evening at the restaurant. Through these commands, he could pick Winterlight up at the airport in a limousine, drive her to a hotel room so that she could shower, and then take her to the restaurant.

> So, I mean, if you have the waiter coming in, you can just kinda get creative. . . . So, I described the menu to her. I found out she didn't like veal, so I asked her if she would mind if I ordered veal . . . because they have really good veal scallopini, . . . and she said that yes, she would mind, so I didn't order veal.
>
> We talked about what her research is. She's working on disease, . . . the biochemistry of coronary artery disease. . . . And so we talked about her research on coronary artery disease, and at the time I was doing nuclear physics and I talked to her about that. We talked for a couple of hours. We talked. And then she had to go to work, so we ended dinner and she left.

This dinner date led to others during which Achilles was tender and romantic, chivalrous and poetic. The intimacy Achilles experienced during his courtship of Winterlight is unknown to Stewart in other contexts. "Winterlight . . . she's a very, she's a good friend. I found out a lot of things —from things about physiology to the color of nail polish she wears." Finally, Achilles asked for Winterlight's hand. When she accepted, they had a formal engagement ceremony on the MUD. In that ceremony, Achilles not only testified to the importance of his relationship with Winterlight, but explained the extent to which the Gargoyle MUD had become his home:

> I have traveled far and wide across these lands. . . . I have met a great deal of people as I wandered. I feel that the friendliest people of all are here at Gargoyle. I consider this place my home. I am proud to be a part of this place. I have had some bad times in the past . . . and the people of Gargoyle were there. I thank the people of Gargoyle for their support. I have recently decided to settle down and be married. I searched far and near for a maiden of beauty with hair of sunshine gold and lips red as the rose. With intelligence to match her beauty . . . Winterlight, you are that woman I seek. You are the beautiful maiden. Winterlight, will you marry me?

Winterlight responded with a "charming smile" and said, "Winterlight knows that her face says all. And then, M'lord . . . I love you from deep in my heart."

At the engagement, Winterlight gave Achilles a rose she had worn in her hair and Achilles gave Winterlight a thousand paper stars. Stewart gave me the transcript of the engagement ceremony. It goes on for twelve single-spaced pages of text. Their wedding was even more elaborate. Achilles prepared for it in advance by creating a sacred clearing in cyberspace, a niche carved out of rock, with fifty seats intricately carved with animal motifs. During their wedding, Achilles and Winterlight recalled their engagement gifts and their love and commitment to each other. They were addressed by the priest Tarniwoof. What follows is excerpted from Stewart's log of the wedding ceremony:

> Tarniwoof says, "At the engagement ceremony you gave one another an item which represents your love, respect and friendship for each other."
> Tarniwoof turns to you.
> Tarniwoof says, "Achilles, do you have any reason to give your item back to Winterlight?"
> Winterlight attends your answer nervously.
> Tarniwoof waits for the groom to answer.
> You would not give up her gift for anything.
> Tarniwoof smiles happily.
> Winterlight smiles at you.
> Tarniwoof turns to the beautiful bride.
> Tarniwoof says, "Winterlight, is there any doubt in your heart about what your item represents?"
> Winterlight looks straightly to Tarniwoof.
> Winterlight would never return the thousand paper stars of Achilles.
> Tarniwoof says, "Do you promise to take Silver Shimmering Winterlight as your mudly wedded wife, in sickness and in health, through timeouts and updates, for richer or poorer, until linkdeath do you part?"

You say, "I do."
Winterlight smiles happily at you.

Although Stewart participated in this ceremony alone in his room with his computer and modem, a group of European players actually traveled to Germany, site of Gargoyle's host computer, and got together for food and champagne. There were twenty-five guests at the German celebration, many of whom brought gifts and dressed specially for the occasion. Stewart felt as though he were throwing a party. This was the first time that he had ever entertained, and he was proud of his success. "When I got married," he told me, "people came in from Sweden, Norway, and Finland, and from the Netherlands to Germany, to be at the wedding ceremony in Germany." In real life, Stewart felt constrained by his health problems, his shyness and social isolation, and his narrow economic straits. In the Gargoyle MUD, he was able to bypass these obstacles, at least temporarily. Faced with the notion that "you are what you pretend to be," Stewart can only hope it is true, for he is playing his ideal self.

PSYCHOTHERAPY OR ADDICTION?

I have suggested that MUDs provided Matthew and Gordon with environments they found therapeutic. Stewart, quite self-consciously, has tried to put MUDding in the service of developing a greater capacity for trust and intimacy, but he is not satisfied with the outcome of his efforts. While MUDding on Gargoyle offered Stewart a safe place to experiment with new ways, he sums up his experience by telling me that it has been "an addicting waste of time."

Stewart's case, in which playing on MUDs led to a net drop in self-esteem, illustrates how complex the psychological effects of life on the screen can be. And it illustrates that a safe place is not all that is needed for personal change. Stewart came to MUDding with serious problems. Since childhood he has been isolated both by his illness and by a fear of death he could not discuss with other people. Stewart's mother, who has always been terribly distressed by his illness, has recurring migraines for which Stewart feels responsible. Stewart has never felt free to talk with her about his own anxieties. Stewart's father protected himself by emotionally withdrawing and losing himself in fix-it projects on lawnmowers and cars, the reassuring things that could be made to work the way a sick little boy could not. Stewart resented his father's periods of withdrawal; he says that too often he was left to be the head of the household. Nevertheless, Stewart now emulates his father's style. Stewart says his ·main defense against depression is "not to feel things." "I'd rather put my problems on the back burner and go on with my life." Before he became involved in MUDs, going on with his life usually meant throwing himself into his schoolwork or into major car repair projects. He fondly remembers a two-week period during which a physics experiment took almost all his waking hours and a school vacation spent tinkering round the clock in his family's garage. He finds comfort in the all-consuming demands of science and with the "reliability of machines." Stewart does not know how to find comfort in the more volatile and unpredictable world of people.

> I have a problem with emotional things. I handle them very badly. I do the things you're not supposed to do. I don't worry about them for a while, and then they come back to haunt me two or three years later. . . . I am not able to talk about my problems while they are happening. I have to wait until they have become just a story.
>   If I have an emotional problem I cannot talk to people about it. I will sit there in a room with them, and I will talk to them about anything else in the entire world except what's bothering me.

Stewart was introduced to MUDs by Carrie, an unhappy classmate whose chief source of solace was talking to people on MUDs. Although Stewart tends to ignore his own troubles, he wanted to connect with Carrie by helping with hers. Carrie had troubles aplenty; she drank too much and had an abusive boyfriend. Yet Carrie rejected Stewart's friendship. Stewart described how, when he visited her in her dorm room, she turned her back to him to talk to "the people in the machine."

> I mean, when you have that type of emotional problem and that kind of pain, it's not an intelligent move to log on to a game and talk to people because they are safe and they won't hurt you. Because that's just not a way out of it. I mean there is a limit to how many years you can spend in front of a computer screen.

Shortly after this incident in Carrie's room, Stewart precipitated the end of their relationship. He took it upon himself to inform Carrie's parents that she had a drinking problem, something that she wanted to sort out by herself. When Carrie confronted him about his "meddling," Stewart could not see her point of view and defended his actions by arguing that morality was on his side. For Carrie, Stewart's intrusions had gone too far. She would no longer speak to him. By the fall of his junior year in college, Stewart was strained to his psychological limit. His friendship with Carrie was over, his mother was seriously ill, and he himself had developed pneumonia.

This bout of pneumonia required that Stewart spend three weeks in the hospital, an experience that brought back the fears of death he had tried to repress. When he finally returned to his dormitory, confined to his room for a fourth week of bed rest, he was seriously depressed and felt utterly alone. Stewart habitually used work to ward off depression, but now he felt too far behind in his schoolwork to try to catch up. In desperation, Stewart tried Carrie's strategy: He turned to MUDs. Within a week, he was spending ten to twelve hours a day on the games. He found them a place where he could talk about his real-life troubles. In particular, he talked to the other players about Carrie, telling his side of the story and complaining that her decision to break off their friendship was unjust.

> I was on the game talking to people about my problems endlessly. . . . I find it a lot easier to talk to people on the game about them because they're not there. I mean, they are there but they're not there. I mean, you could sit there and you could tell them about your problems and you don't have to worry about running into them on the street the next day.

MUDs did help Stewart talk about his troubles while they were still emotionally relevant; nevertheless, he is emphatic that MUDding has ultimately made him feel worse about himself. Despite his MUD socializing, despite the poetry of his MUD romance and the pageantry of his MUD engagement and marriage, MUDding did not alter Stewart's sense of himself as withdrawn, unappealing, and flawed. His experience paralleled that of Cyrano in Rostand's play. Cyrano's success in wooing Roxanne for another never made him feel worthy himself. Stewart says of MUDding:

> The more I do it, the more I feel I need to do it. Every couple of days I'd notice, it's like, "Gee, in the last two days, I've been on this MUD for the total of probably over twenty-eight hours." . . . I mean I'd be on the MUD until I fell asleep at the terminal practically, and then go to sleep, and then I'd wake up and I'd do it again.

Stewart has tried hard to make his MUD self, the "better" Achilles self, part of his real life, but he says he has failed. He says, "I'm not social. I don't like parties. I can't talk to people about my problems." We recall together that these things are easy for him on MUDs and he shrugs and says, "I know." The integration of the social Achilles, who can talk about his troubles, and the asocial Stewart, who can only cope by putting them out of mind, has not occurred. From Stewart's point of view, MUDs have stripped away some of his defenses but have given him nothing in return. In fact, MUDs make Stewart feel vulnerable in a new way. Although he hoped that MUDs would cure him, it is MUDs that now make him feel sick. He feels addicted to MUDs: "When you feel you're stagnating and you feel there's nothing going on in your life and you're stuck in a rut, it's very easy to be on there for a very large amount of time."

In my interviews with people about the possibility of computer psychotherapy, a ventilation model of psychotherapy came up often as a reason why computers could be therapists. In the ventilation model, psychotherapy makes people better by being a safe place for airing problems, expressing anger, and admitting to fears. MUDs may provide a place for people to talk freely—and with other people rather than with a machine —but they also illustrate that therapy has to be more than a safe place to "ventilate." There is considerable disagreement among psychotherapists about what that "more" has to be, but within the psychoanalytic tradition, there is fair consensus that it involves a special relationship with a therapist in which old issues will resurface and be addressed in new ways. When elements from the past are projected onto a relationship with a therapist, they can be used as data for self-understanding. So a psychotherapy is not just a safe place, it is a work space, or more accurately a reworking space.

For Stewart, MUD life gradually became a place not for reworking but for reenacting the kinds of difficulties that plagued him in real life. On the MUD, he declared his moral superiority over other players and lectured them about their faults, the exact pattern he had fallen into with Carrie. He began to violate MUD etiquette, for example by revealing certain players' real-life identities and real-life bad behavior. He took on one prominent player, Ursula, a woman who he thought had taken advantage of her (real-life) husband, and tried to expose her as a bad person to other MUD players. Again, Stewart justified his intrusive actions toward Ursula, as he justified his intrusions on Carrie's privacy, by saying that morality was on his side. When other players pointed out that it was now Stewart who was behaving inappropriately, he became angry and self-righteous. "Ursula deserves to be exposed," he said, "because of her outrageous behavior." A psychotherapist might have helped Stewart reflect on why he needs to be in the position of policeman, judge, and jury. Does he try to protect others because he feels that he has so often been left unprotected? How can he find ways to protect himself? In the context of a relationship with a therapist, Stewart might have been able to address such painful matters. On the MUD, Stewart avoided them by blaming other people and declaring right on his side.

When Matthew and Gordon talked about sharing confidences on MUDs more freely than in real life, they spoke of using anonymity to modulate their exposure. In contrast, Stewart renounced anonymity on MUDs and talked nonstop to anyone who would listen. This wholesale discarding of his most characteristic defenses, withdrawal and reticence, made him feel out of control. He compensated by trying even harder to "put things out of his mind" and by denying that MUDs had been of any value. Again, the comparison with psychotherapy is illuminating. A skillful therapist would have treated Stewart's defenses with greater respect, as tools that might well be helpful if used in modest doses.[12] A little withdrawal can be a good thing. But if a naive psychotherapist had encouraged Stewart to toss away defenses and tell all, that therapist would likely have had an unhappy result similar to what Stewart achieved from his MUD confessions: Stewart's defenses would end up more entrenched than before, but it would be the psychotherapy rather than the MUDs that he would denigrate as a waste of time.

Stewart cannot learn from his character Achilles' experiences and social success because they are too different from the things of which he believes himself capable. Despite his efforts to turn Achilles into Stewart, Stewart has split off his strengths and sees them as possible only for Achilles in the MUD. It is only Achilles who can create the magic and win the girl. In making this split between himself and the achievements of his screen persona, Stewart does not give himself credit for the positive steps he has taken in real life. He has visited other MUD players in America and has had a group of the German players visit him for a weekend of sightseeing. But like an unsuccessful psychotherapy, MUDding has not helped Stewart bring these good experiences inside himself or integrate them into his self-image.

Stewart has used MUDs to "act out" rather than "work through" his difficulties. In acting out we stage our old conflicts in new settings, we reenact our past in fruitless repetition. In contrast, working through usually involves a moratorium on action in order to think about our habitual reactions in a new way. Psychoanalytic theory suggests that it is precisely by not stirring things up at the level of outward action that we are able to effect inner change. The presence of the therapist helps to contain the impulse for action and to encourage an examination of the meaning of the impulse itself. MUDs provide rich spaces for both acting out and working through. There are genuine possibilities for change, and there is room for unproductive repetition. The outcome depends on the emotional challenges the players face and the emotional resources they bring to the game. MUDs can provide occasions for personal growth and change, but not for everyone and not in every circumstance.

Stewart tried and failed to use MUDs for therapeutic purposes. Robert, whom I met after his freshman year in college, presents a contrasting case. Although Robert went through a period during which he looked even more "addicted" to MUDs than Stewart, in the end, his time on MUDs provided him with a context for significant personal growth.

During his final year of high school, Robert had to cope with severe disruptions in his family life. His father lost his job as a fireman because of heavy drinking. The fire department helped him to find a desk job in another state. "My dad was an abusive alcoholic," says Robert. "He lost his good job. They sent him somewhere else. He moved, but my mom stayed in Minnesota with me. She was my security." College in New Jersey took Robert away from his high school friends and his mother. It was his first extended separation from her. Calls to his mother felt unsatisfying. They were too short and too expensive. Robert was lonely during the early days of his freshman year, but then a friend introduced him to MUDs.

For a period of several months, Robert MUDded over eighty hours a week. "The whole second semester, eighty hours a week," he says. During a time of particular stress, when a burst water pipe and a flooded dorm room destroyed all his possessions, Robert was playing for over a hundred and twenty hours a week. He ate at his computer; he generally slept four hours a night. Much of the fun, he says, was being able to put his troubles aside. He liked "just sitting there, not thinking about anything else. Because if you're so involved, you can't think about the problem, your problems."

> When I MUDded with the computer I never got tired. A lot of it was, like, "Oh, whoa, it's this time already." . . . Actually it is very obsessive. I remember up at college, I was once thinking, "Boy, I was just on this too much. I should cut down." But I was not able to. It's like a kind of addiction. . . . It was my life. . . . I was, like, living on the MUD. . . . Most of the time I felt comfortable that this was my life. I'd say I was addicted to it.
>
> I'd keep trying to stop. I'd say, "OK, I'm not going on. I'm going to classes." But something would come up and I wouldn't go to my class. I wouldn't do what I wanted to do.

Much of Robert's play on MUDs was serious work, because he took on responsibilities in cyberspace equivalent to a full-time job. He became a highly placed administrator of a new MUD. Robert told me that he had never before been in charge of anything. Now his MUD responsibilities were enormous.

Building and maintaining a MUD is a large and complicated task. There is technical programming work. New objects made by individual players need to be carefully reviewed to make sure that they do not interfere with the basic technical infrastructure of the MUD. There is management work. People need to be recruited, assigned jobs, taught the rules of the MUD, and put into a chain of command. And there is political work. The MUD is a community of people whose quarrels need to be adjudicated and whose feelings need to be respected. On his MUD, Robert did all of this, and by all accounts, he did it with elegance, diplomacy, and aplomb.

> I had to keep track of each division of the MUD that was being built and its local government, and when the money system came in I had to pay the

local workers. All the officers and enlisted men and women on each ship got paid a certain amount, depending on their rank. I had to make sure they got paid on the same day and the right amount. I had to make sure people had the right building quota, not wasting objects, not building too much.

Matthew and Julee nurtured themselves "in displacement." By helping others they were able to take care of themselves. Robert, too, gave others on MUDs what he most needed himself: a sense of structure and control.

Prior to taking on this job in a MUD, Robert had been known as something of a cut up on MUDs and elsewhere, someone accustomed to thumbing his nose at authority. He had gotten the administrative job on the MUD primarily because of the amount of time he was willing to commit. Robert says his MUD responsibilities gave him new respect for authority ("everyone should get to be a higher-up for a day," he says) and taught him something about himself. Robert discovered that he excels at negotiation and practical administration.

But despite the intensity and gratification of being online, at the end of the school year, Robert's MUDding was essentially over. When I met him in the summer after his freshman year, he was working as a sales clerk, had gotten his first apartment, and had formed a rock band with a few friends. He says that one week he was MUDding "twelve hours a day for seven days," and then the next week he was not MUDding at all.

How had Robert's MUDding ended? For one thing, a practical consideration intervened. At the end of the school year, his college took back the computer they had leased him for his dorm room. But Robert says that by the time his computer was taken away, MUDding had already served its emotional purpose for him.

Robert believes that during the period he was MUDding most intensively, the alternative for him would have been partying and drinking, that is, getting into his father's kind of trouble. He says,"I remember a lot of Friday and Saturday nights turning down parties because I was on the computer. . . . Instead of drinking I had something more fun and safe to do." During his high school years Robert drank to excess and was afraid that he might become an alcoholic like his father. MUDding helped to keep those fears at bay.

MUDding also gave Robert a way to think about his father with some sympathy but reassured him that he was not like his father. Robert's behavior on MUDs reminded him of his father's addictions in a way that increased his feelings of compassion.

It made me feel differently about someone who was addicted. I was a different person on the MUD. I didn't want to be bothered when I was on the MUD about other things like work, school, or classes. . . . I suppose in some way I feel closer to my Dad. I don't think he can stop himself from drinking. . . . maybe with a lot of help he could. But I don't think he can. It's just like I had a hard time stopping MUDs.

Like Stewart, Robert acted out certain of his troubles on the MUDs— the fascination with pushing an addiction to a limit, for example. But unlike Stewart, after he was confident that he could function responsibly and competently on MUDs, Robert wanted to try the same behavior in real life. And unlike Stewart, he was able to use MUDding as an environment in which he could talk about his feelings in a constructive way. In the real world Robert found it painful to talk about himself because he often found himself lying about such simple things as what his father did for a living. Because it was easier to "walk away" from conversations on the MUD, Robert found that it was easier to have them in the first place. While Stewart used MUDs to "tell all," Robert practiced the art of talking about himself in measured doses: "The computer is sort of practice to get into closer relationships with people in real life. . . . If something is bothering me, you don't have to let the person know or you can let the person know."

MUDs provided Robert with what the psychoanalyst Erik Erikson called a psychosocial moratorium.[13] The notion of moratorium was a central aspect of Erikson's theories about adolescent identity development. Although the term implies a time out, what Erikson had in mind was not withdrawal. On the contrary, the adolescent moratorium is a time of intense interaction with people and ideas. It is a time of passionate friendships and experimentation. The moratorium is not on significant experiences but on their consequences. Of course, there are never human actions that are without consequences, so there is no such thing as a pure moratorium. Reckless driving leads to teenage deaths; careless sex to teenage pregnancy. Nevertheless, during the adolescent years, people are generally given permission to try new things. There is a tacit understanding that they will experiment. Though the outcomes of this experimentation can have enormous consequences, the experiences themselves feel removed from the structured surroundings of one's normal life. The moratorium facilitates the development of a core self, a personal sense of what gives life meaning. This is what Erikson called identity.

Erikson developed his ideas about the importance of a moratorium to the development of identity during the late 1950s and early 1960s. At that time, the notion corresponded to a common understanding of what "the college years" were about. Today, thirty years later, the idea of the college years as a "time out" seems remote, even quaint. College is preprofessional and AIDS has made sexual experimentation a potentially deadly game. But if our culture no longer offers an adolescent moratorium, virtual communities do. They offer permission to play, to try things out. This is part of what makes them attractive.

Erikson saw identity in the context of a larger stage theory of development. Identity was one stage, intimacy and generativity were others. Erikson's ideas about stages did not suggest rigid sequences. His stages describe what people need to achieve before they can easily move ahead to another developmental task. For example, Erikson pointed out that successful intimacy in young adulthood is difficult if one does not come to it with a sense of who one is. This is the challenge of adolescent identity building. In real life, however, people frequently move on with incompletely resolved stages, simply doing the best they can. They use whatever materials they have at hand to get as much as they can of what they have missed. MUDs are striking examples of how technology can play a role in these dramas of self-repair. Stewart's case makes it clear that they are not a panacea. But they do present new opportunities as well as new risks.

Once we put aside the idea that Erikson's stages describe rigid sequences, we can look at the stages as modes of experience that people work on throughout their lives. Thus, the adolescent moratorium is not something people pass through but a mode of experience necessary throughout functional and creative adulthoods. We take vacations to escape not only from our work but from our habitual social lives. Vacations are times during which adults are allowed to play. Vacations give a finite structure to periodic adult moratoria. Time in cyberspace reshapes the notion of vacations and moratoria, because they may now exist in always-available windows. Erikson wrote that "the playing adult steps sideward into another reality; the playing child advances forward into new stages of mastery."[14] In MUDs, adults can do both; we enter another reality and have the opportunity to develop new dimensions of self-mastery.

Unlike Stewart, Robert came to his emotional difficulties and his MUDding with a solid relationship with a consistent and competent mother. This good parenting enabled him to identify with other players he met on the MUD who had qualities he wished to emulate. Even more important, unlike Stewart, Robert was able to identify with the better self he played in the game. This constructive strategy is available only to people who are able to take in positive models, to bring other people and images of their better selves inside themselves.

When people like Stewart get stuck or become increasingly self-critical and depressed on MUDs, it is often because deficits in early relationships have made it too hard for them to have relationships that they can turn to constructive purposes. From the earliest days of his life, Stewart's illness and his parents' response to it, his mother's migraines and his father's withdrawals, made him feel unacceptable. In his own words, "I have always felt like damaged goods."

Life in cyberspace, as elsewhere, is not fair. To the question, "Are MUDs good or bad for psychological growth?" the answer is unreassuringly complicated, just like life. If you come to the games with a self that is healthy enough to be able to grow from relationships, MUDs can be very good. If not, you can be in for trouble.

Stewart attended a series of pizza parties I held for MUDders in the Boston area. These were group sessions during which players had a chance to meet face to face and talk about their experiences. There Stewart met a group of people who used the games to role-play characters from diverse cultures and time periods. They played medieval ladies, Japanese warriors, Elizabethan bards. Stewart told me he felt little in common with these players, and he also seemed uncomfortable around them. Perhaps they called into question his desire to see MUDding as a simple extension of his real life. Stewart repeatedly insisted that, despite the fact that his character was "technically" named Achilles, he was in fact playing himself. He reminded the group several times that when he MUDded he actually asked other players to call him Stewart. But during one group session, after insisting for several hours that he plays no role on MUDs, a member of the role-playing contingent casually asked Stewart if he was married. Stewart immediately said, "Yes," and then blushed deeply because he was caught in the contradiction between his insistence that he plays no roles in Gargoyle and his deep investment in his MUD marriage. Paradoxically, Stewart was kept from fully profiting from Achilles' social successes, not because he fully identified with the character as he insisted, but because he ultimately saw Stewart as too unlike Achilles.

In some computer conferences, the subject of the slippage between online personae and one's real-life character has become a focal point of discussion. On the WELL, short for the "Whole Earth 'Lectronic Link," a San Francisco–based virtual community, some contributors have maintained that they enjoy experimenting with personae very different from their RL selves. Others have insisted that maintaining an artificial persona very different from one's sense of oneself in RL is what one called "cheap fuel," a novelty that wears thin fast because of the large amount of "psychic energy" required to maintain it. These people note that they want to reveal themselves to the members of a community that they care about.[15] Yet other contributors take a third position: They stress that cyberspace provides opportunities to play out aspects of oneself that are not total strangers but that may be inhibited in real life. One contributor finds that online experience "seems to interface with the contentious, opinionated, verbal, angry, and snide aspects of my personality beautifully but not with many of the other aspects. My electronic id is given wing here. I'm having a hard time balancing it."[16]

The electronic discussion on the WELL circled around the therapeutic potential of online personae and touched on a point that was very important to many of the people I interviewed: The formats of MUDs, electronic mail, computer conferences, or bulletin boards force one to recognize a highly differentiated (and not always likable) virtual persona, because that persona leaves an electronic trace. In other words, the presence of a record that you can scroll through again and again may alert you to how persistent are your foibles or defensive reactions. One New York City writer told me ruefully, "I would see myself on the screen and say, 'There I go again.' I could see my neuroses in black and white. It was a wake-up call."

INTIMACY AND PROJECTION

Robert had a virtual girlfriend on the MUD, a character played by a West Coast college senior named Kasha. Women are in short supply in MUDs and his friendship with Kasha made Robert the envy of many other male players. In the MUD universe, Kasha built a private planet whose construction took many weeks of programming. On the planet, Kasha built a mansion with flowers in every room. As a special gift to Robert, Kasha designed the most beautiful of these rooms as their bedroom.

Robert traveled cross-country to visit Kasha and, completely smitten, Kasha made plans to move to New Jersey at the end of the academic year. But as that time approached, Robert pulled away.

> I mean, she had a great personality over the computer. We got along pretty well. And then I went to see her. And then—I don't know. Every day I had less and less feeling toward her. And I was thinking of my mom more and more. I'm so confused about what I am doing in college. I just didn't want someone coming to live with me in New Jersey and all. That's what she was talking about. It was all much too fast.

Relationships during adolescence are usually bounded by a mutual understanding that they involve limited commitment. Virtual space is well suited to such relationships; its natural limitations keep things within bounds. So, from one point of view, Robert's pulling back from Kasha was simply a sign that he was ready for commitment in the virtual but not in the real. But Robert and Kasha were also playing out a characteristic pattern on MUDs. As in Thomas Mann's *The Magic Mountain,* which takes place in the isolation of a sanatorium, relationships become intense very quickly because the participants feel isolated in a remote and unfamiliar world with its own rules. MUDs, like other electronic meeting places, can breed a kind of easy intimacy. In a first phase, MUD players feel the excitement of a rapidly deepening relationship and the sense that time itself is speeding up. One player describes the process as follows:

The MUD quickens things. It quickens things so much. You know, you don't think about it when you're doing it, but you meet somebody on the MUD, and within a week you feel like you've been friends forever. It's notorious. One of the notorious things that people who've thought about it will say is that MUDs are both slower because you can't type as fast as you can talk, but they're faster because things seem to move so much faster.

In a second phase, players commonly try to take things from the virtual to the real and are usually disappointed. Peter, a twenty-eight-year-old lecturer in comparative literature, thought he was in love with a MUDding partner who played Beatrice to his Dante (their characters' names). Their relationship was intellectual, emotionally supportive, and erotic. Their virtual sex life was rich and fulfilling. The description of physical actions in their virtual sex (or TinySex) was accompanied by detailed descriptions of each of their thoughts and feelings. It was not just TinySex, it was TinyLovemaking. Peter flew from North Carolina to Oregon to meet the woman behind Beatrice and returned home crushed. "[On the MUD] I saw in her what I wanted to see. Real life gave me too much information."

Since it is not unusual for players to keep logs of their MUD sessions with significant others, Peter had something that participants in real-life relationships never have: a record of every interaction with Beatrice.[17] When he read over his logs, he remarked that he could not find their relationship in them. Where was the warmth? The sense of complicity and empathy?[18]

When everything is in the log and nothing is in the log, people are confronted with the degree to which they construct relationships in their own minds. In Peter's case, as he reflected on it later, his unconscious purpose was to create a love object, someone who reminded him of an idolized and inaccessible older sister.

MUDs encourage projection and the development of transferences for some of the same reasons that a classical Freudian analytic situation does. Analysts sit behind their patients so they can become disembodied voices. Patients are given space to project onto the analyst thoughts and feelings from the past. In MUDs, the lack of information about the real person to whom one is talking, the silence into which one types, the absence of visual cues, all these encourage projection. This situation leads to exaggerated likes and dislikes, to idealization and demonization. So it is natural for people to feel let down or confused when they meet their virtual lovers in person. Those who survive the experience best are the people who try to understand why the persona of a fellow MUDder has evoked such a powerful response. And sometimes, when the feelings evoked in transferences on MUDs are reflected upon, MUD relationships can have a positive effect on self-understanding.

Jeremy, a thirty-two-year-old lawyer, says this about MUDding:

> I dare to be passive. . . . I don't mean in having sex on the MUD. I mean in letting other people take the initiative in friendships, in not feeling when I am in character that I need to control everything. My mother controlled my whole family, well, certainly she controlled me. So I grew up thinking, "Never again!" My real life is exhausting that way. I'm always protecting myself. On MUDs I do something else. . . . I didn't even realize this connection to my mother and the MUDding until [in the game] somebody tried to boss my pretty laid-back character around and I went crazy. . . . I hated her. . . . And then I saw what I was doing. When I looked at the logs I saw that . . . this woman was really doing very little to boss me around. But I hear a woman with an authoritative tone and I go crazy. Food for thought.

To the question, "Are MUDs psychotherapeutic?" one is tempted to say that they stand the greatest chance to be so if the MUDder is also in psychotherapy. Taken by themselves, MUDs are highly evocative and provide much grist for the mill of a psychodynamic therapeutic process. If "acting out" is going to happen, MUDs are relatively safe places, since virtual promiscuity never causes pregnancy or disease. But it is also true that, taken by themselves, virtual communities will only sometimes facilitate psychological growth.

French Sherry

MUDs provide dramatic examples of how one can use experiences in virtual space to play with aspects of the self. Electronic mail and bulletin boards provide more mundane but no less impressive examples. There, role playing may not be as explicit or extravagant, but it goes on all the same.

On America Online, people choose handles by which they are known on the system. One's real name need only be known to the administrators of the service itself. One forty-two-year-old nurse whose real name is Annette calls herself Bette on the system. "Annette," she says, "for all my life that will be sweet, little perky Annette from the *Mickey Mouse Club*. I want to be a Bette. Like Bette Davis. I want to seem mysterious and powerful. There is no such thing as a mysterious and powerful Mouseketeer." On America Online, "Bette" is active on a poetry forum. "I've always wanted to write poetry. I have made little fits and starts through the years. I don't want to say that changing my name made it possible, but I can tell you it made it a whole lot easier. Bette writes poems. Annette just fools around with it. Annette is a nurse. Bette is the name of a writer, more moody, often more morose." When she types at her computer, Annette, who has become a skillful touch typist, says:

> I like to close my eyes and imagine myself speaking as Bette. An authoritative voice. When I type as Bette I imagine her voice. You might ask whether this Bette is real or not. Well, she is real enough to write poetry. I mean it's poetry that I take credit for. Bette gives courage. We sort of do it together.

Annette does not suffer from multiple personality disorder. Bette does not function autonomously. Annette is not dissociated from Bette's behavior. Bette enables aspects of Annette that have not been easy for her to express. As Annette becomes more fluent as Bette, she moves flexibly between the two personae. In a certain sense, Annette is able to relate to Bette with the flexibility that Stewart could not achieve in his relationship with Achilles. Achilles could have social successes but, in Annette's terms, the character could not give courage to the more limited Stewart.

Annette's very positive Bette experience is not unusual in online culture. Experiences like Annette's require only that one use the anonymity of cyberspace to project alternate personae. And, like Annette, people often project underdeveloped aspects of themselves. We can best understand the psychological meaning of this by looking to experiences that do not take place online. These are experiences in which people expand their sense of identity by assuming roles where the boundary between self and role becomes increasingly permeable. When Annette first told me her story, it reminded me of such an experience in my own life.

My mother died when I was nineteen and a college junior. Upset and disoriented, I dropped out of school. I traveled to Europe, ended up in Paris, studied history and political science, and worked at a series of odd jobs from housecleaner to English tutor. The French-speaking Sherry, I was pleased to discover, was somewhat different from the English-speaking one. French-speaking Sherry was not unrecognizable, but she was her own person. In particular, while the English-speaking Sherry had little confidence that she could take care of herself, the French-speaking Sherry simply had to and got on with it.

On trips back home, English-speaking Sherry rediscovered old timidities. I kept returning to France, thirsty for more French speaking. Little by little, I became increasingly fluent in French and confortable with the persona of the resourceful, French-speaking young woman. Now I cycled through the French- and English-speaking Sherrys until the movement seemed natural; I could bend toward one and then the other with increasing flexibility. When English-speaking Sherry finally returned to college in the United States, she was never as brave as French-speaking Sherry. But she could hold her own.

CHAPTER 8

TINYSEX AND GENDER TROUBLE

From my earliest effort to construct an online persona, it occurred to me that being a virtual man might be more comfortable than being a virtual woman.

When I first logged on to a MUD, I named and described a character but forgot to give it a gender. I was struggling with the technical aspects of the MUD universe—the difference between various MUD commands such as "saying" and "emoting," "paging" and "whispering." Gender was the last thing on my mind. This rapidly changed when a male-presenting character named Jiffy asked me if I was "really an it." At his question, I experienced an unpleasurable sense of disorientation which immediately gave way to an unfamiliar sense of freedom.

When Jiffy's question appeared on my screen, I was standing in a room of LambdaMOO filled with characters engaged in sexual banter in the style of the movie *Animal House.* The innuendos, double entendres, and leering invitations were scrolling by at a fast clip; I felt awkward, as though at a party to which I had been invited by mistake. I was reminded of junior high school dances when I wanted to go home or hide behind the punch bowl. I was reminded of kissing games in which it was awful to be chosen and awful not to be chosen. Now, on the MUD, I had a new option. I wondered if playing a male might allow me to feel less out of place. I could stand on the sidelines and people would expect *me* to make the first move. And I could choose not to. I could choose simply to "lurk," to stand by and observe the action. Boys, after all, were not called prudes if they were too cool to play kissing games. They were not categorized as wallflowers if they held back and didn't ask girls to dance. They could simply be shy in a manly way—aloof, above it all.

Two days later I was back in the MUD. After I typed the command that joined me, in Boston, to the computer in California where the MUD resided, I discovered that I had lost the paper on which I had written my MUD password. This meant that I could not play my own character but had to log on as a guest. As such, I was assigned a color: Magenta. As "Magenta_guest" I was again without gender. While I was struggling with basic MUD commands, other players were typing messages for all to see such as "Magenta_guest gazes hot and enraptured at the approach of Fire_Eater." Again I was tempted to hide from the frat party atmosphere by trying to pass as a man.[1] When much later I did try playing a male character, I finally experienced that permission to move freely I had always imagined to be the birthright of men. Not only was I approached less frequently, but I found it easier to respond to an unwanted overture with aplomb, saying something like, "That's flattering, Ribald_Temptress, but I'm otherwise engaged." My sense of freedom didn't just involve a different attitude about sexual advances, which now seemed less threatening. As a woman I have a hard time deflecting a request for conversation by asserting my own agenda. As a MUD male, doing so (nicely) seemed more natural; it never struck me as dismissive or rude. Of course, my reaction said as much about the construction of gender in my own mind as it did about the social construction of gender in the MUD.

Playing in MUDs, whether as a man, a woman, or a neuter character, I quickly fell into the habit of orienting myself to new cyberspace acquaintances by checking out their gender. This was a strange exercise, especially because a significant proportion of the female-presenting characters were RL men, and a good number of the male-presenting characters were RL women. I was not alone in this curiously irrational preoccupation. For many players, guessing the true gender of players behind MUD characters has become something of an art form. Pavel Curtis, the founder of LambdaMOO, has observed that when a female-presenting character is called something like FabulousHotBabe, one can be almost sure there is a man behind the mask.[2] Another experienced MUDder shares the folklore that "if a female-presenting character's description of her beauty goes on for more than two paragraphs, 'she' [the player behind the character] is sure to be an ugly woman."

The preoccupation in MUDs with getting a "fix" on people through "fixing" their gender reminds us of the extent to which we use gender to shape our relationships. Corey, a twenty-two-year-old dental technician, says that her name often causes people to assume that she is male—that is, until she meets them. Corey has long blonde hair, piled high, and admits to "going for the Barbie look."

I'm not sure how it started, but I know that when I was a kid the more people said, "Oh, you have such a cute boy's name," the more I laid on the hairbows. [With my name] they always expected a boy—or at least a tomboy.

Corey says that, for her, part of the fun of being online is that she gets to see "a lot of people having the [same] experience [with their online names that] I've had with my name." She tells me that her girlfriend logged on as Joel instead of Joely, "and she saw people's expectations change real fast." Corey continues:

> I also think the neuter characters [in MUDs] are good. When I play one, I realize how hard it is not to be either a man or a woman. I always find myself trying to be one or the other even when I'm trying to be neither. And all the time I'm talking to a neuter character [she reverses roles here] . . . I'm thinking "So who's behind it?"

In MUDs, the existence of characters other than male or female is disturbing, evocative. Like transgressive gender practices in real life, by breaking the conventions, it dramatizes our attachment to them.

Gender-swapping on MUDs is not a small part of the game action. By some estimates, Habitat, a Japanese MUD, has 1.5 million users. Habitat is a MUD operated for profit. Among the registered members of Habitat, there is a ratio of four real-life men to each real-life woman. But inside the MUD the ratio is only three male characters to one female character. In other words, a significant number of players, many tens of thousands of them, are virtually cross-dressing.[3]

GENDER TROUBLE[4]

What is virtual gender-swapping all about? Some of those who do it claim that it is not particularly significant. "When I play a woman I don't really take it too seriously," said twenty-year-old Andrei. "I do it to improve the ratio of women to men. It's just a game." On one level, virtual gender-swapping is easier than doing it in real life. For a man to present himself as female in a chat room, on an IRC channel, or in a MUD, only requires writing a description. For a man to play a woman on the streets of an American city, he would have to shave various parts of his body; wear makeup, perhaps a wig, a dress, and high heels; perhaps change his voice, walk, and mannerisms. He would have some anxiety about passing, and there might be even more anxiety about not passing, which would pose a risk of violence and possibly arrest. So more men are willing to give virtual cross-dressing a try. But once they are online as female, they soon find that maintaining this fiction is difficult. To pass as a woman for any length of time requires understanding how gender inflects speech, manner, the interpretation of experience. Women attempting to pass as men face the same kind of challenge. One woman said that she "worked hard" to pass in a room on a commercial network service that was advertised as a meeting place for gay men.

> I have always been so curious about what men do with each other. I could never even imagine how they talk to each other. I can't exactly go to a gay bar and eavesdrop inconspicuously. [When online] I don't actually have [virtual] sex with anyone. I get out of that by telling the men there that I'm shy and still unsure. But I like hanging out; it makes gays seem less strange to me. But it is not so easy. You have to think about it, to make up a life, a job, a set of reactions.

Virtual cross-dressing is not as simple as Andrei suggests. Not only can it be technically challenging, it can be psychologically complicated. Taking a virtual role may involve you in ongoing relationships. In this process, you may discover things about yourself that you never knew before. You may discover things about other people's response to you. You are not in danger of being arrested, but you are embarked on an enterprise that is not without some gravity and emotional risk.

In fact, one strong motivation to gender-swap in virtual space is to have TinySex as a creature of another gender, something that suggests more than an emotionally neutral activity. Gender-swapping is an opportunity to explore conflicts raised by one's biological gender. Also, as Corey noted, by enabling people to experience what it "feels" like to be the opposite gender or to have no gender at all, the practice encourages reflection on the way ideas about gender shape our expectations. MUDs and the virtual personae one adopts within them are objects-to-think-with for reflecting on the social construction of gender.

Case, a thirty-four-year-old industrial designer who is happily married to a coworker, is currently MUDding as a female character. In response to my question, "Has MUDding ever caused you any emotional pain?" he says, "Yes, but also the kind of learning that comes from hard times."

> I'm having pain in my playing now. The woman I'm playing in MedievalMUSH [Mairead] is having an interesting relationship with a fellow. Mairead is a lawyer. It costs so much to go to law school that it has to be paid for by a corporation or a noble house. A man she met and fell in love with was a nobleman. He paid for her law school. He bought my [Case slips into referring to Mairead in the first person] contract. Now he wants to marry me although I'm a commoner. I finally said yes. I try to talk to him about the fact that I'm essentially his property. I'm a commoner, I'm basically property and to a certain extent that doesn't bother me. I've grown up with it, that's the way life is. He wants to deny the situation. He says, "Oh no, no, no. . . . We'll pick you up, set you on your feet, the whole world is open to you."

But everytime I behave like I'm now going to be a countess some day, you know, assert myself—as in, "And I never liked this wallpaper anyway"—I get pushed down. The relationship is pull up, push down. It's an incredibly psychologically damaging thing to do to a person. And the very thing that he liked about her—that she was independent, strong, said what was on her mind—it is all being bled out of her.

Case looks at me with a wry smile and sighs, "A woman's life." He continues:

I see her [Mairead] heading for a major psychological problem. What we have is a dysfunctional relationship. But even though it's very painful and stressful, it's very interesting to watch myself cope with this problem. How am I going to dig my persona's self out of this mess? Because I don't want to go on like this. I want to get out of it. . . . You can see that playing this woman lets me see what I have in my psychological repertoire, what is hard and what is easy for me. And I can also see how some of the things that work when you're a man just backfire when you're a woman.

Case has played Mairead for nearly a year, but even a brief experience playing a character of another gender can be evocative. William James said, "Philosophy is the art of imagining alternatives." MUDs are proving grounds for an action-based philosophical practice that can serve as a form of consciousness-raising about gender issues. For example, on many MUDs, offering technical assistance has become a common way in which male characters "purchase" female attention, analogous to picking up the check at an RL dinner. In real life, our expectations about sex roles (who offers help, who buys dinner, who brews the coffee) can become so ingrained that we no longer notice them. On MUDs, however, expectations are expressed in visible textual actions, widely witnessed and openly discussed. When men playing females are plied with unrequested offers of help on MUDs, they often remark that such chivalries communicate a belief in female incompetence. When women play males on MUDs and realize that they are no longer being offered help, some reflect that those offers of help may well have led them to believe they needed it. As a woman, "First you ask for help because you think it will be expedient," says a college sophomore, "then you realize that you aren't developing the skills to figure things out for yourself."

ALL THE WORLD'S A STAGE

Any account of the evocative nature of gender-swapping might well defer to Shakespeare, who used it as a plot device for reframing personal and political choices. As You Like It is a classic example, a comedy that uses gender-swapping to reveal new aspects of identity and to permit greater complexity of relationships.[5] In the play, Rosalind, the Duke's daughter, is exiled from the court of her uncle Frederick, who has usurped her father's throne. Frederick's daughter, Rosalind's cousin Celia, escapes with her. Together they flee to the magical forest of Arden. When the two women first discuss their plan to flee, Rosalind remarks that they might be in danger because "beauty provoketh thieves sooner than gold." In response, Celia suggests that they would travel more easily if they rubbed dirt on their faces and wore drab clothing, thus pointing to a tactic that frequently provides women greater social ease in the world—becoming unattractive. Rosalind then comes up with a second idea—becoming a man: "Were it not better,/Because that I am more than common tall,/That I did suit me all points like a man?"

In the end, Rosalind and Celia both disguise themselves as boys, Ganymede and Aliena. In suggesting this ploy, Rosalind proposes a disguise that will be both physical ("A gallant curtle-axe on my thigh,/A boarspear in my hand") and emotional ("and—in my heart,/Lie there what hidden woman's fear there will"). She goes on, "We'll have a swashbuckling and martial outside,/as many other mannish cowards have/That do outface it with their semblances."[6]

In these lines, Rosalind does not endorse an essential difference between men and women; rather, she suggests that men routinely adopt the same kind of pose she is now choosing. Biological men have to construct male gender just as biological women have to construct female gender. If Rosalind and Celia make themselves unattractive, they will end up less feminine. Their female gender will end up deconstructed. Both strategies —posing as men and deconstructing their femininity—are games that female MUDders play. One player, a woman currently in treatment for anorexia, described her virtual body this way:

> In real life, the control is the thing. I know that it is very scary for me to be a woman. I like making my body disappear. In real life that is. On MUDs, too. On the MUD, I'm sort of a woman, but I'm not someone you would want to see sexually. My MUD description is a combination of smoke and angles. I like that phrase "sort of a woman." I guess that's what I want to be in real life too.

In addition to virtual cross-dressing and creating character descriptions that deconstruct gender, MUDders gender-swap as double agents. That is, in MUDs, men play women pretending to be men, and women play men pretending to be women. Shakespeare's characters play these games as well. In As You Like It, when Rosalind flees Frederick's court she is in love with Orlando. In the forest of Arden, disguised as the boy Ganymede, she encounters Orlando, himself lovesick for Rosalind. As Ganymede, Rosalind says she will try to cure Orlando of his love by playing Rosalind, pointing out the flaws of femininity in the process. In current stagings, Rosalind is usually played by a woman who at this point in the play pretends to be a man who pretends to be a woman. In Shakespeare's time, there was yet another turn because all women's parts were played by boys. So the character of Rosalind was played by a boy playing a girl playing a boy who plays a girl so she can have a flirtatious conversation with a boy. Another twist occurs when Rosalind playing Ganymede playing Rosalind meets Phebe, a shepherdess who falls passionately in love with "him."

As You Like It, with its famous soliloquy that begins "All the world's a stage," is a play that dramatizes the power of the theater as a metaphor for life. The visual pun of Rosalind's role underscores the fact that each of us is an actor playing one part or many parts. But the play has another message that speaks to the power of MUDs as new stages for working on the politics of gender. When Rosalind and Orlando meet "man to man" as Ganymede and Orlando, they are able to speak freely. They are able to have conversations about love quite different from those that would be possible if they followed the courtly conventions that constrain communications between men and women. In this way, the play suggests that donning a mask, adopting a persona, is a step toward reaching a deeper truth about the real, a position many MUDders take regarding their experiences as virtual selves.

Garrett is a twenty-eight-year-old male computer programmer who played a female character on a MUD for nearly a year. The character was a frog named Ribbit. When Ribbit sensed that a new player was floundering, a small sign would materialize in her hand that said, "If you are lost in the MUD, this frog can be a friend."

When talking about why he chose to play Ribbit, Garrett says:

> I wanted to know more about women's experiences, and not just from reading about them. . . . I wanted to see what the difference felt like. I wanted to experiment with the other side. . . . I wanted to be collaborative and helpful, and I thought it would be easier as a female. . . . As a man I was brought up to be territorial and competitive. I wanted to try something new. . . . In some way I really felt that the canonically female way of communicating was more productive than the male—in that all this competition got in the way.

And indeed, Garrett says that as a female frog, he did feel freer to express the helpful side of his nature than he ever had as a man. "My competitive side takes a back seat when I am Ribbit."

Garrett's motivations for his experiment in gender-swapping run deep.

Growing up, competition was thrust upon him and he didn't much like it. Garrett, whose parents divorced when he was an infant, rarely saw his father. His mother offered little protection from his brother's bullying. An older cousin regularly beat him up until Garrett turned fourteen and could inflict some damage of his own. Garrett got the clear idea that male aggression could only be controlled by male force.

In his father's absence, Garrett took on significant family responsibility. His mother ran an office, and Garrett checked in with her every day after school to see if she had any errands for him to run. If so, he would forgo the playground. Garrett recalls these days with great warmth. He felt helpful and close to his mother. When at ten, he won a scholarship to a prestigious private boarding school for boys, a school he describes as being "straight out of Dickens," there were no more opportunities for this kind of collaboration. To Garrett, life now seemed to be one long competition. Of boarding school he says:

> It's competitive from the moment you get up in the morning and you all got to take a shower together and everyone's checking each other out to see who's got pubic hair. It's competitive when you're in class. It's competitive when you're on the sports field. It's competitive when you're in other extra-curricular activities such as speeches. It's competitive all day long, every day.

At school, the older boys had administrative authority over the younger ones. Garrett was not only the youngest student, he was also from the poorest family and the only newcomer to a group that had attended school together for many years. "I was pretty much at the bottom of the food chain," he says. In this hierarchical environment, Garrett learned to detest hierarchy, and the bullies at school reinforced his negative feelings about masculine aggression.

Once out of high school, Garrett committed himself to finding ways to "get back to being the kind of person I was with my mother." But he found it difficult to develop collaborative relationships, particularly at work. When he encouraged a female coworker to take credit for some work they had done together—"something," he says "that women have always done for men"—she accepted his offer, but their friendship and ability to work together were damaged. Garrett sums up the experience by saying that women are free to help men and both can accept the woman's self-sacrifice, "but when a man lets a woman take the credit, the relationship feels too close, too seductive [to the woman]."

From Garrett's point of view, most computer bulletin boards and discussion groups are not collaborative but hostile environments, characterized by "flaming." This is the practice of trading angry and often *ad hominem* remarks on any given topic.

There was a premium on saying something new, which is typically something that disagrees to some extent with what somebody else has said. And
that in itself provides an atmosphere that's ripe for conflict. Another aspect,
I think, is the fact that it takes a certain degree of courage to risk really
annoying someone. But that's not necessarily true on an electronic medium,
because they can't get to you. It's sort of like hiding behind a wall and
throwing stones. You can keep throwing them as long as you want and
you're safe.

Garrett found MUDs different and a lot more comfortable. "On MUDs,"
he says, "people were making a world together. You got no prestige from
being abusive."

Garrett's gender-swapping on MUDs gave him an experience-to-think-
with for thinking about gender. From his point of view, all he had to do
was to replace male with female in a character's description to change
how people saw him and what he felt comfortable expressing. Garrett's
MUD experience, where as a female he could be collaborative without
being stigmatized, left him committed to bringing the helpful frog persona into his life as a male, both on and off the MUD. When I met him,
he had a new girlfriend who was lending him books about the differences
in men's and women's communication styles. He found they reinforced
the lessons he learned in the MUD.

By the time I met Garrett, he was coming to feel that his gender-
swapping experiment had reached its logical endpoint. Indeed, between
the time of our first and second meeting, Garrett decided to blow his
cover on the MUD and tell people that in RL he was really male. He said
that our discussions of his gender-swapping had made him realize that it
had achieved its purpose.

For anthropologists, the experience of *dépaysement* (literally, "decoun-
trifying" oneself) is one of the most powerful elements of fieldwork. One
leaves one's own culture to face something unfamiliar, and upon returning home it has become strange—and can be seen with fresh eyes.
Garrett described his decision to end his gender-swapping in the language of *dépaysement*. He had been playing a woman for so long that it
no longer seemed strange. "I'd gotten used to it to the extent that I was
sort of ignoring it. OK, so I log in and now I'm a woman. And it really
didn't seem odd anymore." But returning to the MUD as a male persona
did feel strange. He struggled for an analogy and came up with this one:

> It would be like going to an interview for a job and acting like I do at a party
> or a volleyball game. Which is not the way you behave at an interview. And
> so it is sort of the same thing. [As a male on the MUD] I'm behaving in a way
> that doesn't feel right for the context, although it is still as much me as it
> ever was.

When Garrett stopped playing the female Ribbit and started playing a helpful male frog named Ron, many of Garrett's MUDding companions interpreted his actions as those of a woman who now wanted to try playing a man. Indeed, a year after his switch, Garrett says that at least one of his MUD friends, Dredlock, remains unconvinced that the same person has actually played both Ribbit and Ron. Dredlock insists that while Ribbit was erratic (he says, "She would sometimes walk out in the middle of a conversation"), Ron is more dependable. Has Garrett's behavior changed? Is Garrett's behavior the same but viewed differently through the filter of gender? Garrett believes that both are probably true. "People on the MUD have . . . seen the change and it hasn't necessarily convinced them that I'm male, but they're also not sure that I'm female. And so, I've sort of gotten into this state where my gender is unknown and people are pretty much resigned to not knowing it." Garrett says that when he helped others as a female frog, it was taken as welcome, natural, and kind. When he now helps as a male frog, people find it unexpected and suspect that it is a seduction ploy. The analogy with his real life is striking. There, too, he found that playing the helping role as a man led to trouble because it was easily misinterpreted as an attempt to create an expectation of intimacy.

Case, the industrial designer who played the female Mairead in MedievalMUSH, further illustrates the complexity of gender swapping as a vehicle for self-reflection. Case describes his RL persona as a nice guy, a "Jimmy Stewart–type like my father." He says that in general he likes his father and he likes himself, but he feels he pays a price for his low-key ways. In particular, he feels at a loss when it comes to confrontation, both at home and in business dealings. While Garrett finds that MUDding as a female makes it easier to be collaborative and helpful, Case likes MUDding as a female because it makes it easier for him to be aggressive and confrontational. Case plays several online "Katharine Hepburn–types," strong, dynamic, "out there" women who remind him of his mother, "who says exactly what's on her mind and is a take-no-prisoners sort." He says:

> For virtual reality to be interesting it has to emulate the real. But you have to be able to do something in the virtual that you couldn't in the real. For me, my female characters are interesting because I can say and do the sorts of things that I mentally want to do, but if I did them as a man, they would be obnoxious. I see a strong woman as admirable. I see a strong man as a problem. Potentially a bully.

In other words, for Case, if you are assertive as a man, it is coded as "being a bastard." If you are assertive as a woman, it is coded as "modern and together."

My wife and I both design logos for small businesses. But do this thought experiment. If I say "I will design this logo for $3,000, take it or leave it," I'm just a typical pushy businessman. If she says it, I think it sounds like she's a "together" woman. There is too much male power-wielding in society, and so if you use power as a man, that turns you into a stereotypical man. Women can do it more easily.

Case's gender-swapping has given him permission to be more assertive within the MUD, and more assertive outside of it as well:

There are aspects of my personality—the more assertive, administrative, bureaucratic ones—that I am able to work on in the MUDs. I've never been good at bureaucratic things, but I'm much better from practicing on MUDs and playing a woman in charge. I am able to do things—in the real, that is —that I couldn't have before because I have played Katharine Hepburn characters.

Case says his Katharine Hepburn personae are "externalizations of a part of myself." In one interview with him, I use the expression "aspects of the self," and he picks it up eagerly, for MUDding reminds him of how Hindu gods could have different aspects or subpersonalities, all the while having a whole self.

You may, for example, have an aspect who is a ruthless business person who can negotiate contracts very, very well, and you may call upon that part of yourself while you are in tense negotiation, to do the negotiation, to actually go through and negotiate a really good contract. But you would have to trust this aspect to say something like, "Of course, I will need my lawyers to look over this," when in fact among your "lawyers" is the integrated self who is going to do an ethics vet over the contract, because you don't want to violate your own ethical standards and this [ruthless] aspect of yourself might do something that you wouldn't feel comfortable with later.

Case's gender-swapping has enabled his inner world of hard-bitten negotiators to find self-expression, but without compromising the values he associates with his "whole person." Role playing has given the negotiators practice; Case says he has come to trust them more. In response to my question, "Do you feel that you call upon your personae in real life?" Case responds:

Yes, an aspect sort of clears its throat and says, "I can do this. You are being so amazingly conflicted over this and I know exactly what to do. Why don't you just let me do it?" MUDs give me balance. In real life, I tend to be extremely diplomatic, nonconfrontational. I don't like to ram my ideas down anyone's throat. On the MUD, I can be, "Take it or leave it." All of my

Hepburn characters are that way. That's probably why I play them. Because they are smart-mouthed, they will not sugarcoat their words.

In some ways, Case's description of his inner world of actors who address him and are capable of taking over negotiations is reminiscent of the language of people with multiple personality. In most cases of multiple personality, it is believed that repeated trauma provokes a massive defense: An "alter" is split off who can handle the trauma and protect the core personality from emotional as well as physical pain. In contrast, Case's inner actors are not split off from his sense of himself. He calls upon their strengths with increasing ease and fluidity. Case experiences himself very much as a collective self, not feeling that he must goad or repress this or that aspect of himself into conformity. To use Marvin Minsky's language, Case feels at ease in his society of mind.

Garrett and Case play female MUD characters for very different reasons. There is a similar diversity in women's motivations for playing male characters. Some share my initial motivation, a desire for invisibility or permission to be more outspoken or aggressive. "I was born in the South and I was taught that girls didn't speak up to disagree with men," says Zoe, a thirty-four-year-old woman who plays male and female characters on four MUDs.

We would sit at dinner and my father would talk and my mother would agree. I thought my father was a god. Once or twice I did disagree with him. I remember one time in particular when I was ten, and he looked at me and said, "Well, well, well, if this little flower grows too many more thorns, she will never catch a man."

Zoe credits MUDs with enabling her to reach a state of mind where she is better able to speak up for herself in her marriage ("to say what's on my mind before things get all blown out of proportion") and to handle her job as the financial officer for a small biotechnology firm.

I played a MUD man for two years. First I did it because I wanted the feeling of an equal playing field in terms of authority, and the only way I could think of to get it was to play a man. But after a while, I got very absorbed by MUDding. I became a wizard on a pretty simple MUD—I called myself Ulysses—and got involved in the system and realized that as a man I could be firm and people would think I was a great wizard. As a woman, drawing the line and standing firm has always made me feel like a bitch and, actually, I feel that people saw me as one, too. As a man I was liberated from all that. I learned from my mistakes. I got better at being firm but not rigid. I practiced, safe from criticism.

Zoe's perceptions of her gender trouble are almost the opposite of Case's. Case sees aggressiveness as acceptable only for women; Zoe sees it as acceptable only for men. Comparison with Garrett is also instructive. Like Case, Garrett associated feminine strength with positive feelings about his mother; Zoe associated feminine strength with loss of her father's love. What these stories have in common is that in all three cases, a virtual gender swap gave people greater emotional range in the real. Zoe says:

> I got really good at playing a man, so good that whoever was on the system would accept me as a man and talk to me as a man. So, other guys talked to Ulysses "guy to guy." It was very validating. All those years I was paranoid about how men talked about women. Or I thought I was paranoid. And then, I got a chance to be a guy and I saw that I wasn't paranoid at all.[7]

Zoe talked to me about her experiences in a face-to-face interview, but there is a great deal of spontaneous discussion of these issues on Internet bulletin boards and discussion groups. In her paper "Gender Swapping on the Internet," Amy Bruckman tracks an ongoing discussion of gender issues on the electronic discussion group rec.games.mud.[8] Individuals may post to it, that is, send a communication to all subscribers. Postings on specific topics frequently start identifiable discussion "threads," which may continue for many months.

On one of these threads, several male participants described how playing female characters had given them newfound empathy with women. One contributor, David, described the trials and tribulations of playing a female character:

> Other players start showering you with money to help you get started, and I had never once gotten a handout when playing a male player. And then they feel they should be allowed to tag along forever, and feel hurt when you leave them to go off and explore by yourself. Then when you give them the knee after they grope you, they wonder what your problem is, reciting that famous saying, "What's your problem? It's only a game."

Carol, an experienced player with much technical expertise about MUDs, concurred. She complained about male players' misconception that "women can't play MUDs, can't work out puzzles, can't even type 'kill monster' without help." Carol noted that men offered help as a way to be ingratiating, but in her case this seduction strategy was ineffectual: "People offering me help to solve puzzles *I* wrote are not going to get very far."

Ellen, another contributor to the rec.games.mud discussion, tried gender-bending on an adventure-style MUD, thinking she would find out:

> if it was true that people would be nasty and kill me on sight and other stuff I'd heard about on r.g.m. [an abbreviation of rec.games.mud]. But, no, everyone was helpful (I was truly clueless and needed the assistance); someone gave me enough money to buy a weapon and armor and someone else showed me where the easy-to-kill newbie [a new player] monsters were. They definitely went out of their way to be nice to a male-presenting newbie. . . . (These were all male-presenting players, btw [by the way].)
>
> One theory is that my male character [named Argyle and described as "a short squat fellow who is looking for his socks"] was pretty innocuous. Maybe people are only nasty if you are "a broad-shouldered perfect specimen of a man" or something of that nature, which can be taken as vaguely attacking.

Ellen concluded that harassment relates most directly to self-presentation: "People are nice if they don't view you as a threat." Short, squat, a bit lost, in search of socks, and thus connoting limpness—Argyle was clearly not a threat to the dominant status of other "men" on the MUD. In the MUD culture Ellen played in, men tended to be competitive and aggressive toward each other; Argyle's nonthreatening self-presentation earned him kind treatment.

For some men and women, gender-bending can be an attempt to understand better or to experiment safely with sexual orientation.[9] But for everyone who tries it, there is the chance to discover, as Rosalind and Orlando did in the Forest of Arden, that for both sexes, gender is constructed.[10]

VIRTUAL SEX

Virtual sex, whether in MUDs or in a private room on a commercial online service, consists of two or more players typing descriptions of physical actions, verbal statements, and emotional reactions for their characters. In cyberspace, this activity is not only common but, for many people, it is the centerpiece of their online experience.

On MUDs, some people have sex as characters of their own gender. Others have sex as characters of the other gender. Some men play female personae to have netsex with men. And in the "fake-lesbian syndrome," men adopt online female personae in order to have netsex with women.[11] Although it does not seem to be as widespread, I have met several women who say they present as male characters in order to have netsex with men. Some people have sex as nonhuman characters, for example, as animals on FurryMUDs. Some enjoy sex with one partner. Some use virtual reality as a place to experiment with group situations. In real life, such behavior (where possible) can create enormous practical and emotional confusion. Virtual adventures may be easier to undertake, but they can also result in significant complications. Different people and different couples deal with them in very different ways.

Martin and Beth, both forty-one, have been married for nineteen years and have four children. Early in their marriage, Martin regretted not having had more time for sexual experimentation and had an extramarital affair. The affair hurt Beth deeply, and Martin decided he never wanted to do it again. When Martin discovered MUDs he was thrilled. "I really am monogamous. I'm really not interested in something outside my marriage. But being able to have, you know, a Tiny romance is kind of cool." Martin decided to tell Beth about his MUD sex life and she decided to tell him that she does not mind. Beth has made a conscious decision to consider Martin's sexual relationships on MUDs as more like his reading an erotic novel than like his having a rendezvous in a motel room. For Martin, his online affairs are a way to fill the gaps of his youth, to broaden his sexual experience without endangering his marriage.

Other partners of virtual adulterers do not share Beth's accepting attitude. Janet, twenty-four, a secretary at a New York law firm, is very upset by her husband Tim's sex life in cyberspace. After Tim's first online affair, he confessed his virtual infidelity. When Janet objected, Tim told her that he would stop "seeing" his online mistress. Janet says that she is not sure that he actually did stop.

> Look, I've got to say the thing that bothers me most is that he wants to do it in the first place. In some ways, I'd have an easier time understanding why he would want to have an affair in real life. At least there, I could say to myself, "Well, it is for someone with a better body, or just for the novelty." It's like the first kiss is always the best kiss. But in MUDding, he is saying that he wants that feeling of intimacy with someone else, the "just talk" part of an encounter with a woman, and to me that comes closer to what is most important about sex.
>
> First I told him he couldn't do it anymore. Then, I panicked and figured that he might do it anyway, because unlike in real life I could never find out. All these thousands of people all over the world with their stupid fake names ... no way I would ever find out. So, I pulled back and said that talking about it was strictly off limits. But now I don't know if that was the right decision. I feel paranoid whenever he is on the computer. I can't get it off my mind, that he is cheating, and he probably is tabulating data for his thesis. It must be clear that this sex thing has really hurt our marriage.

This distressed wife struggles to decide whether her husband is unfaithful when his persona collaborates on writing real-time erotica with man, too easy for her to experience herself as a man, too easy to avoid social consequences of her actions. MUDs provide a situation in which can play out scenarios that otherwise might have remained pure tasy. Yet the status of these fantasies-in-action in cyberspace is unclear. though they involve other people and are no longer pure fantasy, they e not "in the world." Their boundary status offers new possibilities. inySex and virtual gender-bending are part of the larger story of people using virtual spaces to construct identity.

Nowhere is this more dramatic than in the lives of children and adolescents as they come of age in online culture. Online sexual relationships are one thing for those of us who are introduced to them as adults, but quite another for twelve-year-olds who use the Internet to do their homework and then meet some friends to party in a MUD.

CHILDREN AND NETSEX

From around ten years of age, in those circles where computers are readily available, social life involves online flirting, necking, petting, and going all the way. I have already introduced a seventeen-year-old whose virtual affair was causing him to think more about the imaginative, emotional, and conversational aspects of sex. His experience is not unusual. A thirteen-year-old girl informs me that she prefers to do her sexual experimentation online. Her partners are usually the boys in her class at school. In person, she says, it "is mostly grope-y." Online, "They need to talk more." A shy fourteen-year-old, Rob, tells me that he finds online flirting easier than flirting at school or at parties. At parties, there is pressure to dance close, kiss, and touch, all of which he both craves and dreads. He could be rejected or he could get physically excited, and "that's worse," he says. If he has an erection while online, he is the only one who will know about it.

In the grownup world of engineering, there is criticism of text-based virtual reality as "low bandwidth," but Rob says he is able to get "more information" online than he would in person.

Face to face, a girl doesn't always feel comfortable either. Like about not saying "Stop" until they really mean *"Stop there! Now!"* But it would be less embarrassing if you got more signals like about more or less when to stop. I think girls online are more communicative.

And online, he adds, "I am able to talk with a girl all afternoon—and not even try anything [sexual] and it does not seem weird. It [online conversation] lends itself to telling stories, gossiping; much more so than when you are trying to talk at a party."

another persona in cyberspace. And beyond this, sh
ence if unbeknownst to the husband his cyberspace
be a nineteen-year-old male college freshman? What i
eighty-year-old man in a nursing home? And even mor
if she is a twelve-year-old girl? Or a twelve-year-old boy?

TinySex poses the question of what is at the heart of se.
it the physical action? Is it the feeling of emotional intimacy
other than one's primary partner? Is infidelity in the head or
Is it in the desire or in the action? What constitutes the violat
And to what extent and in what ways should it matter who
sexual partner is in the real world? The fact that the physical
been factored out of the situation makes these issues both sub
harder to resolve than before.

Janet feels her trust has been violated by Tim's "talk intimacy" w.
other woman. Beth, the wife who gave her husband Martin permissi
have TinySex, feels that he violated her trust when he chose to play a fer
character having a sexual encounter with a "man." When Beth read the
of one of these sessions, she became angry that Martin had drawn on h
knowledge of her sexual responses to play his female character.

For Rudy, thirty-six, what was most threatening about his girlfriend's
TinySex was the very fact that she wanted to play a character of the
opposite sex at all. He discovered that she habitually plays men and has
sex with female characters in chat rooms on America Online (like MUDs
in that people can choose their identities). This discovery led him to
break off the relationship. Rudy struggles to express what bothers him
about his ex-girlfriend's gender-bending in cyberspace. He is not sure of
himself, he is unhappy, hesitant, and confused. He says, "We are not ready
for the psychological confusion this technology can bring." He explains:

> It's not the infidelity. It's the gnawing feeling that my girlfriend—I mean, I
> was thinking of marrying her—is a dyke. I know that everyone is bisexual,
> I know, I know . . . but that is one of those things that I knew but it never
> had anything to do with me. . . . It was just intellectual.
>
> What I hate about the rooms on America Online is that it makes it so easy
> for this sort of thing to become real. Well, in the sense that the rooms are
> real. I mean, the rooms, real or not, make it too easy for people to explore
> these things. If she explored it in real life, well, it would be hard on me, but
> it would have been hard for her. If she really wanted to do it, she would do
> it, but it would have meant her going out and doing it. It seems like more
> of a statement. And if she had really done it, I would know what to make of
> it. Now, I hate her for what she does online, but I don't know if I'm being
> crazy to break up with her about something that, after all, is only words.

Rudy complained that virtual reality made it too easy for his girlfriend
to explore what it might be like to have a sexual relationship with another

A thirteen-year-old girl says that she finds it easier to establish relationships online and then pursue them offline. She has a boyfriend and feels closer to him when they send electronic mail or talk in a chat room than when they see each other in person. Their online caresses make real ones seem less strained. Such testimony supports Rob's descriptions of online adolescent sexual life as less pressured than that in RL. But here, as in other aspects of cyberlife, things can cut both ways. A twelve-year-old girl files this mixed report on junior high school cyberromance:

> Usually, the boys are gross. Because you can't see them, they think they can say whatever they want. But other times, we just talk, or it's just [virtual] kissing and asking if they can touch your breast or put their tongue in your mouth.

I ask her if she thinks that online sexual activity has changed things for her. She says that she has learned more from "older kids" whom she wouldn't normally have been able to hang out with. I ask her if she has ever been approached by someone she believes to be an adult. She says no, but then adds: "Well, now I sometimes go online and say that I am eighteen, so if I do that more it will probably happen." I ask her if she is concerned about this. She makes it very clear that she feels safe because she can always just "disconnect."

There is no question that the Internet, like other environments where children congregate—playgrounds, scout troops, schools, shopping malls —is a place where they can be harassed or psychologically abused by each other and by adults. But parental panic about the dangers of cyberspace is often linked to their unfamiliarity with it. As one parent put it, "I sign up for the [Internet] account, but I can barely use e-mail while my [fourteen-year-old] daughter tells me that she is finding neat home pages [on the World Wide Web] in Australia."

Many of the fears we have for our children—the unsafe neighborhoods, the drugs on the street, the violence in the schools, our inability to spend as much time with them as we wish to—are displaced onto those unknowns we feel we can control. Fifteen years ago, when children ran to personal computers with arms outstretched while parents approached with hands behind their backs, there was much talk about computers as addicting and hypnotic. These days, the Internet is the new unknown.

Parents need to be able to talk to their children about where they are going and what they are doing. This same commonsense rule applies to their children's lives on the screen. Parents don't have to become technical experts, but they do need to learn enough about computer networks to discuss with their children what and who is out there and lay down some basic safety rules. The children who do best after a bad experience on the Internet (who are harassed, perhaps even propositioned) are those who can talk to their parents, just as children who best handle bad experiences in real life are those who can talk to a trusted elder without shame or fear of blame.

DECEPTION

Life on the screen makes it very easy to present oneself as other than one is in real life. And although some people think that representing oneself as other than one is is always a deception, many people turn to online life with the intention of playing it in precisely this way. They insist that a certain amount of shape-shifting is part of the online game. When people become intimate, they are particularly vulnerable; it is easy to get hurt in online relationships. But since the rules of conduct are unclear, it is also easy to believe that one does not have the right to feel wounded. For what can we hold ourselves and others accountable?

In cyberculture, a story that became known as the "case of the electronic lover" has taken on near-legendary status. Like many stories that become legends, it has several versions. There were real events, but some tellings of the legend conflate several similar incidents. In all the versions, a male psychiatrist usually called Alex becomes an active member of a CompuServe chat line using the name of a woman, usually Joan. In one version of the story, his deception began inadvertently when Alex, using the computer nickname Shrink, Inc., found that he was conversing with a woman who assumed he was a female psychiatrist. Alex was stunned by the power and intimacy of this conversation. He found that the woman was more open with him than were his female patients and friends in real life.[12] Alex wanted more and soon began regularly logging on as Joan, a severely handicapped and disfigured Manhattan resident. (Joan said it was her embarrassment about her disfigurement that made her prefer not to meet her cyberfriends face to face.) As Alex expected, Joan was able to have relationships of great intimacy with "other" women on the computer service. Alex came to believe that it was as Joan that he could best help these women. He was encouraged in this belief by his online female friends. They were devoted to Joan and told her how central she had become to their lives.

In most versions of the story, Joan's handicap plays an important role. Not only did it provide her with an alibi for restricting her contacts to online communication, but it gave focus to her way of helping other people. Joan's fighting spirit and ability to surmount her handicaps served as an inspiration. She was married to a policeman and their relationship gave other disabled women hope that they, too, could be loved. Despite her handicaps, Joan was lusty, funny, a woman of appetites.

As time went on and relationships deepened, several of Joan's grateful online friends wanted to meet her in person, and Alex realized that his game was getting out of control. He decided that Joan had to die. Joan's "husband" got online and informed the community that Joan was ill and in the hospital. Alex was overwhelmed by the outpouring of sympathy and love for Joan. Joan's friends told her husband how important Joan was to them. They offered moral support, financial assistance, names of specialists who might help. Alex was in a panic. He could not decide whether to kill Joan off. In one account of the story, "For four long days Joan hovered between life and death."[13] Finally, Alex had Joan recover. But the virtual had bled into the real. Joan's "husband" had been pressed for the name of the hospital where Joan was staying so that cards and flowers could be sent. Alex gave the name of the hospital where he worked as a psychiatrist. One member of the bulletin board called the hospital to confirm its address and discovered that Joan was not there as a patient. The ruse began to unravel.

All the versions of the story have one more thing in common: The discovery of Alex's deception led to shock and outrage. In some versions of the story, the anger erupts because of the initial deception—that a man had posed as a woman, that a man had won confidences as a woman. The case presents an electronic version of the movie *Tootsie,* in which a man posing as a woman wins the confidence of another woman and then, when he is found out, her fury. In other versions, the anger centers on the fact that Joan had introduced some of her online women friends to lesbian netsex, and the women involved felt violated by Joan's virtual actions. These women believed they were making love with a woman, but in fact they were sharing intimacies with a man. In other accounts, Joan introduced online friends to Alex, a Manhattan psychiatrist, who had real-life affairs with several of them.[14] In these versions, the story of the electronic lover becomes a tale of real-life transgression.

The con artist is a stock character who may be appreciated for his charm in fictional presentations, but in real life is more often reviled for his duplicity and exploitiveness. In this sense, Alex was operating as part of a long tradition. But when familiar phenomena appear in virtual form, they provoke new questions. Was the reclusive, inhibited Alex only pretending to have the personality of the sunny, outgoing, lusty Joan? What was his real personality? Did Joan help her many disabled online friends who became more active because of her inspiration? When and how did Alex cross the line from virtual friend and helper to con artist? Was it when he dated Joan's friends? Was it when he had sexual relations with them? Or was it from the moment that Alex decided to pose as a woman?

At a certain point, traditional categories for sorting things out seem inadequate.

In the past fifteen years, I have noticed a distinct shift in people's way of talking about the case of the electronic lover. In the early 1980s, close to the time when the events first took place, people were most disturbed by the idea that a man had posed as a woman. By 1990, I began to hear more complaints about Joan's online lesbian sex. What most shocks today's audience is that Alex used Joan to pimp for him. The shock value of online gender-bending has faded. Today what disturbs us is when the shifting norms of the virtual world bleed into real life.

In 1993, the WELL computer network was torn apart by controversy over another electronic lover where the focus was on these shifting norms and the confusion of the real and the virtual. The WELL has a "Women's Only" forum where several women compared notes on their love lives in cyberspace. They realized that they had been seduced and abandoned (some only virtually, some also in the flesh) by the same man, whom one called a "cyber-cad." As they discussed the matter with more and more women, they found out that Mr. X's activities were far more extensive and had a certain consistency. He romanced women via electronic mail and telephone calls, swore them to secrecy about their relationship, and even flew across the country to visit one of them in Sausalito, California. But then he dropped them. One of the women created a topic (area for discussion) on the WELL entitled "Do You Know this Cyber ScamArtist?" Within ten days, nearly one thousand messages had been posted about the "outing" of Mr. X. Some supported the women, some observed that the whole topic seemed like a "high-tech lynching."[15]

At the time of the incident and its widespread reporting in the popular media, I was interviewing people about online romance. The story frequently came up. For those who saw a transgression it was that Mr. X had confused cyberworld and RL. It was not just that he used the relationships formed in the cyberworld to misbehave in RL. It was that he treated the relationships in the cyberworld as though they were RL relationships. A complex typology of relationships began to emerge from these conversations: real-life relationships, virtual relationships with the "real" person, and virtual relationships with a virtual other. A thirty-five-year-old woman real estate broker tried hard to make clear how these things needed to be kept distinct.

In a MUD, or chat room, or on IRC, it might be OK to have different flings with other people hiding behind other handles. But this man was coming on to these women as though he was interested in them really—I mean he said he was falling in love with them, with the real women. And he even did meet—and dump—some. Do you see the difference, from the beginning he didn't respect that online is its own place.

Mr. X himself did not agree that he had done anything wrong. He told the computer network that although he had been involved in multiple, simultaneous consensual relationships, he believed that the rules of cyberspace permitted that. Perhaps they do. But even if they do, the boundaries between the virtual and real are staunchly protected. Having sex with several characters on MUDs is one thing, but in a virtual community such as the WELL, most people are creating an electronic persona that they experience as coextensive with their physically embodied one. There, promiscuity can be another thing altogether.

Once we take virtuality seriously as a way of life, we need a new language for talking about the simplest things. Each individual must ask: What is the nature of my relationships? What are the limits of my responsibility? And even more basic: Who and what am I? What is the connection between my physical and virtual bodies? And is it different in different cyberspaces? These questions are framed to interrogate an individual, but with minor modifications, they are equally central for thinking about community. What is the nature of our social ties? What kind of accountability do we have for our actions in real life and in cyberspace? What kind of society or societies are we creating, both on and off the screen?

BEING DIGITAL

In the last two chapters we have seen people doing what they have always done: trying to understand themselves and improve their lives by using the materials they have at hand. Although this practice is familiar, the fact that these materials now include the ability to live through virtual personae means two fundamental changes have occurred in our situation. We can easily move through multiple identities, and we can embrace—or be trapped by—cyberspace as a way of life.

As more and more people have logged on to this new way of life and have experienced the effects of virtuality, a genre of cultural criticism is emerging to interpret these phenomena. An article in *The New York Times* described new books on the subject by dividing them into three categories: utopian, utilitarian, and apocalyptic.[16] Utilitarian writers emphasize the practical side of the new way of life. Apocalyptic writers warn us of increasing social and personal fragmentation, more widespread surveillance, and loss of direct knowledge of the world. To date, however, the utopian approaches have dominated the field. They share the technological optimism that has dominated post-war culture, an optimism captured in the advertising slogans of my youth: "Better living through chemistry," "Progress is our most important product." In our current situation, technological optimism tends to represent urban decay, social alienation, and economic polarization as out-of-date formulations of a problem that could be solved if appropriate technology were applied in sufficient doses, for example, technology that would link everyone to the "information superhighway." We all want to believe in some quick and relatively inexpensive solution to our difficulties. We are tempted to believe with the utopians that the Internet is a field for the flowering of participatory democracy and a medium for the transformation of education. We are tempted to share in the utopians' excitement at the prospect of virtual pleasures: sex with a distant partner, travel minus the risks and inconvenience of actually having to go anywhere.

In the next two chapters I try to capture some of what is most challenging about the new way of life, what Nicholas Negroponte, the director of the MIT Media Lab, refers to as being digital.[17] The new practice of entering virtual worlds raises fundamental questions about our communities and ourselves. My account challenges any simple utilitarian story. For every step forward in the instrumental use of a technology (what the technology can do for us), there are subjective effects. The technology changes us as people, changes our relationships and sense of ourselves. My account also calls into question the apocalyptic and utopian views. The issues raised by the new way of life are difficult and painful, because they strike at the heart of our most complex and intransigent social problems: problems of community, identity, governance, equity, and values. There is no simple good news or bad news.

Although it provides us with no easy answers, life online does provide new lenses through which to examine current complexities. Unless we take advantage of these new lenses and carefully analyze our situation, we shall cede the future to those who want to believe that simple fixes can solve complicated problems. Given the history of the last century, thoughts of such a future are hardly inspiring.

Chapter 9

VIRTUALITY AND ITS DISCONTENTS

The anthropologist Ray Oldenberg has written about the "great good place," a place where members of a community can gather for the pleasure of easy company, conversation, and a sense of belonging.[1] He considers these places—the local bar, bistro, and coffee shop—to be at the heart of individual social integration and community vitality. Today, we see a resurgence of interest in coffee shops and bistros, but most often the new structures are merely nostalgic because they have not grown out of coherent communities or neighborhoods. Some people are trying to fill the gap with neighborhoods in cyberspace. Take Dred's Bar, for example, a watering hole on the MUD LambdaMOO. It is described as having a "castle decor" and a polished oak dance floor. Recently I (here represented by my character or persona "ST") visited Dred's Bar with Tony, a persona I had met on another MUD. After passing the bouncer, Tony and I encountered a man asking for a $5 cover charge, and once we paid it our hands were stamped.

> The crowd opens up momentarily to reveal one corner of the club. A couple
> is there, making out madly. Friendly place. . . .
> You sit down at the table. The waitress sees you and indicates that she will
> be there in a minute.
> The waitress comes up to the table, "Can I get anyone anything from the
> bar?" she says as she puts down a few cocktail napkins.
> Tony says, "When the waitress comes up, type order name of drink."
> Abigail [a character at the bar] dries off a spot where some drink spilled on
> her dress.
> The waitress nods to Tony and writes on her notepad.
> Order margarita [I type this line, following Tony's directions].
> You order a margarita. [This is the result of the line I typed, causing this line

to appear on my screen and "ST orders a margarita" to appear on the screens of everyone else in the room.]

The waitress nods to ST and writes on her notepad.

Tony sprinkles some salt on the back of his hand.

Tony remembers he ordered a margarita, not tequila, and brushes the salt off.

. . .

You say, "I like salt on my margarita too."

The DJ makes a smooth transition from The Cure into a song by 10,000 Maniacs.

After the arrival of the drinks comes the following interchange:

You say, "L'chaim."

Tony says, "Excuse me?"

After some explanations, Tony says, "Ah . . . ," smiles, and introduces me to several of his friends. Tony and I take briefly to the dance floor to try out some MUD features that allow us to waltz and tango,[2] then we go to a private booth to continue our conversation.

MAIN STREET, MALL, AND VIRTUAL CAFÉ

What changes when we move from Oldenberg's great good places to something like Dred's Bar on LambdaMOO? To answer this question, it helps to consider some intermediate steps, for example, the steps implied in moving from a sidewalk café to a food court in a suburban shopping mall[3] or from Main Street in an American small town to Disneyland's Main Street USA.

"Disneyland," writes the French social theorist Jean Baudrillard, "is there to conceal the fact that it is the 'real' country, all of 'real' America, which is Disneyland."[4] Baudrillard means that once we experience the re-creations of Disneyland, Los Angeles will strike us as real. Once Disneyland's Main Street, USA, is the standard for artifice, Los Angeles's shopping malls seem authentic, even though they, too, are re-creations. The shopping malls enclose another dream: a golden age that never was of idyllic small-town life. What we have are dreams within dreams.

But as the shopping malls try to recreate the Main Streets of yesteryear, critical elements change in the process. Main Street is a public place; the shopping mall is planned to maximize purchasing. On Main Street you are a citizen; in the shopping mall, you are customer as citizen. Main Street had a certain disarray: there was a drunk, a panhandler, a traveling snake-oil salesman. In the mall, you are in a relatively controlled space; the street theater is planned and paid for in advance; the appearance of serendipity is part of the simulation.

Disneyland and shopping malls are elements of a way of life I have called the culture of simulation. Television is a major element as well. On any given evening, nearly eighty million people in the United States are watching television. The average American household has a television turned on more than six hours a day, reducing eye contact and conversation to a minimum.[5] Computers and the virtual worlds they provide are adding another dimension of mediated experience. Perhaps computers and virtuality in its various forms feel so natural because of their similarity to watching TV, our dominant media experience for the past forty years.

The bar featured for a decade in the television series *Cheers* no doubt figures so prominently in the American imagination at least partly because most of us don't have a neighborhood place where "everybody knows your name." Instead, we identify with the place on the screen, and most recently have given it some life off the screen as well. Bars designed to look like the one on *Cheers* have sprung up all over the country, most poignantly in airports, our most anonymous of locales. Here, no one will know your name, but you can always buy a drink or a souvenir sweatshirt.

In the postwar atomization of American social life, the rise of middle-class suburbs created communities of neighbors who often remained strangers. Meanwhile, as the industrial and economic base of urban life declined, downtown social spaces such as the neighborhood theater or diner were replaced by malls and cinema complexes in the outlying suburbs. In the recent past, we left our communities to commute to these distant entertainments; increasingly, we want entertainment (such as video on demand) that commutes right into our homes. In both cases, the neighborhood is bypassed. We seem to be in the process of retreating further into our homes, shopping for merchandise in catalogues or on television channels, shopping for companionship via personals ads.

Technological optimists think that computers will reverse some of this social atomization, touting virtual experience and virtual community as ways for people to widen their horizons. But is it really sensible to suggest that the way to revitalize community is to sit alone in our rooms, typing at our networked computers and filling our lives with virtual friends?[6]

THE LOSS OF THE REAL

Which would you rather see—a Disney crocodile robot or a real croco- dile? The Disney version has a certain vividness. It rolls its eyes, it moves from side to side, it disappears beneath the surface and rises again. It is designed to thrill us, to command our attention at all times. None of these qualities is necessarily visible in a real crocodile in a zoo, which seems to spend most of its time sleeping. And you may have neither the means nor the inclination to observe a real crocodile in the Nile or the River Gambia.

Compare a rafting trip down the Colorado River to an adolescent girl using an interactive CD-ROM to explore the same territory. In the physical rafting trip, there is likely to be physical danger and with it, a sense of real consequences. One may need to strain one's resources to survive. There might be a rite of passage. What might await the girl who picks up an interactive CD-ROM called "Adventures on the Colorado"? A touch-sensitive screen lets her explore the virtual Colorado and its shoreline. Clicking a mouse brings up pictures and descriptions of local flora and fauna. She can have all the maps and literary references she wants. All this might be fun, perhaps useful. But it is hard to imagine it marking a transition to adulthood. But why not have both—the virtual Colorado and the real one? Not every exploration need be a rite of passage. The virtual and the real may provide different things. Why make them compete?

This question recalls the controversy about simulation that divided the MIT faculty during Project Athena. Those who wanted to keep their students away from simulations argued that once students have seen an experiment unfold perfectly in a simulation, the messiness of a real experiment—the imperfections of measurement, the crack in the equipment that means you have to repeat it, the rough edges on a hand sketch of a building site—all these come to seem like a waste of time. The seductiveness of simulation does not mean that it is a bad thing or something to be avoided at all cost, but it does mean that simulation carries certain risks. It is not retrograde to say that if we value certain aspects of life *off* the screen, we may need to do something to protect them.[7]

Searching for an easy fix, we are eager to believe that the Internet will provide an effective substitute for face-to-face interaction. But the move toward virtuality tends to skew our experience of the real in several ways. One way is to make denatured and artificial experiences seem real. Let's call it the Disneyland effect. After a brunch on Disneyland's Royal Street, a cappuccino at a restaurant chain called Bonjour Café may seem real by comparison. After playing a video game in which your opponent is a computer program, the social world of MUDs may seem real as well. At least there are real people playing most of the parts and the play space is relatively open. One player compares the roles he was able to play on video games and MUDs. "Nintendo has a good one [game] where you can play four characters. But even though they are very cool," he says, "they are written up for you." They seem artificial. In contrast, on the MUDs, he says, "There is nothing written up." He says he feels free. MUDs are "for real" because you make them up yourself.

Such sentiments remind me of a comment by a high-school junior who was upset by what she described as the flight of her friends to the Internet. She complained, "Now they just want to talk online. It used to be that things weren't so artificial. We phoned each other every afternoon." To this young woman, phone calls represented the natural, intimate, and immediate. We build our ideas about what is real and what is natural with the cultural materials available. When I was in college and living in Paris, I stayed with a family who avoided the telephone for everything but emergency communications. An intimate communication would go by a *pneumatique*. One brought (or had delivered) a handwritten message to the local post office. There, it was placed in a cannister and sent through a series of underground tubes to another post office. It would then be hand delivered to its destination. I was taught that the *pneumatique* was the favored medium for love letters, significant apologies, or requests for an important meeting. Although mediated by significant amounts of technology, the handwritten *pneumatique* bore the trace of the physical body of the person who sent it; it was physically taken from that person's hand and put into the hand of the person to whom it was sent. The *pneumatique*'s insistence on physical presence may have ill-prepared me for the lessons of postmodernism, but it has made e-mail seem oddly natural.

Another effect of simulation, which I'll call the artificial crocodile effect, makes the fake seem more compelling than the real. In *The Future Does Not Compute: Warnings from the Internet,* Stephen L. Talbott quotes educators who say that years of exciting nature programming have compromised wildlife experiences for children. The animals in the woods are unlikely to perform as dramatically as those captured on the camera.[8] The world of direct, unmediated experience is thus devalued. I have a clear memory of a Brownie Scout field trip to the Brooklyn Botanical Gardens where I asked an attendant if she could make the flowers "open fast." For a long while, no one understood what I was talking about. Then they figured it out: I was hoping that the attendant could make the flowers behave as they did in the time-lapse photography I had seen on *Walt Disney.*

I was reminded of this incident when several years ago I interviewed children about their experiences role-playing in the game Dungeons and Dragons. One ten-year-old boy explained that Dungeons and Dragons was like history, except that Dungeons and Dragons "is more complicated. . . . There are hundreds and hundreds of books about Dungeons and Dragons." As far as this boy knew, there was only one book about history, his textbook.

A similar point about the devaluation of direct experience is familiar to those who have followed the discussion about the effect of television on our sensibilities, including the developing sensibilities of children. Media critics have suggested that quick cuts, rapid transitions, changing camera angles, all heighten stimulation through editing, a hyperactive style that is shared by *Sesame Street* and MTV. For some, this rapid cycling of events spoils us for the real: "One can only guess at the effect upon viewers of these hyperactive images, aside from fixating attention on the television set. . . . They must surely . . . contribute to the . . . inability to absorb information that comes muddling along at natural, real-life speed."[9] Direct experience is often messy; its meaning is never exactly clear. Interactive multimedia comes already interpreted. It is already someone else's version of reality.

A third effect is that a virtual experience may be so compelling that we believe that within it we've achieved more than we have. Many of the people I interviewed claimed that virtual gender-swapping enabled them to understand what it's like to be a person of the other gender, and I have no doubt that this is true, at least in part. But as I listened to this boast, my mind often traveled to my own experiences of living in a woman's body. These include worry about physical vulnerability, fears of unwanted pregnancy and of infertility, fine-tuned decisions about how much make-up to wear to a job interview, and the difficulty of giving a professional seminar while doubled over with monthly cramps. To a certain extent, knowledge is inherently experiential, based on a physicality that we each experience differently.

Pavel Curtis, the founder of LambdaMOO, began his paper on its social dimensions with a quote from E. M. Forster: "The Machine did not transmit nuances of expression. It only gave a general idea of people—an idea that was good enough for all practical purposes."[10] But what are practical purposes? And what about impractical purposes? To the question, "Why must virtuality and real life compete—why can't we have both?" the answer is of course that we *will* have both. The more important question is "How can we get the best of both?"

THE POLITICS OF VIRTUALITY

When I began exploring the world of MUDs in 1992, the Internet was open to a limited group of users, chiefly academics and researchers in affiliated commercial enterprises. The MUDders were mostly middle-class college students. They chiefly spoke of using MUDs as places of play and escape, though some used MUDs to address personal difficulties.

By late 1993, network access could easily be purchased commercially, and the number and diversity of people on the Internet had expanded dramatically. With many more people drawn to social virtual reality (MUDs, chat lines, bulletin boards, etc.) conversations with MUDders began to touch on new themes. Earlier interviews with participants in MUDs had touched on them as recreation and as an escape from RL experiences of broken homes, parental alcoholism, and physical and sexual abuse. While those earlier themes were still present, I heard about a new one. This was RL as a place of economic insecurity for young people trying to find meaningful work and trying to hold on to the middle-class status they had grown up in. Socially speaking, there was nowhere to go but down in RL, whereas MUDs were seen as a vehicle of virtual social mobility.

Josh is a twenty-three-year-old college graduate who lives in a small studio apartment in Chicago. After six months of looking for a job in marketing, the field in which he recently received his college degree, Josh has had to settle for a job working on the computer system that maintains inventory records at a large discount store. He considers this a dead end. When a friend told him about MUDs, he thought the games sounded diverting enough to give them a try. Josh talked the friend into letting him borrow his computer account for one evening, and then for another. Within a week, MUDs had become more than a diversion. Josh had stepped into a new life.

Now, Josh spends as much time on MUDs as he can. He belongs to a class of players who sometimes call themselves Internet hobos. They solicit time on computer accounts the way panhandlers go after spare change. In contrast to his life in RL, which he sees as boring and without prospects, Josh's life inside MUDs seems rich and filled with promise. It has friends, safety, and *space*. "I live in a terrible part of town. I see a rat hole of an apartment, I see a dead-end job, I see AIDS. Down here [in the MUD] I see friends, I have something to offer, I see safe sex." His job taking inventory has him using computers in ways he finds boring. His programming on MUDs is intellectually challenging. Josh has worked on three MUDs, building large, elaborate living quarters in each. In addition, he has become a specialist at building virtual cafés in which bots serve as waiters and bartenders. Within MUDs, Josh serves as a programming consultant to many less-experienced players and has even become something of an entrepreneur. He "rents" ready-built rooms to people who are not as skilled in programming as he is. He has been granted wizard privileges on various MUDs in exchange for building food-service software. He dreams that such virtual commerce will someday lead to more —that someday, as MUDs become commercial enterprises, he will build them for a living. MUDs offer Josh a sense of participation in the American dream.

MUDs play a similar role for Thomas, twenty-four, whom I met after giving a public lecture in Washington, D.C. As I collected my notes, Thomas came up to the lectern, introduced himself as a dedicated MUD player and asked if we could talk. After graduating from college, Thomas entered a training program at a large department store. When he discovered that he didn't like retailing, he quit the program, thinking that he would look for something in a different area of business. But things did not go well for him:

> My grades had not been fantastic. Quitting the training program looked bad to people. . . . I would apply for a job and two hundred other people would be there. You better bet that in two hundred people there was someone who had made better grades and hadn't quit his first job.

Finally, Thomas took a job as a bellhop in the hotel where I had just given my lecture. "I thought that working evening hours would let me continue looking for something that would get me back into the middle class," Thomas says. "I haven't found that job yet. But MUDs got me back into the middle class."

Thomas sees himself as someone who should be headed for a desk job, a nice car, and life in the suburbs. "My family is like that," he says, "and they spent a lot of money sending me to college. It wasn't to see me bellhop, I can promise you that." During the day Thomas carries luggage, but at night on MUDs he feels that he is with and recognized by his own kind. Thomas has a group of MUD friends who write well, program, and read science fiction. "I'm interested in MUD politics. Can there be democracy in cyberspace? Should MUDs be ruled by wizards or should they be democracies? I majored in political science in college. These are important questions for the future. I talk about these things with my friends. On MUDs."

Thomas moves on to what has become an obvious conclusion. He says, "MUDs make me more what I really am. Off the MUD, I am not as much me." Tanya, also twenty-four, a college graduate working as a nanny in rural Connecticut, expresses a similar aspiration for upward mobility. She says of the MUD on which she has built Japanese-style rooms and a bot to offer her guests a kimono, slippers, and tea, "I feel like I have more stuff on the MUD than I have off it."

Josh, Thomas, and Tanya belong to a generation whose college years were marked by economic recession and a deadly sexually transmitted disease. They scramble for work; finances force them to live in neighborhoods they don't consider safe; they may end up back home living with parents. These young people are looking for a way back into the middle class. MUDs provide them with the sense of a middle-class peer group. So it is really not that surprising that it is in virtual social life they feel most like themselves.

If a patient on the antidepressant medication Prozac tells his therapist he feels more like himself with the drug than without it, what does this do to our standard notions of a real self?[11] Where does a medication end and a person begin? Where does real life end and a game begin? Is the real self always the naturally occurring one? Is the real self always the one in the physical world? As more and more real business gets done in cyberspace, could the real self be the one who functions best in that realm? Is the real Tanya the frustrated nanny or the energetic programmer on the MUD? The stories of these MUDders point to a whole set of issues about the political and social dimension of virtual community. These young people feel they have no political voice, and they look to cyberspace to help them find one.

ESCAPE OR RESISTANCE

In *Reading the Romance,* the literary scholar Janice Radaway argues that when women read romance novels they are not escaping but building realities less limited than their own.[12] Romance reading becomes a form of resistance, a challenge to the stultifying categories of everyday life. This perspective, sensitive to the ways people find to resist constraints of race, class, and gender, is widely shared in contemporary cultural studies. In a similar spirit, the media researcher Henry Jenkins has analyzed the cultures built by television fans as a form of resistance and as enriching for people whose possibilities for fulfillment in real life are seriously limited.

Jenkins quotes a song written by a science fiction fan, which describes how her "Weekend-Only World" at science fiction conventions has more reality for her than her impoverished "real-time life."

> In an hour of make-believe
> In these warm convention halls
> My mind is free to think
> And feel so deeply
> An intimacy never found
> Inside their silent walls
> In a year or more
> Of what they call reality.[13]

Jenkins writes that this song "expresses the fans' recognition that fandom offers not so much an escape from reality as an alternative reality whose values may be more humane and democratic than those held by mundane society." The author of the song, in Jenkins's view, "gains power and identity from the time she spends within fan culture; fandom allows her to maintain her sanity in the face of the indignity and alienation of everyday life."[14] A similar perspective can be heard in the many online discussions of addictions to virtuality.

On an Internet mailing list discussing MUDding, one player reported on a role-playing conference in Finland that debated (among other things) whether "the time spent in [the] computer (yeah, IN it, not in front of it)" was

> a bad thing or what; the conclusion was, that it is at least better than watching "The Bald [sic] & The Beautiful" for 24H a day—and here we talked about MUDs or such mostly where people communicate with real people through the machine. . . . Well, hasty judging people might say that the escapists are weak and can't stand the reality—the truly wise see also the other side of the coin: there must be something wrong with Reality, if so many people want to escape from it. If we cannot change the reality, what can we do?[15]

One of the things that people I've interviewed have decided to do is change the reality of virtual reality. An example of this can be found in the history of the MUD LambdaMOO. LambdaMOO has recently undergone a major change in its form of governance. Instead of the MUD wizards (or system administrators) making policy decisions, there is a complex system of grass-roots petitions and collective voting. Thomas, the Washington, D.C., bellhop who sees himself as a yuppie manqué, says he is very involved in this experiment. He goes on at length about the political factions with which he must contend to "do politics" on LambdaMOO. Our conversation is taking place in the fall of 1994. His home state has an upcoming senatorial race, hotly contested, ideologically charged, but he hasn't registered to vote and doesn't plan to. I bring up the Senate race. He shrugs it off: "I'm not voting. Doesn't make a difference. Politicians are liars."

One might say that MUDs compensate individuals like Thomas for their sense of political impotence. Or, if we take the perspective sketched by Radaway and Jenkins, we can look at MUDs as places of resistance to many forms of alienation and to the silences they impose. Chat lines, e-mail, bulletin boards, and MUDs are like a weekend-only world in which people can participate every day. Are these activities best understood in terms of compensation or resistance? The logic of compensation suggests that the goal of virtual experience is to feel better; the logic of resistance suggests that it is political empowerment.

The question of how to situate users of seductive technology on a continuum between psychological escape and political empowerment is reminiscent of a similar question posed by the enthusiasm of personal computer hobbyists of the late 1970s. MUDders like Josh, Thomas, and

Tanya—out of college and not yet in satisfying work—have much in common with these early computer hobbyists. Both groups express unfulfilled intellectual and political aspirations within computer microworlds. In the case of the home hobbyists, programmers who no longer had a sense of working with a whole problem on the job demanded a sense of the whole in their recreational computing. In the case of the MUDders, people who feel a loss of middle-class status find reassurance in virtual space. Although the MUDs' extravagant settings—spaceships and medieval towns—may not seem likely places to reconstruct a sense of middle-class community, that is exactly the function they serve for some people who live in them. Many have commented that one appeal of LambdaMOO may be due to its being built as a home, modeled after the large, rambling house where its designer actually lives.

There is a special irony in bringing together the stories of pioneer personal computer owners and pioneer MUDders. The politics of the hobbyists had a grass-roots flavor. To their way of thinking, personal computers were a path to a new populism. Networks would allow citizens to band together to run decentralized schools and local governments. They thought that personal computers would create a more participatory political system because "people will get used to understanding things, to being in control of things, and they will demand more." Hobbyists took what was most characteristic of their relationships with the computer —building safe microworlds of transparent understanding—and turned it into a political metaphor. When nearly twenty years later, another group of people has turned to computation as a resource for community building, the communities they are thinking of exist on and through the computer. When Thomas talks to me about his passion for politics, about his undergraduate political science major, and how being politically involved makes him feel more like himself, he is talking about the MUD, not about life in Washington, D.C.[16]

Yet the Internet has become a potent symbol and organizational tool for current grass-roots movements—of both right and left. The hobbyists dreamed that the early personal computers would carry a political message about the importance of understanding how a system worked. The Internet carries a political message about the importance of direct, immediate action and interest-group mobilization. It is the symbol and tool of a postmodern politics.

The hobbyists I interviewed nearly two decades ago were excited, enthusiastic, and satisfied with what they were doing with their machines. As an ethnographer I thought it appropriate to report this enthusiasm and try to capture a sense of the pleasures and satisfactions that these individuals were deriving from their "non-alienated" relationships with their computers. In the same sense, it seems appropriate to report the enthusiasm

of most MUD users. They take pleasure in building their virtual friendships and virtual spaces and taking on responsibility for virtual jobs.
However, fifteen years ago, when reflecting on hobbyists' deeply-felt populism, I also worried about a darker side:

> Will the individual satisfactions of personal computation (which seem to
> derive some of their power from the fact that they are at least in part
> responsive to political dissatisfactions) take the individual away from collec-
> tive politics? People will not change unresponsive political systems or intel-
> lectually deadening work environments by building machines that are
> responsive, fun, and intellectually challenging. They will not change the
> world of human relations by retreating into a world of things. It would
> certainly be inappropriate to rejoice at the holistic and humanistic relation-
> ships that personal computers offer if it turns out that, when widespread,
> they replace religion as an opiate of the masses.[17]

These words can easily be transposed into the current context, substituting MUDs for personal computers. My misgivings are similar: Instead
of solving real problems—both personal and social—are we choosing to
live in unreal places? Women and men tell me that the rooms and mazes
on MUDs are safer than city streets, virtual sex is safer than sex anywhere,
MUD friendships are more intense than real ones, and when things don't
work out you can always leave.[18]

It is not hard to agree that MUDs provide an outlet for people to work
through personal issues in a productive way; they offer a moratorium that
can be turned to constructive purpose, and not only for adolescents. One
can also respect the sense in which political activities in a MUD demonstrate resistance to what is unsatisfying about political life more generally.
And yet, it is sobering that the personal computer revolution, once conceptualized as a tool to rebuild community, now tends to concentrate on
building community inside a machine.

If the politics of virtuality means democracy online and apathy offline,
there is reason for concern. There is also reason for concern when access
to the new technology breaks down along traditional class lines.[19] Although some inner-city communities have used computer-mediated communication as a tool for real-life community building,[20] the overall trend
seems to be the creation of an information elite at the same time that the
walls around our society's traditional underclass are maintained. Perhaps
people are being even more surely excluded from participation, privilege, and responsibility in the information society than they have been
from the dominant groups of the past.

Today many are looking to computers and virtual reality to counter
social fragmentation and atomization; to extend democracy; to break

down divisions of gender, race, and class; and to lead to a renaissance of learning. Others are convinced that these technologies will have negative effects. Dramatic stories supporting both points of view are always enticing, but most people who have tried to use computer-mediated communication to change their conditions of life and work have found things more complex. They have found themselves both tantalized and frustrated.

Vanessa, thirty-four, is one of the founding wizards of a large and successful MUD. She is a skilled computer programmer whose talents and energy have always enabled her to earn good money. But she has never been happy in the computer industry because she found little support for her creative style, which she characterizes as "thinking along with people." She is the kind of person whose creativity emerges in conversation. Things went from bad to worse for her when she was forced to telework from home for a period of time. "I was going crazy. Now there was no one. I was so lonely I couldn't get myself to work." But then, a MUD-like chat window gave Vanessa some of what she wanted:

> There was one woman I was working with . . . on a project and we would always have a chat on a talk window on our machines. There we could talk about the project and the testing we were doing and say, "OK, type this," "OK, see if it works," "OK, you know I've changed this file now." . . . That talk window was an important piece of support to me.

That project and the chat sessions are over. When I meet her, Vanessa has no such intellectual companionship in her job. But she has it when she collaborates with others in MUDs. She comments, "So I think that's why I spend so much time on the MUD. . . . I am looking for environments with that sort of support." Vanessa has not yet been able to take the work style she has carved out in a virtual world and use it to enlarge her real-world job. She does not find room for "that sort of support" in the company where she works. There, she describes the highly productive people as driven individualists while her preferred work style is seen as time-wasting.

Vanessa's story would not read as an escape into MUDs if she had found a way to use her experiences there to model a more fulfilling style of RL work. Her story points toward new possibilities for using MUDs to foster collaboration in work settings. These are early days for such experiments, but they are beginning. For example, Pavel Curtis, the designer of LambdaMOO, is creating a new virtual space—enhanced with audio and video—for Xerox PARC in Palo Alto, California, a research facility funded by the Xerox Corporation. The MUD is called Jupiter, and Xerox PARC employees will step in and out of it depending on whom they want to talk to and what tools they want to use. Jupiter is meant to pick up where the physical workplace leaves off. Smooth transitions back and forth are a key design principle.

Xerox PARC is not the only workplace where MUDs are either in operation or being planned. The MIT Media Lab has MediaMOO, a MUD built and maintained by Amy Bruckman and dedicated to collaboration and community building among media researchers all over the world. Some veteran MUDders are building similar environments for members of international corporations, to make it easier for them to participate in meetings with their colleagues. What these situations have in common is the permeable border between the real and the virtual.

On a more widespread level, chat windows in which collaborators "talk" while editing shared documents, take notes on shared "white boards," and manipulate shared data are becoming increasingly common. Three doctors at three different physical locations, all looking at the same CAT scan images on their screens, consult together about a young child with a tumor, but the subsequent conversation about the recommended treatment will take place at the child's bedside with the family members present. Similarly permeable are virtual communities such as the WELL. In *The Virtual Community*, Howard Rheingold describes how WELL members have been able to support one another in real life. They have elicited information and contacts that saved lives (for example, of an American Buddhist nun in Katmandu who developed an amoebic liver infection). They have brought electronic consolation and personal visits in times of grief (for example, to a WELL member dying of cancer). Rheingold himself believes that this permeability is essential for the word "community" to be applied to our virtual social worlds. To make a community work "at least some of the people [must] reach out through that screen and affect each other's lives."[21]

PANOPTICON

Much of the conversation about electronic mail, bulletin boards, and the information superhighway in general is steeped in a language of liberation and utopian possibility. It is easy to see why. I write these words in 1995. To date, a user's experience of the Internet is of a dizzyingly free zone. On it information is easily accessible. One can say anything to anyone. Bulletin boards and information utilities are run by interested and motivated people—a graduate student in comparative literature here, an unemployed philosopher there, as well as insurance salesmen, housewives, and bellhops. These people obviously have something in common, access to the Internet and enough money or connections to buy or borrow a computer and modem, but they are a diverse enough group to foster fantasies about a new kind of social power. People who usually think of the world in materialist terms play with the idea that the somehow immaterial world of computer networks has created a new space for power without traditional forms of ownership. People who think of the world in bureaucratic terms play with the ways in which electronic communities undermine traditional forms of organization and status. Such musings are no longer restricted to professional social theorists. The August 1995 issue of *NetGuide,* a monthly magazine written for beginning Internet users, carries the cover story "Take Charge: Create Your Own Online Service."[22]

I am talking with Ray, an MIT freshman who is discussing his first Internet experiences (an Internet account comes as part of MIT's registration package). Ray quickly turns the conversation to the issue of power and access. He is thrilled with how much there is to explore and about being able to connect with people who would otherwise be inaccessible. He says he would never dare to make an appointment to see one of his professors without something very specific to say, but would send off an e-mail to inquire about a difficult assignment. Ray is on an electronic mailing list with one of his intellectual heroes. "They say Marvin Minsky is actually on this list they let me join. He hasn't posted anything yet. But as soon as he does, I feel like I could comment on something he said." Ray comments that the idea that he and Marvin Minsky are receiving the same e-mail makes him feel like "the two of us are sharing a *New York Times* over coffee and bagels on Sunday." Ray has also discovered LambdaMOO and is impressed with its efforts at self-governance. He says, "This is what American democracy should be."

Despite many people's good intentions, there is much in recent social thought that casts a sobering light on such enthusiasms. Michel Foucault's work, for example, elaborates a perspective on information, communication, and power that undermines any easy links between electronic communication and freedom.[23] He argues that power in modern society is imposed not by the personal presence and brute force of an elite caste but by the way each individual learns the art of self-surveillance. Modern society must control the bodies and behaviors of large numbers of people. Force could never be sufficiently distributed. Discourse substitutes and does a more effective job.

The social philosopher Jeremy Bentham, best known for his espousal of utilitarianism, proposed a device called the Panopticon, which enabled a prison guard to see all prisoners without being seen. At any given moment, any one prisoner was perhaps being observed, perhaps not. Prisoners would have to assume they were being observed and would therefore behave according to the norms that the guard would impose, if watching. Individuals learn to look at themselves through the eyes of the prison guard. Foucault has pointed out that this same kind of self-surveillance has extended from the technologies of imprisonment to those of education and psychotherapy. We learn to see ourselves from a teacher's or a therapist's point of view, even in their absence.

In our day, increasingly centralized databases provide a material basis for a vastly extended Panopticon that could include the Internet. Even now there is talk of network censorship, in part through (artificially) intelligent agents capable of surveillance. From Foucault's perspective, the most important factor would not be how frequently the agents are used or censorship is enforced. Like the threat of a tax audit, what matters most is that people know that the possibility is always present.

Ray's attitude about being online is totally positive. But Andy and Daniella, two other MIT freshmen, express reservations about computer-mediated communication. Neither knows about Foucault's work, but their ideas resonate with his on the way social control operates through learned self-surveillance. Andy hangs out on a MUD on which wizards have the power to enter any room without being seen. This means that they can "overhear" private conversations. He is organizing a petition to put a stop to this practice. Although he has been successful in marshaling support, he does not think his efforts will succeed. "We need the wizards. They are the ones willing to do the work. Without them, there would be no MUD." His comment provokes the following remark from Daniella: "Do you know that if you type the finger command, you can see the last time someone got online? So you are responsible for your e-mail if you log on to your computer, because everybody can know that you got your messages. But you don't know who asked about [that is, who has fingered] you." Andy nods his assent and replies, "I don't think that's the worst." He continues:

I subscribed to a list about cyberpunk and I wrote every day. It was such a release. My ideas were pretty wild. Then I found out that the list is archived in three places. E-mail makes you feel as though you are just talking. Like it will evaporate. And then what you say is archived. It won't evaporate. It's like somebody's always putting it on your permanent record. You learn to watch yourself.

Such considerations about power, discourse, and domination have been the province of social theorists. The experience of the Internet, that most ephemeral of objects, has made these considerations more concrete. Of course, people have known for decades that each time they place an order from a mail-order catalogue or contribute to a political cause, they are adding information to a database. New catalogues and new requests for political contributions arrive that are more and more finely tuned to the profile of the electronic personae they have created through their transactions. But people are isolated in their reflections about their electronic personae. On the Internet, such matters are more likely to find a collective voice.

In discussing the parallels between hobbyists and MUDders, I have balanced a language of psychological compensation (people without power and resources in the real find a compensatory experience in the virtual) with a language of political criticism (the satisfactions that people experience in virtual communities underscore the failures of our real ones). Both approaches give precedence to events in the real world. Do MUDs oblige us to find a new language that does not judge virtual experiences purely in terms of how far they facilitate or encumber "real" ones? Perhaps the virtual experiences are "real enough."

When people pursue relationships through letter writing, we are not concerned that they are abandoning their real lives. Relationships via correspondence seem romantic to people for whom MUDding seems vaguely unsavory. Some envisage letter writing as a step toward physical presence rather than as an alternative to it. Some imagine the letter writers speaking in their own voice rather than role playing. But neither of these ideas is necessarily true of letter writing or untrue of MUDding. In MUDs it is hard not to play an aspect of oneself, and virtual encounters often lead to physical ones. What makes an eighty-hour work week in investment banking something a parent can be proud of while Robert's eighty hours a week building and administering his MUD raises fears of addiction? Would Robert seem less addicted to his MUD activities if he were being paid for them? Would they have a different feel if his relationship with Kasha—the fellow-MUDder he traveled cross-country to meet —had blossomed, as some MUD friendships do, into marriage and family? In an electronic discussion group on virtual community, Barry Kort, one of the founders of a MUD for children, argued in a similar vein: "I don't think anyone would have said that Socrates, Plato, and Aristotle were addicted to the Agora. The computer nets are the modern Agora, serving a role similar to talk radio and tabloid journalism, but with more participation, less sensationalism, and more thinking between remarks."[24]

What are the social implications of spinning off virtual personae that can run around with names and genders of our choosing, unhindered by the weight and physicality of embodiment? From their earliest days, MUDs have been evocative objects for thinking about virtuality and accountability.

Habitat was an early MUD, initially built to run on Commodore 64 personal computers in the early 1980s. It had a short run in the United States before it was bought and transferred to Japan.[25] Its designers, Chip

Morningstar and F. Randall Farmer, have written about how its players struggled to establish the rights and responsibilities of virtual selves. On Habitat, players were originally allowed to have guns and other weapons. Morningstar and Farmer say that they "included these because we felt that players should be able to 'materially' affect each other in ways that went beyond simply talking, ways that required real moral choices to be made by the participants."[26] However, death in Habitat had little in common with the RL variety. "When an Avatar is killed, he or she is teleported back home, head in hands (literally), pockets empty, and any object in hand at the time dropped on the ground at the scene of the crime."[27] This eventuality was more like a setback in a game of Chutes and Ladders than real mortality, and for some players thievery and murder became the highlights of the game. For others, these activities were a violent intrusion on their peaceful world. An intense debate ensued.

Some players argued that guns should be eliminated, for in a virtual world a few lines of code can translate into an absolute gun ban. Others argued that what was dangerous in virtual reality was not violence but its trivialization. These individuals maintained that guns should be allowed, but their consequences should be made more serious; when you are killed, your character should cease to exist and not simply be sent home. Still others believed that since Habitat was just a game and playing assassin was part of the fun, there could be no harm in a little virtual violence.

As the debate continued, a player who was a Greek Orthodox priest in real life founded the first Habitat church, the "Order of the Holy Walnut," whose members pledged not to carry guns, steal, or engage in virtual violence of any kind. In the end, the game designers divided the world into two parts. In town, violence was prohibited. In the wilds outside of town, it was allowed. Eventually a democratic voting process was installed and a sheriff elected. Participants then took up discussion on the nature of Habitat laws and the proper balance between law and order and individual freedom. It was a remarkable situation. Participants in Habitat were seeing themselves as citizens; and they were spending their leisure time debating pacifism, the nature of good government, and the relationship between representations and reality. In the nineteenth century, utopians built communities in which political thought could be lived out as practice. On the cusp of the twenty-first century, we are building MUDs, possible worlds that can provoke a new critical discourse about the real.

MUD RAPE: ONLY WORDS?

Consider the first moments of a consensual sexual encounter between the characters Backslash and Targa. The player behind Backslash, Ronald, a mathematics graduate student in Memphis, types "emote fondles Targa's breast" and "say You are beautiful Targa." Elizabeth, Targa's player, sees on her screen:

> Backslash fondles Targa's breast.
> You are beautiful Targa.

Elizabeth responds with "say Touch me again, and harder. Please. Now. That's how I like it." Ronald's screen shows:

> Targa says, "Touch me again, and harder. Please. Now. That's how I like it."

But consensual relationships are only one facet of virtual sex. Virtual rape can occur within a MUD if one player finds a way to control the actions of another player's character and can thus "force" that character to have sex. The coercion depends on being able to direct the actions and reactions of characters independent of the desire of their players. So if Ronald were such a culprit, he would be the only one typing, having gained control of Targa's character. In this case Elizabeth, who plays Targa, would sit at her computer, shocked to find herself, or rather her "self," begging Backslash for more urgent caresses and ultimately violent intercourse.

In March 1992, a character calling himself Mr. Bungle, "an oleaginous, Bisquick-faced clown dressed in cum-stained harlequin garb and girdled with a mistletoe-and-hemlock belt whose buckle bore the inscription 'KISS ME UNDER THIS, BITCH!' " appeared in the LambdaMOO living room. Creating a phantom that masquerades as another player's character is a MUD programming trick often referred to as creating a voodoo doll. The "doll" is said to possess the character, so that the character must do whatever the doll does. Bungle used such a voodoo doll to force one and then another of the room's occupants to perform sexual acts on him. Bungle's first victim in cyberspace was legba, a character described as "a Haitian trickster spirit of indeterminate gender, brown-skinned and wearing an expensive pearl gray suit, top hat, and dark glasses." Even when ejected from the room, Bungle was able to continue his sexual assaults. He forced various players to have sex with each other and then forced legba to swallow his (or her?) own pubic hair and made a character called Starsinger attack herself sexually with a knife. Finally, Bungle was immobilized by a MOO wizard who "toaded" the perpetrator (erased the character from the system).

The next day, legba took the matter up on a widely read mailing list within LambdaMOO about social issues. Legba called both for "civility" and for "virtual castration." When chronicling this event, the journalist

Julian Dibbell contrasted the cyberspace description of the event with what was going on in real life. The woman who played the character of legba told Dibbell that she cried as she wrote those words, but he points out their "precise tenor," mingling "murderous rage and eyeball-rolling annoyance was a curious amalgam that neither the RL nor the VR facts alone can quite account for."

> Where virtual reality and its conventions would have us believe that legba and Starsinger were brutally raped in their own living room, here was the victim legba scolding Mr. Bungle for a breach of "civility." Where real life, on the other hand, insists the incident was only an episode in a free-form version of Dungeons and Dragons, confined to the realm of the symbolic and at no point threatening any player's life, limb, or material well-being, here now was the player legba issuing aggrieved and heartfelt calls for Mr. Bungle's dismemberment. Ludicrously excessive by RL's lights, woefully understated by VR's, the tone of legba's response made sense only in the buzzing, dissonant gap between them.[28]

Dibbell points out that although the RL and the VR description of the event may seem to "march in straight, tandem lines separated neatly into the virtual and the real, its meaning lies always in that gap." He describes the way MUD players tend to learn this lesson during their early sexual encounters in MUDs.

> Amid flurries of even the most cursorily described caresses, sighs, and penetrations, the glands do engage, and often as throbbingly as they would in a real-life assignation—sometimes even more so, given the combined power of anonymity and textual suggestiveness to unshackle deep-seated fantasies. And if the virtual setting and the interplayer vibes are right, who knows? The heart may engage as well, stirring up passions as strong as many that bind lovers who observe the formality of trysting in the flesh.

The issue of MUD rape and violence has become a focal point of conversation on discussion lists, bulletin boards, and newsgroups to which MUD players regularly post. In these forums one has the opportunity to hear from those who believe that MUDs should be considered as games and that therefore virtual rape should be allowed. One posting defending MUD rape was from someone who admitted to being a MUD rapist.

> MUDs are Fantasy. MUDs are somewhere you can have fun and let your "hidden" self out. Just to let you all in on what happened, here is the story:
>
> On a MUD (who's [sic] name I will not release, like I said, you know who you are) a friend of mine and myself were reprimanded for actions we took.

We have a little thing we do, he uses emote . . . to emote "<his name> holds <victim's name> down for <my name> to rape." Then I use emote and type "<my name> rapes the held down <victim's name>."

Now this may be an odd thing to do, but it is done in a free non-meaningful manner. We don't do it to make people feel victimized, (like this GOD said we were doing) we do it for fun. OK, it is plain out sick, but that isn't the point. On this MUD the victim isn't the one who complained. It was several other PCs who complained about what we did. Let the victim complain about it. It happened to the victim, not the bystanders. The victim didn't actually mind, she thought it was somewhat humorous. Well, along comes Mr. GOD saying "raping the Player's character, is the same as raping the player."

BULL SHIT

This is a GAME, nothing more. This particular GOD needs to chill out and stop being so serious. MUDs are supposed to be fun, not uptight. I will never return to this MUD of my own choice. There are other MUDs where we have done the same thing and even though the victim didn't like it, the GODs told the victim "too bad. it's not like they Pkilled you." [This refers to "player killing," in which one player kills another player. It is often considered different in nature from being "toaded."][29]

There was a postscript after the signature on this communication. The author asks his readers to "Please excuse my grammer [sic] as I am a Computer Science Major, not an English Major. Also excuse the no indenting as our netnews poster eats posts that have indents in them. Argh." Rape was not all that was on this MUDder's mind. Grammar and the limitations of his text formatting system also loomed large.

Discussion of the MUD rape occupied LambdaMOO for some time. In one of a series of online meetings that followed it, one character asked, "Where does the body end and the mind begin? Is not the mind a part of the body?" Another answered, "In MOO, the body is the mind." MUD rape occurred without physical contact. Is rape, then, a crime against the mind? If we say that it is, we begin to approach the position taken by the feminist legal theorist Catharine MacKinnon, who has argued that some words describing acts of violence toward women are social actions. Thus some pornography should be banned because it is not "only words." Dibbell says that he began his research on MUDs with little sympathy for this point of view but admits that "the more seriously I took the notion of virtual rape, the less seriously I was able to take the notion of freedom of speech, with its tidy division of the world into the symbolic and the real."[30] While legal scholars might disagree that any such "tidy division" is to be found in American law, no one can doubt that examples of MUD rape—of which the incident on LambdaMOO was only one—raise the question of accountability for the actions of virtual personae who have only words at their command.

Similar issues of accountability arise in the case of virtual murder. If your MUD character erases the computer database on which I have over many months built up a richly described character and goes on to announce to the community that my character is deceased, what exactly have *you,* the you that exists in real life, done? What if my virtual apartment is destroyed along with all its furniture, VCR, kitchen equipment, and stereo system? What if you kidnap my virtual dog (the bot Rover, which I have trained to perform tricks on demand)? What if you destroy him and leave his dismembered body in the MUD?

In the physically embodied world, we have no choice but to assume responsibility for our body's actions. The rare exceptions simply prove the rule as when someone with multiple personality disorder claims that a crime was committed by an "alter" personality over which he or she has no control or we rule someone mentally incompetent to stand trial. The possibilities inherent in virtuality, on the other hand, may provide some people with an excuse for irresponsibility, just as they may enable creative expressions that would otherwise have been repressed. When society supported people in unitary experiences of self, it often maintained a narrowness of outlook that could be suffocating. There were perhaps great good places, but there was also a tendency to exclude difference as deviance. Turning back provides no solutions. The challenge is to integrate some meaningful personal responsibility in virtual environments. Virtual environments are valuable as places where we can acknowledge our inner diversity. But we still want an authentic experience of self.

One's fear is, of course, that in the culture of simulation, a word like authenticity can no longer apply. So even as we try to make the most of virtual environments, a haunting question remains. For me, that question is raised every time I use the MUD command for taking an action. The command is "emote." If while at Dred's bar on LambdaMOO, I type "emote waves," the screens of all players in the MUD room will flash "ST waves." If I type "emote feels a complicated mixture of desire and expectation," all screens will flash "ST feels a complicated mixture of desire and expectation." But what exactly do I *feel?* Or, what exactly do *I* feel? When we get our MUD persona to "emote" something and observe the effect, do we gain a better understanding of our real emotions, which can't be switched on and off so easily, and which we may not even be able to describe? Or is the emote command and all that it stands for a reflection of what Fredric Jameson called the flattening of affect in postmodern life?

Chapter 10

IDENTITY CRISIS

Every era constructs its own metaphors for psychological well-being. Not so long ago, stability was socially valued and culturally reinforced. Rigid gender roles, repetitive labor, the expectation of being in one kind of job or remaining in one town over a lifetime, all of these made consistency central to definitions of health. But these stable social worlds have broken down. In our time, health is described in terms of fluidity rather than stability. What matters most now is the ability to adapt and change—to new jobs, new career directions, new gender roles, new technologies.

In *Flexible Bodies,* the anthropologist Emily Martin argues that the language of the immune system provides us with metaphors for the self and its boundaries.[1] In the past, the immune system was described as a private fortress, a firm, stable wall that protected within from without. Now we talk about the immune system as flexible and permeable. It can only be healthy if adaptable.

The new metaphors of health as flexibility apply not only to human mental and physical spheres, but also to the bodies of corporations, governments, and businesses. These institutions function in rapidly changing circumstances; they too are coming to view their fitness in terms of their flexibility. Martin describes the cultural spaces where we learn the new virtues of change over solidity. In addition to advertising, entertainment, and education, her examples include corporate workshops where people learn wilderness camping, high-wire walking, and zip-line jumping. She refers to all of these as flexibility practicums.

In her study of the culture of flexibility, Martin does not discuss virtual communities, but these provide excellent examples of what she is talking about. In these environments, people either explicitly play roles (as in

MUDs) or more subtly shape their online selves. Adults learn about being multiple and fluid—and so do children. "I don't play so many different people online—only three," says June, an eleven-year-old who uses her mother's Internet account to play in MUDs. During our conversation, I learn that in the course of a year in RL, she moves among three house-holds—that of her biological mother and stepfather, her biological father and stepmother, and a much-loved "first stepfather," her mother's second husband. She refers to her mother's third and current husband as "second stepfather." June recounts that in each of these three households the rules are somewhat different and so is she. Online switches among perso-nae seem quite natural. Indeed, for her, they are a kind of practice. Martin would call them practicums.

"LOGINS R US"

On a WELL discussion group about online personae (subtitled "boon or bête-noire") participants shared a sense that their virtual identities were evocative objects for thinking about the self. For several, experiences in virtual space compelled them to pay greater attention to what they take for granted in the real. "The persona thing intrigues me," said one. "It's a chance for all of us who aren't actors to play [with] masks. And think about the masks we wear every day."[2]

In this way, online personae have something in common with the self that emerges in a psychoanalytic encounter. It, too, is significantly virtual, constructed within the space of the analysis, where its slightest shifts can come under the most intense scrutiny.[3]

What most characterized the WELL discussion about online personae was the way many of the participants expressed the belief that life on the WELL introduced them to the many within themselves. One person wrote that through participating in an electronic bulletin board and letting the many sides of ourselves show, "We start to resemble little corporations, 'Logins R Us,' and like any company, we each have within us the bean-counter, the visionary, the heart-throb, the fundamentalist, and the wild child. Long may they wave."[4] Other participants responded to this com-ment with enthusiasm. One, echoing the social psychologist Kenneth Gergen,[5] described identity as a "pastiche of personalities" in which "the test of competence is not so much the integrity of the whole but the apparent correct representation appearing at the right time, in the right context, not to the detriment of the rest of the internal 'collective.'"[6] Another said that he thought of his ego "as a hollow tube, through which, one at a time, the 'many' speak through at the appropriate moment. . . . I'd like to hear more . . . about the possibilities surrounding the notion that what we perceive as 'one' in any context is, perhaps, a conglomerate of 'ones.' " This writer went on:

> Hindu culture is rooted in the "many" as the root of spiritual experience. A person's momentary behavior reflects some influence from one of hundreds of gods and/or goddesses. I am interested in . . . how this natural assumption of the "many" creates an alternative psychology.[7]

Another writer concurred:

> Did you ever see that cartoon by R. Crumb about "Which is the real R. Crumb?" He goes through four pages of incarnations, from successful businessman to street beggar, from media celebrity to gut-gnawing recluse, etc. etc. Then at the end he says, "Which is the real one?" . . . "It all depends on what mood I'm in!"
> We're all like that on-line.[8]

Howard Rheingold, the member of the WELL who began the discussion topic, also referred to Gergen's notion of a "saturated self," the idea that communication technologies have caused us to "colonize each other's brains." Gergen describes us as saturated with the many "voices of humankind—both harmonious and alien." He believes that as "we absorb their varied rhymes and reasons, they become part of us and we of them. Social saturation furnishes us with a multiplicity of incoherent and unrelated languages of the self." With our relationships spread across the globe and our knowledge of other cultures relativizing our attitudes and depriving us of any norm, we "exist in a state of continuous construction and reconstruction; it is a world where anything goes that can be negotiated. Each reality of self gives way to reflexive questioning, irony, and ultimately the playful probing of yet another reality. The center fails to hold."[9]

Although people may at first feel anguish at what they sense as a breakdown of identity, Gergen believes they may come to embrace the new possibilities. Individual notions of self vanish "into a stage of relatedness. One ceases to believe in a self independent of the relations in which he or she is embedded."[10] "We live in each other's brains, as voices, images, words on screens," said Rheingold in the online discussion. "We are multiple personalities and we include each other."[11]

Rheingold's evocation of what Gergen calls the "raptures of multiplicitous being" met with support on the WELL. One participant insisted that all pejorative associations be removed from the notion of a saturated self. "Howard, I *like* being a saturated self, in a community of similarly saturated selves. I grew up on TV and pop music, but it just ain't enough.

Virtual communities are, among other things, the co-saturation of selves who have been, all their lives, saturated in isolation."[12] To which Rheingold could only reply, "I like being a saturated self too."[13] The cybersociety of the WELL is an object-to-think-with for reflecting on the positive aspects of identity as multiplicity.

IDENTITY AND MULTIPLICITY

Without any principle of coherence, the self spins off in all directions. Multiplicity is not viable if it means shifting among personalities that cannot communicate. Multiplicity is not acceptable if it means being confused to a point of immobility.[14] How can we be multiple and coherent at the same time? In *The Protean Self,* Robert Jay Lifton tries to resolve this seeming contradiction. He begins by assuming that a unitary view of self corresponded to a traditional culture with stable symbols, institutions, and relationships. He finds the old unitary notion no longer viable because traditional culture has broken down and identifies a range of responses. One is a dogmatic insistence on unity. Another is to return to systems of belief, such as religious fundamentalism, that enforce conformity. A third is to embrace the idea of a fragmented self.[15] Lifton says this is a dangerous option that may result in a "fluidity lacking in moral content and sustainable inner form." But Lifton sees another possibility, a healthy protean self. It is capable, like Proteus, of fluid transformations but is grounded in coherence and a moral outlook. It is multiple but integrated.[16] You can have a sense of self without being one self.

Lifton's language is theoretical. Experiences in MUDs, on the WELL, on local bulletin boards, on commercial network services, and on the World Wide Web are bringing his theory down to earth. On the Web, the idiom for constructing a "home" identity is to assemble a "home page" of virtual objects that correspond to one's interests. One constructs a home page by composing or "pasting" on it words, images, and sounds, and by making connections between it and other sites on the Internet or the Web. Like the agents in emergent AI, one's identity emerges from whom one knows, one's associations and connections. People link their home page to pages about such things as music, paintings, television shows, cities, books, photographs, comic strips, and fashion models. As I write this book I am in the process of constructing my own home page. It now contains links to the text of my curriculum vitae, to drafts of recent papers (one about MUDs, one about French psychoanalysis), and to the reading lists for the two courses I shall teach next fall. A "visitor" to my home page can also click a highlighted word and watch images of Michel Foucault and Power Rangers "morph," one into the other, a visual play on my contention that children's toys bring postmodernism down to earth.

This display, affectionately referred to as "The Mighty Morphin' Michel Foucault," was a present from my assistant at MIT, Cynthia Col. A virtual home, like a real one, is furnished with objects you buy, build, or receive as gifts.

My future plans for my home page include linking to Paris (the city has a home page), the bot Julia, resources on women's studies, Imari china, and recent research on migraines. I am not limited in the number of links I can create. If we take the home page as a real estate metaphor for the self, its decor is postmodern. Its different rooms with different styles are located on computers all over the world. But through one's efforts, they are brought together to be of a piece.

Home pages on the Web are one recent and dramatic illustration of new notions of identity as multiple yet coherent; in this book we have met others. Recall Case, the industrial designer who plays the female lawyer Mairead in MedievalMUSH. He does not experience himself as a unitary self, yet says that he feels in control of "himselves" and "herselves." He says that he feels fulfilled by his real and virtual work, marriage, and friendships. While conventional thinking tends to characterize multiple personae in pathological terms, this does not seem to capture what is most meaningful about Case playing Mairead or Garrett (introduced in Chapter 8) playing Ribbit.

Within the psychoanalytic tradition, there have been schools that departed from the standard unitary view of identity. As we have seen, the object-relations theorists invented a language for talking about the many voices that we bring inside ourselves in the course of development. Jungian psychology encouraged the individual to become acquainted with a whole range of personae and to understand them as manifestations of universal archetypes, such as innocent virgins, mothers and crones, eternal youths and old men.[17] Jung believed that for each of us, it is potentially most liberating to become acquainted with our dark side, as well as the other-gendered self called anima in men and animus in women. Jung was banished from the ranks of orthodox Freudians for such suggestions. The object-relations school, too, was relegated to the margins. As America became the center of psychoanalytic politics in the mid-twentieth century, ideas about a robust executive ego became the psychoanalytic mainstream.

Through the fragmented selves presented by patients and through theories that stress the decentered subject, contemporary psychology confronts what is left out of theories of the unitary self. Now it must ask, What is the self when it functions as a society?[18] What is the self when it divides its labors among its constituent "alters"?[19] Those burdened by post-traumatic dissociative disorders suffer these questions; here I have suggested that inhabitants of virtual communities play with them.

Ideas about mind can become a vital cultural presence when they are carried by evocative objects-to-think-with.[20] I said earlier that these objects need not be material. For example, dreams and slips of the tongue were objects-to-think-with that brought psychoanalytic ideas into everyday life. People could play with their own and others' dreams and slips. Today, people are being helped to develop ideas about identity as multiplicity by a new practice of identity as multiplicity in online life. Virtual personae are objects-to-think-with.

When people adopt an online persona they cross a boundary into highly-charged territory. Some feel an uncomfortable sense of fragmentation, some a sense of relief. Some sense the possibilities for self-discovery, even self-transformation. Serena, a twenty-six-year-old graduate student in history, says, "When I log on to a new MUD and I create a character and know I have to start typing my description, I always feel a sense of panic. Like I could find out something I don't want to know." Arlie, a twenty-year-old undergraduate, says, "I am always very self-conscious when I create a new character. Usually, I end up creating someone I wouldn't want my parents to know about. It takes me, like, three hours. But that someone is part of me." In these ways and others, many more of us are experimenting with multiplicity than ever before.

With this last comment, I am not implying that MUDs or computer bulletin boards are causally implicated in the dramatic increase of people who exhibit symptoms of multiple personality disorder (MPD), or that people on MUDs have MPD, or that MUDding is like having MPD. What I am saying is that the many manifestations of multiplicity in our culture, including the adoption of online personae, are contributing to a general reconsideration of traditional, unitary notions of identity.

The history of a psychiatric symptom is inextricably tied up with the history of the culture that surrounds it. When I was in graduate school in psychology in the 1970s, clinical psychology texts regarded multiple personality as so rare (perhaps one in a million) as to be barely worthy of mention. In these rare cases, there was typically one alter personality in addition to the host personality.[21] Today, cases of multiple personality are much more frequent and typically involve up to sixteen alters of different ages, races, genders, and sexual orientations.[22] In multiple personality disorder, it is widely believed that traumatic events have caused various aspects of the self to congeal into virtual personalities, the "ones" often hiding from the "others" and hiding too from that special alter, the host personality. Sometimes, the alters are known to each other and to the host; some alters may see their roles as actively helping others. Such differences led the philosopher Ian Hacking to write about a "continuum of dissociation."[23] These differences also suggest a way of thinking about the self in terms of a continuum of how accessible its parts are to each other.

At one extreme, the unitary self maintains its oneness by repressing all that does not fit. Thus censored, the illegitimate parts of the self are not accessible. This model would of course function best within a fairly rigid social structure with clearly defined rules and roles. At the other extreme is the MPD sufferer whose multiplicity exists in the context of an equally repressive rigidity. The parts of the self are not in easy communication. Communication is highly stylized; one personality must speak to another personality. In fact, the term "multiple personality" is misleading, because the different parts of the self are not full personalities. They are split-off, disconnected fragments. But if the disorder in multiple personality disorder is the need for the rigid walls between the selves (blocking the secrets those selves protect), then the study of MPD may begin to furnish ways of thinking about healthy selves as nonunitary but with fluid access among their many aspects. Thus, in addition to the extremes of unitary self and MPD, we can imagine a flexible self.

The essence of this self is not unitary, nor are its parts stable entities. It is easy to cycle through its aspects and these are themselves changing through constant communication with each other. The philosopher Daniel Dennett speaks to the flexible self in his multiple drafts theory of consciousness.[24] Dennett's notion of multiple drafts is analogous to the experience of having several versions of a document open on a computer screen where the user is able to move between them at will. The presence of the drafts encourages a respect for the many different versions while it imposes a certain distance from them. No one aspect can be claimed as the absolute, true self. When I got to know French Sherry I no longer saw the less confident English-speaking Sherry as my one authentic self. What most characterizes the model of a flexible self is that the lines of communication between its various aspects are open. The open communication encourages an attitude of respect for the many within us and the many within others.

As we sense our inner diversity we come to know our limitations. We understand that we do not and cannot know things completely, not the outside world and not ourselves. Today's heightened consciousness of incompleteness may predispose us to join with others. The historian of science Donna Haraway equates a "split and contradictory self" with a "knowing self." She is optimistic about its possibilities: "The knowing self is partial in all its guises, never finished, whole, simply there and original; it is always constructed and stitched together imperfectly; and *therefore* able to join with another, to see together without claiming to be another."[25]

When identity was defined as unitary and solid it was relatively easy to recognize and censure deviation from a norm. A more fluid sense of self allows a greater capacity for acknowledging diversity. It makes it easier to accept the array of our (and others') inconsistent personae—perhaps with humor, perhaps with irony. We do not feel compelled to rank or judge the elements of our multiplicity. We do not feel compelled to exclude what does not fit.

VIRTUALITY AS TRANSITIONAL SPACE

In a journal published on the Internet, Leslie Harris speculates on how virtual experiences become part of the perceptual and emotional background "that changes the way we see things."[26] Harris describes an episode of *Star Trek: The Next Generation* in which Captain Picard plays Caiman, an inhabitant of the virtual world Catanh. On Catanh, Picard lives the experiences he had to forgo in order to make a career in Starfleet. He has a virtual experience of love, marriage, and fatherhood. He develops relationships with his community that are not possible for him as a Starfleet commander. "On" Catanh, the character Caiman "learns" to play the Ressiccan flute. Harris says, "He can eventually fall in love with a fellow crew member in his 'real life' because he experienced the feelings of love, commitment, and intimacy 'on' Catanh."[27] When in his real life Picard plays the flute with a fellow Starfleet officer he realizes that he is in love with her. Picard is aware that he has projected his desire for music and sensuality onto his virtual world. It is this awareness that lets him use music to link the "real" Captain Picard to the emotional growth he was able to experience as the virtual Caiman.

Here, virtuality is powerful but transitional. Ultimately, it is put in the service of Picard's embodied self. Picard's virtual Catanh, like the space created within psychoanalysis, operates in a time out of normal time and according to its own rules. In a successful psychoanalysis, the meetings between analyst and analysand come to an end, although the analytic process goes on forever. It is internalized within the person, just as Picard brought Catanh inside himself. Buddhists speak of their practice as a raft to get to the other shore, liberation. But the raft, like an analytic treatment, is thought of as a tool that must be set aside, even though the process of crossing the river is conceived of as never-ending. Wittgenstein takes up a similar idea in *The Tractatus,* when he compares his work to a ladder that is to be discarded after the reader has used it to reach a new level of understanding.

In April 1995, a town meeting was held at MIT on the subject "Doing Gender on the Net." As the discussion turned to using virtual personae to try out new experiences, a thirty-year-old graduate student, Ava, told her story. She had used a MUD to try out being comfortable with a disability. Several years earlier, Ava had been in an automobile accident that left her without a right leg. During her recuperation, she began to MUD. "Without giving it a lot of advance thought," Ava found herself creating a one-legged character on a MUD. Her character had a removable prosthetic limb. The character's disability featured plainly in her description, and the friends she made on the MUD found a way to deal with her handicap. When Ava's character became romantically involved, she and her virtual lover acknowledged the "physical" as well as the emotional aspects of the virtual amputation and prosthesis. They became comfortable with making virtual love, and Ava found a way to love her own virtual body. Ava told the group at the town meeting that this experience enabled her to take a further step toward accepting her real body. "After the accident, I made love in the MUD before I made love again in real life," she said. "I think that the first made the second possible. I began to think of myself as whole again." For her, the Internet had been a place of healing.

Virtual reality gave Ava choices. She could have tried being one of this MUD's many FabulousHotBabes. If so, she might have never felt safe leaving the anonymity of the virtual world. But instead she was able to reimagine herself not as whole but as whole-in-her-incompleteness. Each of us in our own way is incomplete. Virtual spaces may provide the safety for us to expose what we are missing so that we can begin to accept ourselves as we are.

Virtuality need not be a prison. It can be the raft, the ladder, the transitional space, the moratorium, that is discarded after reaching greater freedom. We don't have to reject life on the screen, but we don't have to treat it as an alternative life either. We can use it as a space for growth. Having literally written our online personae into existence, we are in a position to be more aware of what we project into everyday life. Like the anthropologist returning home from a foreign culture, the voyager in virtuality can return to a real world better equipped to understand its artifices.

CYBORG DREAMS

I have argued that Internet experiences help us to develop models of psychological well-being that are in a meaningful sense postmodern: They admit multiplicity and flexibility. They acknowledge the constructed nature of reality, self, and other. The Internet is not alone in encouraging such models. There are many places within our culture that do so. What they have in common is that they all suggest the value of approaching one's "story" in several ways and with fluid access to one's different aspects. We are encouraged to think of ourselves as fluid, emergent, decentralized, multiplicitous, flexible, and ever in process.[28] The metaphors travel freely among computer science, psychology, children's games, cultural studies, artificial intelligence, literary criticism, advertising, molecular biology, self-help, and artificial life. They reach deep into the popular culture. The ability of the Internet to change popular understandings of identity is heightened by the presence of these metaphors.

For example, a recent *Newsweek* article reports on a new narrative movement in psychotherapy, describing the trend as consistent with the "postmodernist idea that we don't so much perceive the world as interpret it." "The psyche," says *Newsweek,* "is not a fixed objective entity, but a fluid, social construct—a story that is subject to revision."[29] The new therapeutic movement described by *Newsweek* draws on deconstructionist literary criticism and on recent currents of psychoanalytic thought that emphasize conflicting narratives as a way of thinking about the analytic experience.[30] But its breezy and accessible newsmagazine coverage makes it clear that psychotherapy, too, can bring postmodernism down to earth.

The literary scholar Katherine Hayles, writing on the cultural resonances of chaos theory, has made the circulation of dominant metaphors a central theme of her work. She suggests that similarities arise in diverse scholarly disciplines and within popular culture "because of broadly based movements within the culture which made the deep assumptions underlying the new paradigms thinkable, perhaps inevitable, thoughts."[31] These assumptions carry a sense of the times that manifests itself in one place and then another, here as developmental psychology and there as a style of engineering, here as a description of our bodies and there as a template for corporate organization, here as a way to build a computer network and there as a manifesto of political ideals.

We are all dreaming cyborg dreams. While our children imagine "morphing" humans into metallic cyber-reptiles, our computer scientists dream themselves immortal. They imagine themselves thinking forever, downloaded onto machines. The AI researcher W. Daniel Hillis says,

> I have the same nostalgic love of human metabolism that everybody else does, but if I can go into an improved body and last for 10,000 years I would do it in an instant, no second thoughts. I actually don't think I'm going to have that option, but maybe my children will.[32]

Hillis's musings exemplify the mythic side of cybernetics, apparent from its earliest days. Norbert Wiener, a pioneer in the field, once wrote, "This is an idea with which I have toyed before—that it is conceptually possible for a human being to be sent over a telegraph line."[33] Today, the cyborg, in which human and machine are one, has become a postmodern myth.[34] The myth is fed by the extravagances of Robocop, the Terminator, and Power Rangers as well as by the everyday reality of children plugged into video games. When William Gibson was asked about his sources of inspiration for *Neuromancer,* he described the merging of human and machine as he watched a teenager playing a video game in a downtown arcade.

> Video games weren't something I'd done much, and I'd have been embarrassed to actually go into these arcades because everyone was so much younger than I was, but when I looked into one, I could see in the physical intensity of their postures how *rapt* these kids were. It was like one of those closed systems out of a Pynchon novel: you had this feedback loop, with photons coming off the screen into the kids' eyes, the neurons moving through their bodies, electrons moving through the computer. And these kids clearly *believed* in the space these games projected.
>
> Everyone who works with computers seems to develop an intuitive faith that there's some kind of *actual* space behind the screen.[35]

Thus, for Gibson, the video game player has already merged with the computer. The video game player is already a cyborg, an insight Gibson spun into a postmodern mythology. Over the past decade, such mythologies have been recasting our sense of collective identity.

For Will, a thirty-seven-year-old writer who has recently gone online, the Internet inspires a personal mythology in which he feels part of something far larger than himself: "The Internet is like a giant brain. . . . It's developing on its own. And people and computers are its neural net." This view puts human brains and computers in a provocative symmetry and together they contribute to a larger evolving structure. Will tells me that his new idea about the Internet as a brain made up of human and computer parts "felt like an epiphany." In an age where we feel fragmented as individuals, it is not surprising to see the emergence of popular mythologies that try to put the world back together again.

Will creates his unity by treating both people and machines as Internet nodes, sites through which information travels. Like the fantasies of Wiener and Hillis, his epiphany depends on a central notion of artificial intelligence and artificial life: Emergent or not, when reduced to our most basic elements, we are made up, mind and body, of information. Some believe that thinking about people as information carries the possibility for leveling the playing field.[36] For example, if all people are ultimately thought to be information, then such categories as race, class, and gender may be stripped of their cultural charge. But thinking about people as information also carries the serious risk of impoverishing our sense of the human. Even as we recognize the risks of reducing people to strings of code, we must remember that we are peculiarly vulnerable to the message (whether from scientists, futurists, novelists, or filmmakers) that we and machines are kin. In this book we've seen many examples of people treating even a very primitive computer as an other, worthy of relationship.

At the MIT Artificial Intelligence Laboratory, Rodney Brooks has embarked on a project to build an artificial two-year-old. Brooks calls his new "creature" Cog in order to evoke both the mechanical nature of this two-year-old (and perhaps others) and its aspiration to cognition. Brooks's artificial life research, inspired by Herbert Simon's description of the ant walking across a sand dune, takes as a starting assumption that much of what we see as complex behavior is actually simple responses to a complex environment. After over fifteen years of using this strategy to build robots that aspired to insect-level intelligence, Brooks decided, in his words, "to go for the whole enchilada." Cog is being designed to "learn" from its interaction with its environment—most particularly from its interaction with the many researchers who will dedicate themselves to its education. Cog is controversial: for some a noble experiment that takes seriously the notion of embodied, emergent intelligence, for others a grandiose fantasy. When I heard about Cog, I was extremely skeptical. I decided to pay a visit.

Cog's mobile torso, neck, and head stand on a pedestal. Trained to track the largest moving object in its field (because this will usually be a human being) Cog "noticed" me soon after I entered its room. Its head turned to follow me and I was embarrassed to note that this made me happy. I found myself competing with another visitor for its attention. At one point, I felt sure that Cog's eyes had "caught" my own. My visit left me shaken—not by anything that Cog was able to accomplish but by my own reaction to "him." For years, whenever I had heard Rodney Brooks speak about his robotic "creatures," I had always been careful to mentally put quotation marks around the word. But now, with Cog, I had found that the quotation marks disappeared. Despite myself and despite my continuing skepticism about this research project, I had behaved as though in the presence of another being.

In the introduction to this book I quoted Ralph Waldo Emerson: "Dreams and beasts are two keys by which we are to find out the secrets of our nature. . . . They are test objects."[37] And I said that if he lived today, Emerson would have added computers to his list. But computers are more than a simple addition. Through virtual reality they enable us to spend more of our time in our dreams. And through "beings" like Cog they revise what we understand as "beast." Not only are computers evocative in their own right but they have transformed the nature of the test objects that have come before.

DWELLERS ON A THRESHOLD

In the past decade, the computer culture has been the site of a series of battles over contested terrains. There have been struggles between formal logic and bricolage, about profound disruptions in our traditional ways of categorizing people and things, and about the nature of the real in a culture of simulation. These struggles marked the three sections of this book, in which we have seen the computer as tool, as mirror, and as gateway to a world through the looking glass of the screen. In each of these domains we are experiencing a complex interweaving of modern and postmodern, calculation and simulation. The tensions are palpable.

In the struggle of epistemologies, the computer is caught between its natural pluralism and the fact that certain styles of computing are more culturally resonant than others. On one hand, the computer encourages a natural diversity of responses. Different people make the computer their own in their own ways. On the other hand, computers are increasingly expressing a constellation of ideas associated with postmodernism, which has been called our new cultural dominant.[38] We have moved in the direction of accepting the postmodern values of opacity, playful experimentation, and navigation of surface as privileged ways of knowing.

In the contest over where the computer fits into categories such as what is or is not intelligent, alive, or person-like, the game is still very much in play. Here, too, we saw tension. In one context, people treat the machine as sentient, an other; in a different context, they insist on its difference from us, its "other-*ness*." As people have become more comfortable psychologizing computers and have come to grant them a certain capacity for intelligence, the boundary dispute between people and machines now falls on the question of life.

The final contest concerns the notion of the real. In simulated science experiments, virtual chemicals are poured from virtual beakers, and virtual light bounces off virtual walls. In financial transactions, virtual money changes hands. In film and photography, realistic-looking images depict scenes that never took place between people who never met. And on the networked computers of our everyday lives, people have compelling interactions that are entirely dependent on their online self-representations. In cyberspace, hundreds of thousands, perhaps already millions, of users create online personae who live in a diverse group of virtual communities where the routine formation of multiple identities undermines any notion of a real and unitary self. Yet the notion of the real fights back. People who live parallel lives on the screen are nevertheless bound by the desires, pain, and mortality of their physical selves. Virtual communities offer a dramatic new context in which to think about human identity in the age of the Internet. They are spaces for learning about the lived meaning of a culture of simulation. Will it be a separate world where people get lost in the surfaces or will we learn to see how the real and the virtual can be made permeable, each having the potential for enriching and expanding the other? The citizens of MUDs are our pioneers.

As we stand on the boundary between the real and the virtual, our experience recalls what the anthropologist Victor Turner termed a liminal moment, a moment of passage when new cultural symbols and meanings can emerge.[39] Liminal moments are times of tension, extreme reactions, and great opportunity. In our time, we are simultaneously flooded with predictions of doom and predictions of imminent utopia. We live in a crucible of contradictory experience. When Turner talked about liminality, he understood it as a transitional state—but living with flux may no longer be temporary. Donna Haraway's characterization of irony illuminates our situation: "Irony is about contradictions that do not resolve into larger wholes . . . about the tension of holding incompatible things together because both or all are necessary and true."[40] It is fitting that the story of the technology that is bringing postmodernism down to earth itself refuses modernist resolutions and requires an openness to multiple viewpoints.

Multiple viewpoints call forth a new moral discourse. I have said that the culture of simulation may help us achieve a vision of a multiple but integrated identity whose flexibility, resilience, and capacity for joy comes from having access to our many selves. But if we have lost reality in the process, we shall have struck a poor bargain. In Wim Wenders's film *Until the End of the World,* a scientist develops a device that translates the electrochemical activity of the brain into digital images. He gives this technology to his family and closest friends, who are now able to hold small battery-driven monitors and watch their dreams. At first, they are charmed. They see their treasured fantasies, their secret selves. They see the images they otherwise would forget, the scenes they otherwise would repress. As with the personae one can play in a MUD, watching dreams on a screen opens up new aspects of the self.

However, the story soon turns dark. The images seduce. They are richer and more compelling than the real life around them. Wenders's characters fall in love with their dreams, become addicted to them. People wander about with blankets over their heads the better to see the monitors from which they cannot bear to be parted. They are imprisoned by the screens, imprisoned by the keys to their past that the screens seem to hold.

We, too, are vulnerable to using our screens in these ways. People can get lost in virtual worlds. Some are tempted to think of life in cyberspace as insignificant, as escape or meaningless diversion. It is not. Our experiences there are serious play. We belittle them at our risk. We must understand the dynamics of virtual experience both to foresee who might be in danger and to put these experiences to best use. Without a deep understanding of the many selves that we express in the virtual we cannot use our experiences there to enrich the real. If we cultivate our awareness of what stands behind our screen personae, we are more likely to succeed in using virtual experience for personal transformation.

The imperative to self-knowledge has always been at the heart of philosophical inquiry. In the twentieth century, it found expression in the psychoanalytic culture as well. One might say that it constitutes the ethic of psychoanalysis. From the perspective of this ethic, we work to know ourselves in order to improve not only our own lives, but those of our families and society. I have said that psychoanalysis is a survivor discourse. Born of a modernist worldview, it has evolved into forms relevant to postmodern times. With mechanistic roots in the culture of calculation, psychoanalytic ideas become newly relevant in the culture of simulation. Some believe that we are at the end of the Freudian century. But the reality is more complex. Our need for a practical philosophy of self-knowledge has never been greater as we struggle to make meaning from our lives on the screen.

NOTES

INTRODUCTION

1. William Gibson, *Neuromancer* (New York: Ace, 1984).
2. For a general introduction to LambdaMOO and MUDding, see Pavel Curtis, "Mudding: Social Phenomena in Text-Based Virtual Realities," available via anonymous ftp://parcftp.xerox.com/pub/MOO/papers/DIAC92.*; Amy Bruckman, "Identity Workshop: Emergent Social and Psychological Phenomena in Text-Based Virtual Reality," unpub. ms., March 1992, available via anonymous ftp://media.mit.edu/pub/asb/papers/identity-workshop.*; and the chapter on MUDs in Howard Rheingold's *Virtual Community: Homesteading on the Electronic Frontier* (New York: Addison-Wesley, 1993). On virtual community in general, see Allucquere Rosanne Stone, "Will the Real Body Please Stand Up?: Boundary Stories about Virtual Cultures," in *Cyberspace: First Steps,* ed. Michael Benedikt (Cambridge, Mass.: MIT Press, 1992), pp. 81–118. The asterisk in a net address indicates that the document is available in several formats.
3. The number of MUDs is changing rapidly. Most estimates place it at over five hundred, but an increasing number are private and so without any official "listing." The software on which they are based (and which gives them their names as MOOs, MUSHes, MUSEs, etc.) determines several things about the game; among these is the general layout of the game space. For example, in the class of MUDs known as AberMUDs, the center of town is similar from one game to another, but the mountains, castles, and forests that surround the town are different in different games, because these have been built specifically for that game by its resident "wizards." MUDs also differ in their governance. In MUD parlance, wizards are administrators; they usually achieve this status through virtuosity in the game. In AberMUDs only wizards have the right to build onto the game. In other kinds of MUDs, all players are invited to build. Who has the right to build and how building is monitored (for example, whether the MUD government should allow a player to build a machine that would destroy other players' property or characters) is an important feature that distinguishes types of MUDs. Although it may be technically correct to refer to being in a MUD (as in a dungeon), it is also common to speak of being on a MUD (as in logging on to a program). To

me, the dual usage reflects the ambiguity of cyberspace as both space and program. I (and my informants) use both in this book.

4. A flame war is computer culture jargon for an incendiary expression of differences of opinion. In flame wars, participants give themselves permission to state their positions in strong, even outrageous terms with little room for compromise.

5. I promised Doug anonymity, a promise I made to all the people I interviewed in researching this book. Doug has been told that his name will be changed, his identity disguised, and the names and distinguishing features of his MUD characters altered. It is striking that even given these reassurances, which enable him to have an open conversation with me about his social and sexual activities on MUDs, he wants to protect his FurryMUD character.

6. I immersed myself in these "French lessons," first in the aftermath of the May 1968 student revolt, a revolt in which Lacan and Foucault became intellectual heroes. Later, in 1973–74, the immersion continued while I studied the intellectual fallout of that failed revolution. That fallout included a love affair with things Freudian and an attack on unitary models of self. While followers of Lacan relied on reinterpretations of Freud that challenged models of a centralized ego, Deleuze and Guattari proposed more radical views that described the self as a multiplicity of desiring machines. See Gilles Deleuze and Félix Guattari, *Anti-Oedipus: Capitalism and Schizophrenia,* trans. Robert Hurley, Mark Seem, and Helen R. Lane (Minneapolis: University of Minnesota Press, 1983).

7. Jill Serpentelli, "Conversational Structure and Personality Correlates of Electronic Communication," unpub. ms., 1992.

8. The student's association of Derrida and hypertext may be unsophisticated, but it is far from outlandish. See, for example, George P. Landow, *Hypertext: The Convergence of Critical Theory and Technology* (Baltimore: Johns Hopkins, 1992), pp. 1–34; and in George P. Landow and Paul Delany, eds., *Hypermedia and Literary Studies* (Cambridge, Mass.: MIT Press, 1991).

9. Richard A. Lanham, *The Electronic Word: Democracy, Technology, and the Arts* (Chicago: The University of Chicago Press, 1993), p. 51. George Landow sees critical theory and technology in the midst of a "convergence." See Landow, *Hypertext.*

10. I say almost unthinkable because a small number of postmodern writers had begun to associate their work with the possibilities of computer technology. See, in particular, Jean-François Lyotard, *The Postmodern Condition: A Report on Knowledge,* trans. Geoff Bennington and Brian Massumi (Minneapolis: University of Minnesota Press, ·1984).

11. *The Wall Street Journal,* 3 January 1995: A3, A4, and *The Wall Street Journal,* 10 January 1995: B1, B3.

12. Here I have changed the name of the MUD (there is to my knowledge no DinoMUD) to protect the confidentiality I promise all informants. I use the real name of a MUD when it is important to my account and will not compromise confidentiality. See "A Note on Method."

13. See, for example, Donna Haraway, "A Manifesto for Cyborgs: Science, Technology, and Socialist Feminism in the 1980s," *Socialist Review* 80 (March–April 1985): 65–107.

14. The quotation is from a journal entry by Emerson in January 1832. The passage reads in full, "Dreams and beasts are two keys by which we are to find out the secrets of our nature. All mystics use them. They are like comparative anatomy. They are our test objects." See Joel Porte, ed., *Emerson in His Journals* (Cambridge, Mass.: Belknap Press, 1982), p. 81.

15. See "A Note on Method."

16. In a recent review of the place of genetics in contemporary popular culture, Dorothy Nelkin and Susan Lindee have said: "DNA has taken on the social and cultural functions of the soul." See their *The DNA Mystique: The Gene as a Cultural Icon* (San Francisco and New York: W. H. Freeman, 1995), p. 42.

17. Peter Kramer, *Listening to Prozac: A Psychiatrist Explores Mood-Altering Drugs and the New Meaning of the Self* (New York: Viking, 1993).

18. Nelkin and Lindee's *The DNA Mystique* documents the degree to which genetic essentialism dominates American popular culture today.

19. Evelyn Fox Keller, "The Body of a New Machine: Situating the Organism Between Telegraphs and Computers," *Perspectives on Science* 2, no. 3 (1994): 302–23.

20. For a view of this matter from the perspective of the 1980s, see Sherry Turkle, *The Second Self: Computers and the Human Spirit* (New York: Simon & Schuster, 1984).

CHAPTER 1   A TALE OF TWO AESTHETICS

1. When people talk about the computer as though it were a part of them as well as of the outside world, their words evoke the power of what the psychoanalyst D. W. Winnicott called transitional objects. See D. W. Winnicott, *Playing and Reality* (New York: Basic Books, 1971). These are the objects, like Linus's baby blanket or the tattered rag doll or bit of silk from the first pillow, to which children remain attached even as they embark on the exploration of the world beyond the nursery. For Winnicott, they are mediators between the child's earliest bonds with the mother, whom the infant experiences as inseparable from the self, and the growing capacity to develop relationships with other people who will be experienced as separate beings. The infant knows transitional objects as both almost-inseparable parts of the self and, at the same time, as the first not-me possessions. As the child grows, the actual objects are discarded, but the experience of them remains. The experience is diffused in the intense experiencing in later life of a highly charged intermediate space between the self and certain objects. This experience has traditionally been associated with religion, spirituality, notions of beauty, sexual intimacy, and the sense of connection with nature. Now it is associated with using computers.

2. When children grow up in the computer culture, the machine's holding power depends not only on children's ability to express and discover themselves through computational media but on the computer's profound developmental vocation. From their earliest years, computers enter into the way children grow up.

In an early developmental stage, when children are between about three and eight years old, the computer, reactive and interactive, provokes children to find new reasons for what makes things alive. It is a philosophical stage.

At around the age of eight, however, children's preoccupations in the presence of computers turn from philosophizing to competing. Now what is most compelling is not reflection but action: Children want to beat the game or master the machine. At this stage in development, mastery takes on a privileged role. It becomes the key to autonomy, to the growth of confidence in one's ability to move beyond parents to peers, to move into relationships of competition and collaboration. Later, when adolescence begins, with its new sexual pressures and social demands, mastery can provide respite. The microworlds of sports, chess, or books become places of escape and safe platforms from which to test the difficult waters of adolescence. Involvement with computers can also provide a safe place. Ideally, from this strong, secure place, children move out at their own pace to less secure terrain.

For some, the issues that arise during adolescence are so threatening that the safe place is never abandoned. Sexuality is too frightening to be embraced. Intimacy with other people is unpredictable to the point of being intolerable. As we grow up, we tend to forge our identities by building on the last place in psychological development where we felt safe. As a result, some people come to define themselves in terms of the issues raised by the mastery phase, in terms of competence and what can be controlled. Of course, if the sense of self becomes defined in terms of those things over which one can exert perfect control, the world of safe things becomes severely limited, because those things tend to be things, not people.

It is in this circumstance that the experiences computers offer become most compelling and in some cases are able to exert their hold to the exclusion of other pleasures. Children may be drawn to the worlds of simulation that make them feel like "masters of the universe." The rules of a program, a simulation, a spreadsheet, a game, need not respect the physical laws that constrain the real world. You can build a world that never existed and that could never exist. A virtuoso programmer described to me how as a very young child he took clocks apart and tried "to put them together in new ways—to make new kinds of clocks." But while there were limits to how far he could make the materials of a clock into something new, the computer he was given on his ninth birthday presented him with none. Many are seduced by the computer's promise of perfect mastery. Of course, the need to compensate for a vulnerable identity by establishing a sense of control is not a strategy limited to adolescents. People often try to control what feels chaotic inside through action on the outside. They diet, or they swear off cigarettes or alcohol. With such healthy efforts can come an enhanced sense of autonomy, of being an actor (an "act-or") in one's life. The computer which offers an environment for a new level of control and mastery becomes a key player in this kind of drama.

This way of thinking about how computers enter into development is elaborated in Sherry Turkle, *The Second Self: Computers and the Human Spirit* (New York: Simon & Schuster, 1984), pp. 29–164.

3. Sherry Turkle, "The Subjective Computer: A Study in the Psychology of Personal Computation," *Social Studies of Science* 12 (1982): 173–205.

4. Turkle, "The Subjective Computer."
5. Microsoft Windows was announced on 10 November 1983, and shipped on 20 November 1985.
6. William Gibson, *Neuromancer* (New York: Ace, 1984).
7. Turkle, "The Subjective Computer."
8. Fredric Jameson, "Postmodernism, or the Cultural Logic of Late Capitalism," *New Left Review* 146 (July–August 1984): 53–92. See also Jean Baudrillard, "On Seduction," in *Jean Baudrillard: Selected Writings,* ed. Mark Poster, trans. Jacques Mourrain (Stanford: Stanford University Press, 1988), pp. 149–65.
9. In "Postmodernism, or the Cultural Logic of Late Capitalism," Jameson wrote about how "one fundamental feature" of the many variants of theories we think of as postmodern is "the effacement... of the older... frontier between high culture and so-called mass or commercial culture" (p. 54). This effacement is a central element in the contemporary experience of computer hypertext. Hypertext software enables texts to be linked and commented upon—a scene from Shakespeare and a student essay about Elizabethan England. The comments can be commented upon, new texts can be added, new links can be made between texts (the Shakespeare) and comments (the student essay), between texts and texts, comments and comments. In the end, in Jameson's words when writing about postmodernism, "the past as 'referent' finds itself gradually bracketed and then effaced altogether, leaving us with nothing but texts" (p. 66). The hypertext system has created a flat textual landscape where the boundary between texts and comments starts to break down. And the juxtapositions of texts from different centuries, levels of expertise, and forms of expression provide an experience of the eroding frontier between high and mass culture that Jameson wrote about.
0. Theoretically, this was altogether predictable. Postmodernism undermines the epistemologies of depth that stood behind traditional representation; the manifest does not refer back to the latent, existence does not refer back to essence, signifier does not refer back to signified. So the only objects that could represent this world of surface would be precisely those that do not allow traditional representation. Put otherwise, postmodernism could only be represented by objects that challenge representation itself.
11. Jameson, "Postmodernism, or the Cultural Logic of Late Capitalism": 89.
12. I have taken this perspective on the problem of the appropriability of ideas in other of my writings. For example, I have argued that the interest in, indeed the cultural infatuation with the ideas of Jacques Lacan in France after 1968 can be explained by the way even a relatively superficial understanding of these ideas allowed large numbers of people to think through their social circumstances. See Sherry Turkle, *Psychoanalytic Politics: Jacques Lacan and Freud's French Revolution,* 2nd rev. ed. (New York: Guilford Press, 1992 [1978]).
13. Susan Sontag, *On Photography* (New York: Farrar, Straus, & Giroux, 1977).
14. Jean Baudrillard, *Simulations,* trans. Paul Foss, Paul Patton, and Philip Beitchman (New York: Sémiotext(e), 1983).
15. Jameson, "Postmodernism, or the Cultural Logic of Late Capitalism": 58.
16. Claude Lévi-Strauss, *The Savage Mind* (Chicago: University of Chicago Press, 1968).

17. And of course, the traffic does not flow in only one direction. In our current situation, science fiction informs social criticism; theme parks such as Disneyland become not simply objects of analysis but exemplars of theory. The notion of ideas moving out may be heuristically useful but is too simple. Postmodern theory has underscored the traffic between diverse and seemingly separate realms. With it, high culture comes to contemplate advertising, science fiction fandom, and the romance novel.

18. Mary Douglas, *Purity and Danger: An Analysis of the Concepts of Pollution and Taboo* (London: Routledge, ARK, 1966). Douglas's analysis begins with Genesis and the story of the creation, in which a three-fold classification unfolds. There is earth, water, and sky. Leviticus, where the kosher rules are set out, takes up this scheme, says Douglas, and "allots to each element its proper kind of animal life." She states, "Holiness is exemplified by completeness. . . . Holiness means keeping distinct the categories of creation" (p. 53). It follows that "any class of creatures which is not equipped for the right kind of locomotion in its element is contrary to holiness" (p. 55).

> If the proposed interpretation of the forbidden animals is correct, the dietary laws would have been like signs which at every turn inspired meditation on the oneness, purity, and completeness of God. By rules of avoidance holiness was given a physical expression in every encounter with the animal kingdom and at every meal. [p. 57]

19. See Turkle, *Psychoanalytic Politics*, pp. 227–40.

20. I see the Macintosh as a concrete emissary for significant elements of postmodern thought, most dramatically for the willingness to accept opacity and dialogue with machines. And it would not be an exaggeration to say that, to date, the Macintosh style of simulated desktop has been our most widely disseminated cultural introduction to virtual reality. As we shall see, the sociologist of science, Bruno Latour, stresses the importance of such concrete emissaries, or "footsoldiers," in *The Pasteurization of France,* trans. Alan Sheridan and John Law (Cambridge, Mass.: Harvard University Press, 1988).

21. First, the relationship with the object may carry theory as an unintended side effect. But as things evolve, people seek out the object to have the pleasure of contact with the ideas it carries.

CHAPTER 2   THE TRIUMPH OF TINKERING

1. See Sherry Turkle, *The Second Self: Computers and the Human Spirit* (New York: Simon & Schuster, 1984); Sherry Turkle and Seymour Papert, "Epistemological Pluralism: Styles and Voices Within the Computer Culture," *Signs* 16 (1990): 128–57; and Sherry Turkle and Seymour Papert, "Epistemological Pluralism and the Revaluation of the Concrete," *Journal of Mathematical Behavior* 11 (1992): 3–33.

2. See Claude Lévi-Strauss, *The Savage Mind* (Chicago: University of Chicago Press, 1968), pp. 16–33.

3. See Seymour Papert, *Mindstorms: Children, Computers, and Powerful Ideas,*

2nd rev. ed. (New York: Basic Books, 1993 [1980]). The visibility and near-tangibility of computational objects allow a sense of direct manipulation that encultured mathematicians can feel in traditional formal systems. Many people are not comfortable with mathematical experiences that manipulate symbols on quadrille-lined paper. The ambivalent nature of computational objects—at once concrete and abstract—can permit such people a first access to mathematics.

4. My reference is to the work of Carol Gilligan, who traced the development of a different voice in the realm of moral reasoning. Carol Gilligan, *In a Different Voice: Psychological Theory and Women's Development* (Cambridge, Mass.: Harvard University Press, 1982).

5. See Jean Piaget, *The Child's Conception of Number,* trans. C. Gattegno and F. M. Hodgson (London: Routledge & Kegan Paul, 1952); *The Child's Conception of Space,* trans. F. J. Langdon and J. L. Lunzer (London: Routledge & Kegan Paul, 1956); *The Child's Conception of Physical Causality,* trans. Marjorie Gabain (London: Routledge & Kegan Paul, 1930).

6. Lévi-Strauss, *The Savage Mind.* Also see note 10, below, on work in the social studies of science that takes seriously the presence of bricolage in Western science.

7. Feminist thought offered two interpretations of why this might be. One focused on the different ways our culture socializes boys and girls into relational and hierarchical values. The other borrowed from psychoanalytic theory to examine the earliest relationships between infants and their parents.

Among the most influential writings in this second tradition, which borrowed heavily from what has come to be known as the object-relations perspective in psychoanalysis, are Evelyn Fox Keller, *Reflections on Gender and Science* (New Haven, Conn.: Yale University Press, 1985) and Nancy Chodorow, *The Reproduction of Mothering: Psychoanalysis and the Sociology of Gender* (Berkeley: University of California Press, 1978). For an excellent review of the object relations perspective, see Jay R. Greenberg and Stephen A. Mitchell, *Object Relations in Psychoanalytic Theory* (Cambridge, Mass.: Harvard University Press, 1983). These writings noted that in our culture, boys are generally more comfortable with boundaries and girls with attachments, because in boys, identity formation puts a greater premium on the ability to tolerate separation and discontinuity in relationships.

Psychoanalytic theory portrays the infant beginning life in a closely bonded relationship with the mother, one in which the boundaries between self and other are not clear. Nor does the child experience a clear separation between the self and the outer world in general. On this point see D. W. Winnicott, *Playing and Reality* (New York: Basic Books, 1971). The child's gradual development of a consciousness of separate existence begins with a separation from the mother. Slowly, the infant develops the sense of an objective reality "out there," separate from his or her self. As this process occurs, it takes on gender meanings. Differentiation and delineation, first brought into play in the effort to separate from the mother, become marked as not-mother, not-female. And since, in our culture, our earliest and most compelling experiences of merging are usually with the mother, later experiences where boundaries are not clear become marked as female.

Up to this point, the gender meanings of events are common to girls and boys. But at around age three or four, at what is known as the Oedipal stage, there is a fork in the road. The boy becomes involved in a fantasized romance with the mother. The father steps in to break it up, and in doing so strikes a second blow against closely bonded relationships with a woman. The father's intervention is also another time when the boy sees the pressure for separation as male, a perception reinforced by the fact that this time the boy gives up the idea of a romance with the mother by identifying himself with his father. The path to someday having a woman like his mother is through becoming like his father. Thus, for the boy, separation from the mother is more brutal, because in a certain sense it happens twice, first in the loss of the original bonded relationship with her and then again at the point of the Oedipal struggle.

Since separation from the mother made possible the first experiences of the world as "out there," we might call it the discovery of the objective. Because the boy goes through this separation twice, for him objectivity becomes more highly charged. The boy feels a greater desire for it; the objective, distanced relationship feels like safe, approved ground. In the boy's development, there is more of a taboo on the fusional, along with a correspondingly greater fear of giving in to its forbidden pleasures. The contemptuous comment of one fourth-grade boy who overheard a classmate talking about putting himself in the place of a computational object when he programmed can be interpreted from this point of view. (In the Logo computer language, a "sprite," like a "turtle," is a computational object that can be manipulated on the screen.) "That's baby talk," the fourth-grader said. "I am not in the computer. I'm just making things happen there." The remark reflects an insistence on boundaries and the development of a world view that will fall easily into line with the objective scientific stance, culturally coded as male.

Girls are less driven to objectivity because they are allowed to maintain more elements of their original relationship with the mother. She is positioned to become both an intimate friend and an object with which to identify. Correspondingly, it is more natural for girls to play with the pleasures of closeness with other objects as well, objects that include people and things, spouses and—in a way that ultimately can become fraught with tension—computers. See Evelyn Fox Keller, "Gender and Science," *Psychoanalysis and Contemporary Thought* 1 (1978): 409–33. Also see Keller, *Reflections on Gender and Science;* Chodorow, *The Reproduction of Mothering;* Gilligan, *In a Different Voice,* and Turkle and Papert, "Epistemological Pluralism and the Revaluation of the Concrete."

8. Of course, hackers like Greenblatt were technically able to write programs with well-delineated subprocedures, but their way of working had little in common with the techniques of the hard master. They did not write a program in sections that could be assembled into a product. They wrote simple working programs and shaped them gradually by successive modifications. If

a particular small change did not work, they undid it with another small change. Hackers saw themselves as artists. They were proud of sculpting their programs.

9. The term "situated learning" has become associated with this approach. A sample of studies on situated learning in everyday life is contained in Barbara Rogoff and Jean Lave, eds., *Everyday Cognition: Its Development in Social Context* (Cambridge, Mass.: Harvard University Press, 1984). See also Jean Lave, *Cognition in Practice: Mind, Mathematics and Culture in Everyday Life* (Cambridge, England: Cambridge University Press, 1988).

10. A sample of relevant work in the social studies of science is provided in Karin D. Knorr-Cetina and Michael Mulkay, eds., *Science Observed: Perspectives on the Social Studies of Science* (London: Sage Publications, 1983). See also Karin Knorr-Cetina, *The Manufacture of Knowledge: An Essay on the Constructivist and Contextual Nature of Science* (New York: Pergamon Press, 1981); Bruno Latour and Stephen Woolgar, *Laboratory Life: The Social Construction of Scientific Facts* (Beverly Hills, Calif.: Sage, 1979); and Sharon Traweek, *Beamtimes and Lifetimes: The World of High Energy Physicists* (Cambridge, Mass.: Harvard University Press, 1988).

11. See for example, Mary Field Belenky et al., *Women's Ways of Knowing: The Development of Self, Voice, and Mind* (New York: Basic Books, 1986); Evelyn Fox Keller, *A Feeling for the Organism: The Life and Work of Barbara McClintock* (San Francisco: W. H. Freeman, 1983); and Keller, *Reflections on Gender and Science*. Edited collections that focus on approaches to knowing in science include: Ruth Bleier, ed., *Feminist Approaches to Science* (New York: Pergamon, 1986), and Sandra Harding and Merrill B. Hintikka, eds., *Discovering Reality: Feminist Perspectives on Epistemology, Metaphysics, Methodology, and Philosophy of Science* (London: Reidel, 1983). An overview that highlights many issues relevant to this chapter is provided by Elizabeth Fee, "Critiques of Modern Science: The Relationship of Feminism to Other Radical Epistemologies," in Bleier, ed., *Feminist Approaches to Science*.

12. Gilligan, *In a Different Voice*, p. 26.

13. Keller, *A Feeling for the Organism*, p. 198. Here, Keller describes McClintock's approach as dependent on a capacity to "forget herself," immerse herself in observation, and "hear what the material has to say."

14. Keller, *A Feeling for the Organism*, p. 117.

15. Lisa's experience gives us reason to take a critical look at the characterization of large numbers of people (and in particular large numbers of women) as suffering from computer phobia, a popular diagnosis in the 1980s. Before one falls back on the idea that people respond to computing with a pathological phobia, why not attend to all those people who wanted to approach computing in the noncanonical style but were rarely given the opportunity to do so? Like Lisa and Robin, they could pass a course or pass a test. They were not computer phobic. They didn't need to stay away because of fear or panic. But they were computer reticent. They wanted to stay away, because the computer came to represent an alien way of thinking. They learned to get by. And they learned to keep a certain distance. One symptom of their alienation was the language with which they tended to neutralize the com-

puter, like Lisa dismissing it as just a tool, when they were denied the possibility of using it in an authentic way.

Through the mid-1980s, discrimination in the computer culture took the form of discrimination against epistemological orientations, most strikingly against the approach preferred by Lisa and Robin. Like Lisa and Robin, some people changed their style or "faked it." Some went underground. Recall that the graduate student who admitted that she thought of programming code as potting clay also told me, "Keep this anonymous. It makes me sound crazy." My French lessons taught me to see my kind of writing as cheating. Through the mid-1980s, the ideology of the dominant computer culture taught soft masters that their kind of programming was cheating.

16. How can those who wanted transparent understanding when they programmed relate to the Macintosh operating system that presents them with opaque icons? And yet, bricoleurs do enjoy the Macintosh. Most, in fact, embraced it as soon as they met it. The seeming contradiction is resolved by looking at the issue of authorship. There is a crucial difference in how people behave when they are programming and when they are using programs, software and operating systems, written by others. When programming, bricoleurs prefer the transparent style and planners the opaque. For example, when the physicist George said the Macintosh was a comedown after his transparent Apple II because he wanted his computer "to be as clear to me as my Swiss Army knife," he was talking about the pleasures of transparent systems programming. Planners want to bring their own programs to a point where they can be black-boxed and made opaque, while bricoleurs prefer to keep them transparent. But when dealing with simulation software or an operating system built by others, the situation is reversed. Now, planners are often frustrated when they don't know how the program or system works, and bricoleurs are happy to get to know a new environment by interacting with it.

17. Keller, *Reflections on Gender and Science,* pp. 33ff.; and Carolyn Merchant, *The Death of Nature: A Feminist Reappraisal of the Scientific Revolution* (New York: Harper and Row, 1980).

18. For a discussion of the shift to emergent and biological metaphors in artificial intelligence see Stephen Graubard, ed., *The Artificial Intelligence Debate: False Starts, Real Foundations* (Cambridge, Mass.: MIT Press, 1988). This collection was originally published as a special issue on artificial intelligence in *Daedalus* 117, no. 1 (Winter 1988).

19. Project Athena was MIT's experiment in using state-of-the-art educational software in undergraduate education. See Sherry Turkle, Donald Schön, Brenda Nielson, M. Stella Orsini, and Wim Overmeer, "Project Athena at MIT," unpub. ms., May 1988; and Sherry Turkle, "Paradoxical Reactions and Powerful Ideas: Educational Computing in a Department of Physics," in *Sociomedia: Multimedia, Hypermedia, and the Social Construction of Knowledge,* ed. Edward Barrett (Cambridge, Mass.: MIT Press, 1992), pp. 547–78.

20. He then illustrated his argument by telling the story of one student who was using computer-aided design to plan a road on a nearby site. The student put the road on a slope that was far too steep to support it. When the professor

asked the student to justify his design, the student replied, "Well, it's only one contour," referring to the fact that on the computer screen the slope of the land where the road was laid out was represented by one contour line. The professor pointed out that the way the program had been set up meant that one contour was twenty-five feet. "The computer had led to a distortion of the site in the student's mind." The computer screen was too small for the student to put in more contour lines. Twenty-five feet was the only practical scale that the student could use to prevent the lines from running into each other. Since the scale could not be adjusted on the digitized site plan, the student had become disoriented. To the professor it was clear that the student would not have run into this trouble if he had been using tracing paper to draw and redraw imprecise (but to scale) sketches of possible roads. "That's how you get to know a terrain—by tracing and retracing it, not by letting the computer 'regenerate' it for you."

21. His students described him as fanatical about making sure that they knew how every program they used in the laboratory worked, "just as you should know how every piece of equipment you use works." He told them, "You may not actually build your equipment yourself, but you should feel that you could if you wanted to."

22. This striking example of rules in the game culture of eleven-year-olds was cited in Eugene F. Provenzo, Jr., *Video Kids: Making Sense of Nintendo* (Cambridge, Mass.: Harvard University Press, 1991), pp. 45–46.

23. *The Journal of Myst* (Novato, Calif.: Broderbund Software, 1993).

24. Paul Starr, writing about simulation games, has associated this form of learning with the cycle of moving from a traditional assembly line to a computerized one as described by Shoshana Zuboff, *In The Age of the Smart Machine* (New York: Basic Books, 1988). See Paul Starr, "Seductions of Sim: Policy as a Simulation Game," *The American Prospect* 17 (Spring 1994): 19–29.

25. Starr, "Seductions of Sim": 20.

26. Starr, "Seductions of Sim": 20.

27. Starr, "Seductions of Sim": 25.

28. Cited in Starr, "Seductions of Sim": 25.

29. Eugene Provenzo has summed up the video games' moral philosophy by quoting J. David Bolter's description of "Turing's man." "Turing's man, while on one level empowered by the new technology, is, on another level, ignorant of history, unaware of the natural world that surrounds him, insensitive to deeper human motive, and guided by an instrumental type of reason." (See Provenzo, *Video Kids,* p. 121.) Provenzo supports his argument by comparing the Nintendo game Platoon to the director Oliver Stone's film on which it is loosely based. In a final soliloquy, the film's hero tells us what Stone intends as its message. It is a statement about the futility of war.

Looking back, we did not fight the enemy, we fought ourselves. And the enemy was in us. . . . Those of us who did make it have an obligation to build again, to teach to others what we know, and that the trial of what's left of our life is to find a goodness and meaning in this life. [p. 122]

Platoon, the game, reduces all of this to an instrumental goal: You win if you kill enough other people without being killed. All deaths are factored into the set of algorithms that are the game. "Killing innocent Vietnamese peasants is portrayed ... as undesirable because it will lower one's morale rating" (pp. 122–23). After reviewing Platoon and Blades of Steel, a hockey video game which penalizes players if they do *not* physically fight their opponents (exactly the opposite from what would be the case in "real" hockey), Provenzo says, "Through the computer it is possible to create a decontextualized microworld that conforms in its simulation to the philosophy of individualism and to a decontextualized sense of self. What *are* we teaching our children?"

30. Julian Bleeker, "Urban Crisis: Past, Present, and Virtual," *Socialist Review,* 24 (1995): 189–221.

31. Bleeker, "Urban Crisis." The work Bleeker cites on the King trial is "Real Time/Real Justice" by Kimberlé Crenshaw and Gary Peller in *Reading Rodney King, Reading Urban Uprising,* ed. Robert Goding-Williams (New York: Routledge, 1993).

CHAPTER 3   MAKING A PASS AT A ROBOT

1. Sherry Turkle, *The Second Self: Computers and the Human Spirit* (New York: Simon & Schuster, 1984), pp. 29–30.

2. These studies are reported in Jean Piaget, *The Child's Conception of the World,* trans. Joan and Andrew Tomlinson (Totowa, N.J.: Littlefield, Adams, 1960). See especially Chapters 5, 6, and 7: "Consciousness Attributed to Things," "The Concept of Life," and "The Origins of Child Animism, Moral Necessity and Physical Determinism," pp. 169–252.

3. For details of the earlier studies of children and computers see Turkle, *The Second Self.* There I stress that children used psychology to talk about the aliveness of inanimate things other than computers. One five-year-old told me that the sun is alive "because it has smiles." Another said that a cloud is alive "because it gets sad. It cries when it rains." But if an eight-year-old argued that clouds or the sun are alive, the reasons given are almost always related to their motion—their way of moving across the sky or the fact that they seem to act on their own accord. In contrast, as children grew older and more sophisticated about computers, their arguments about the computer's aliveness or lack of aliveness became focused on increasingly refined psychological distinctions. The machines are "sort of alive" because they think but do not feel, because they learn but do not decide what to learn, because they cheat but do not know they are cheating.

Piaget told a relatively simple story that accounted for increasing sophistication about the question of aliveness through the development of notions of physical causality. My observations of children discussing the aliveness of computers agreed with those of many investigators who stress the greater complexity of children's animistic judgments. For example, in *Conceptual Change in Childhood* (Cambridge, Mass.: MIT Press, 1985), Susan Carey demonstrates that alongside the development Piaget traces, something else is going on: the development of biological knowledge. Children's encounters with computers underscore the importance of a third area of relevant knowledge, psychological knowledge and the ability to make psychological distinctions. Carey discusses how children develop an intuitive psychology that is prebiological, a way of interpreting biological aspects of life as aspects of human behavior (as when a child answers the question "Why do animals eat?" by saying "they *want* to" rather than "they must eat to live"). When I refer to children's responses about the nature of the computer's aliveness as psychological, I am talking about something else—aspects of psychology that will not be replaced by a biological discourse but will grow into mature psychological distinctions, most significantly, the distinction between thought and feeling.

The computer evokes an increasingly sophisticated psychological discourse, just as Carey points out that involvement with and discussion about animals evoke a more developed biological discourse. Noncomputational machines—for example, cars, telephones, and, as we have seen, even simple tube radios—entered into children's thinking but in essence did not disturb traditional patterns of thinking about life and mind. Cognitive psychologists Rochel Gelman and Elizabeth Spelke say the reason is the great difference between traditional machines and computers. "A machine may undergo complex transformations of states that are internal and unseen. But it lacks the capacity for mental transformations and processes." In their work on children's animistic thinking they say that, "for purposes of exposition," they "disregard or set aside modern manmade machines that mimic in one way or more the characteristics of man." Their intuition that the computer would be disruptive to their expository categories is correct. The marginal computer does upset the child's constructed boundaries between thing and person. Rochel Gelman and Elizabeth Spelke, "The Development of Thoughts About Animate and Inanimate Objects: Implications for Research on Social Cognition," in *Social Cognitive Development: Frontiers and Possible Futures,* eds. John H. Flavell and Lee Ross (Cambridge, England: Cambridge University Press, 1981).

4. Among the children I have studied was a thirteen-year-old named Alex who played daily with a chess computer named Boris, which allowed its human user to set its level of play. Although Alex always lost if he asked the computer to play its best game, Alex claimed that "it doesn't feel like I'm *really* losing." Why? Because as Alex saw it, chess with Boris is like chess with a "cheater." Boris has "all the most famous, all the best chess games right there to look at. I mean, they are inside of him." Alex knew he could study up on his game, but Boris would always have an unfair advantage. "It's like in between every move Boris could read all the chess books in the world." Here, Alex defined what was special about being a person not in terms of strengths but of a certain frailty. For this child, honest chess was chess played within the boundaries of human limitations. His heroes were the great chess masters whose skill depends not only on "memorizing games" but on "having a spark," a uniquely human creativity.

5. The evocative phrase an "animistic trace" is of Jim Berkley's coinage.

6. Alan Turing, "Computing Machinery and Intelligence," *Mind* 59 (1950): 434–460. Turing's article has been widely reprinted; for example, see Edward Feigenbaum and Julian Feldman, eds., *Computers and Thought* (New York: McGraw-Hill, 1963), pp. 11–35.

7. John Searle, "The Myth of the Computer," *The New York Review of Books,* 29 April 1982: 5.

8. Searle, "The Myth of the Computer": 5.

9. Nor are they particularly surprised to learn that during restricted Turing tests, ones in which conversation is limited to prescribed domains, machines are regularly mistaken for people and people are regularly mistaken for machines.

10. Indeed, some explications of the film make the point that the hero may in fact also be a machine, a replicant. This idea has been a subject of spirited discussion on several Internet newsgroups, among them alt.cyberpunk and alt.fan.bladerunner. In 1993, alt.cyberpunk carried the Blade Runner FAQ (Frequently Asked Questions) compiled by Murray Chapman in which he noted:

> The Director's Cut includes a scene not in the original release. It is a dream sequence, showing Deckard's dream of a white unicorn. Given this, one can argue that Gaff [another detective Deckard works with] left the unicorn outside Deckard's apartment because he knew that Deckard dreamt of a unicorn. If Gaff knew what Deckard was dreaming, then we can assume that Deckard was a replicant himself, and Gaff knew he would be dreaming of a unicorn. [Murray Chapman, *Blade Runner FAQ,* version 1.0, January 1993.]

In a later section of the FAQ, fans continue a list of factors that make the case that Deckard is a replicant. These range from the fact that Ridley Scott and Harrison Ford have stated that this is the case to the surmise that only a replicant could survive the beating Deckard took from the replicant Roy. Other fans argued the case against Deckard's being a replicant. Their position took off from the idea that Deckard is in the film to embody "man against machine." One fan put it this way:

> A major point of the film was to show Deckard (ie [sic] The Common Man) how good it is to be alive, and how much we take it for granted. "What's it like to live in fear?" If all the main characters become replicants, the contrast between humans and replicants is lost.

11. In the discussion that follows I allow myself to share in the common practice of taking Julia at interface value and referring to the program as a person and a "she." This is certainly something that the program insists on:

> You say, "Julia, what sex are you?"
> Julia says, "Can't you tell?"
> You say, "Julia, what sex are you?"
> Julia says, "Don't you know?"
> You say, "Julia, what sex are you?"
> Julia says, "I'm female."

12. Julia is also able to initiate certain actions. For example, if she is not paged by a player who requests her presence somewhere else in the MUD, Julia will remain in a given room for several minutes. But if during that time, no one speaks to her, she leaves, giving the appearance of deciding that wandering off to explore the maze is a better use of her time.

13. For example, Julia has a set of responses to patterns that begin with "Are you a" and include the characters "robot," "bot," "computer," "autom," "machine," "program," "simulati."

14. Mauldin, "ChatterBots, TinyMUDs, and the Turing Test: Entering the Loebner Prize Competition," *Proceedings of the Twelfth National Conference on Artificial Intelligence,* vol. 1 (Menlo Park: AAAI Press; Cambridge, Mass.: MIT Press, 1994), pp. 16–21. Mauldin has been working on a family of such bots, known as ChatterBots, since November 1989. Julia has been running on the Internet since January 1990. Different versions of the program can be running on different MUDs at the same time. The excerpts of Julia dialogue that I use in this paper are drawn from Michael Mauldin's "ChatterBots" and Leonard Foner, "What's an Agent Anyway?" unpub. ms., 1993. The Mauldin paper and other Julia materials (including the possibility of talking directly to Julia) are available on the World Wide Web at http://fuzine.mt.cs.cmu.edu/mlm/julia.html). The Foner paper is available via anonymous ftp://media.mit.edu/pub/Foner/Papers/Julia/What's-an-Agent-Anyway--Julia.*.

Unlike Foner and Mauldin, I have, in general, corrected MUD players' spelling and punctuation for the sake of clarity. In the spirit of equal treatment of bots and humans, this means that I have also corrected Julia's typos. Julia's typos are made on purpose to make the program seem more human. In Turing test mode, Julia is programmed so that her supposedly typed sentences appear on the screen with realistic delays between the characters, given the layout of the QWERTY keyboard. This stylistic decision was a close call for me, and I can argue a good case for making the opposite decision. Where I have departed from this rule I have noted the text [sic].

15. Mauldin, "ChatterBots."

16. See Foner, "What's an Agent Anyway?"

17. This dialogue, reported in Foner, "What's an Agent Anyway?" is copyright © 1990 by Ashne.

18. The dialogues are published in Foner, "What's an Agent Anyway" and Mauldin, "ChatterBots." Both authors disguised Barry's true identity. I have changed some further details in the interest of protecting his privacy.

19. Julia banters by creating slang nicknames according to a convention popularized on *Saturday Night Live,* a popular television show. Bob becomes Bobmeister or Bobdude. Barry in this case becomes Barrymeister. He might also become Barrydude.

20. See previous note. Now it is Barry who is imitating a *Saturday Night Live* convention. Laura would become Laurarama.

21. Julia gave this computer-like response in TinyMUD mode. Julia also has a contest mode, in which she avoids such gaffes. The following gives a sense of the difference between Julia's contest mode and TinyMUD mode:

You say, "Julia, how old are you?"
Julia says, "I am 4 years and 6 months old."
You say, "Julia contest mode on."
Julia says, "contest mode is now on."
You say, "Julia, how old are you?"
Julia says, "I am 21 years old."

22. Foner, "What's an Agent Anyway?"
23. Foner, "What's an Agent Anyway?"
24. Turing, "Computing Machinery and Intelligence."
25. The citations from the 1993 Loebner competition are drawn from Mauldin, "ChatterBots."
26. Mauldin, "ChatterBots."
27. Joseph Bates, "The Role of Emotion in Believable Agents," *Communications of the ACM* 37, no. 7 (July 1994): 122–25. Bates cites Frank Thomas and Ollie Johnston, *Disney Animation: The Illusion of Life* (New York: Abbeville Press, 1981).
28. Bates cites foundational work by Rodney Brooks and Pattie Maes of MIT. Rodney Brooks, "A Robust Layered Control System for a Mobile Robot," *IEEE Journal of Robotics and Automation* RA-2 (1986): 14–23; and Pattie Maes, "How to Do the Right Thing," *Connection Science* 1, no. 3 (1989): 291–323. "The emphasis in alternative AI on reactivity," says Bates, "could be seen as choosing one of the believability requirements and elaborating it to a position of importance, while downgrading other qualities, such as those related to our idealized view of intelligence." See Bates, "The Role of Emotion in Believable Agents."
29. Rodney A. Brooks, "Intelligence Without Representation," unpub. ms., 1987, p. 7.
30. Herbert Simon, *The Sciences of the Artificial* (Cambridge, Mass.: The MIT Press, 1969), p. 24.
31. Rodney Brooks, *Fast, Cheap, and Out of Control* (Cambridge, Mass.: MIT Artificial Intelligence Laboratory, 1989); Rodney Brooks, "Elephants Don't Play Chess," *Robotics and Autonomous Systems* 6, nos. 1–2 (June 1990): 3–15.
32. On the technique of combining evolution with learning, see David Ackley and Michael Littman, "Interactions between Learning and Evolution," in *Artificial Life II,* eds. Christopher Langton et al., Santa Fe Institute Studies in the Sciences of Complexity (Redwood City, Calif.: Addison-Wesley, 1992), pp. 487–510.
33. Beerud Sheth and Pattie Maes, "Evolving Agents for Personalized Information Filtering," *Proceedings of the Ninth IEEE Conference on Artificial Intelligence for Applications* (Los Alamitos, Calif.: IEEE Computer Society Press, 1993), pp. 345–52. This project has agents search through USENET netnews messages. USENET is an information service on the Internet. A similar project could search other information resources, for example, the archives of *The New York Times.*
34. Yezdi Lashkari, Max Metral, and Pattie Maes, "Collaborative Interface Agents,"

Proceedings of the Twelfth National Conference on Artificial Intelligence, vol. 1 (Menlo Park: AAAI Press; Cambridge, Mass.: MIT Press, 1994): 444–49.

35. Robyn Kozierok and Pattie Maes, "A Learning Interface Agent for Scheduling Meetings," *Proceedings of the ACM-SIGCHIL International Workshop on Intelligent User Interfaces,* Florida, January 1993. The agents acquire their competence from observing the user, from noting which of its suggestions are ignored, from noting which of its suggestions are taken or explicitly rejected, from explicit examples provided by the user to train the agent. See Pattie Maes, "Agents That Reduce Work and Information Overload," *Communications of the ACM* 37, no. 7 (July 1994): 31–40, 146.

36. Pattie Maes, "Agents That Reduce Work and Information Overload": 32.

37. The intellectual historian Bruce Mazlish has written about these challenges as assaults on people's sense of fundamental discontinuity with the rest of the cosmos. The first such assault, the Copernican revolution, unseated the Earth and people on it from their place at the center of the universe. Humankind was not set apart from nature. A second, the Darwinian, made it clear that people were not the first of God's creations, nor were they really so far removed from animals. And a third, the Freudian, decentered each individual from his or her sense of being a conscious, intentional actor. Freud showed us that there was another stage, albeit unconscious, on which some of the most compelling and significant dramas of the human mind were played. Free will—for some people, emblematic of the divine in humanity—had to share the stage with aspects of the unconscious in the determination of intention and action. Information processing artificial intelligence took Freudian decentering a step further. If computer programs could think, the mind might well be such a program, and if so, where was intention, where was self? Furthermore, artificial intelligence challenged people's sense that their intelligence set them apart from artifacts, just as the theory of nature selection challenged their sense that God had set them apart from animals. It was a fourth discontinuity. Bruce Mazlish, *The Fourth Discontinuity: The Co-Evolution of Humans and Machines* (New Haven, Conn.: Yale University Press, 1993).

38. Douglas R. Hofstadter and the Fluid Analogies Research Group, *Fluid Concepts and Creative Analogies: Computer Models of the Fundamental Mechanisms of Thought* (New York: Basic Books, 1995), p. 157.

CHAPTER 4   TAKING THINGS AT INTERFACE VALUE

1. Clifford Nass, Jonathan Steuer, and Ellen R. Tauber, "Computers Are Social Actors," *Social Responses to Communications Technologies Paper #109,* Stanford University (submitted to CHI '94, Conference of the ACM/SIGCHI, Boston, Mass., April 1994).

2. Fredric Jameson, "Postmodernism, or the Cultural Logic of Late Capitalism," *New Left Review* 146 (July–August 1984): 53–94.

3. The observations and interviews that I draw on in this chapter span nearly twenty years. They are based on the experiences of nearly three hundred

people, about half of them college and graduate students in the Boston area, about half of them older and outside academic settings. See "A Note on Method."

4. In June 1984, my local Cape Cod drive-in was showing *Revenge of the Nerds*. By June 1985, it was showing *Perfect*, in which John Travolta played a *Rolling Stone* reporter with a laptop computer.

5. This shift in attitude is reflected in research that has recorded patients' pleasure and displeasure at being asked to use computer programs in psychiatric settings. A typical report from the first period is in Jon H. Greist, Marjorie H. Klein, Lawrence J. Van Cura, "A Computer Interview for Target Psychiatric Symptoms," *Archives of General Psychiatry* 29 (August 1973): 247–51. Twenty-seven patients interacted with a computer program that asked them to specify their symptoms. Sixteen did not like using the computer. Another four liked it only moderately. Twenty-two of the twenty-seven reported that they would have preferred to give such information to a doctor. Consistent with my findings about the fascination with the computer as a blank slate for self-expression, all patients were willing to reveal deeply personal concerns to the machine. And consistent with my findings that part of the holding power of the machine is the human inclination to project life onto it, is the fact that after their sessions with the computer, patients continued to make comments about the machine. These ranged from "Machine broke down, made me angry" to "Fix the cord on the machine so it doesn't die" (from a very angry, obsessional young man) to "I'm hungry" and "Good-bye."

In the mid-1980s, the research literature regularly reports that people are comfortable talking to computers or prefer talking to computers rather than to people. Indeed, the computer psychotherapy movement was given a great boost by the finding that people tend to be more candid with a computer than with a human therapist in disclosing sensitive or embarrassing material. See Harold P. Erdman, Marjorie H. Klein, and John H. Greist, "Direct Patient Computer Interviewing," *Journal of Consulting and Clinical Psychology* 53 (1985): 760–73.

By the late 1980s and early 1990s, the reports of people preferring dialogue with a machine on sensitive issues was widely reported in the popular press. See, for example, Christopher Joyce, "This Machine Wants to Help You," *Psychology Today*, February 1988, 44–50; and Kathleen Murray, "When the Therapist Is a Computer," *The New York Times*, 9 May 1993: C25.

6. Weizenbaum had just published a critique of artificial intelligence as an exemplar of "instrumental reason." Joseph Weizenbaum, *Computer Power and Human Reason: From Judgment to Calculation* (San Francisco: W. H. Freeman, 1976).

7. The program also had a vocabulary of cue words, such as "mother" or "father," that triggered preset responses. So, for example, the words "miserable," "unhappy," "sad," and "depressed" might trigger the stock phrase, "I AM SORRY TO HEAR THAT" from ELIZA. ELIZA's trigger words were arranged by priority. In Weizenbaum's version, for example, the word "computer" would always be picked up for conversation, taking precedence over any other topic, because Weizenbaum originally believed that anyone talking with

ELIZA would be thinking about computers. Other versions of the program have given precedence to words about feelings, such as "love," "hate," and "loneliness."

8. Weizenbaum first wrote about ELIZA in a 1966 article, "A Computer Program for the Study of Natural Language Communication Between Man and Machine." It appeared in *Communications of the Association of Computing Machinery* 9 (1966): 36–45. The technical title reflected his view that ELIZA's contribution lay in computer science, not psychotherapy. Unlike Weizenbaum, who had published his account of ELIZA in a computer science journal, Colby published in a journal for clinicians and announced a therapeutic rather than a computer science breakthrough. "We have written a computer program which can conduct psychotherapeutic dialogue." Kenneth Mark Colby, James B. Watt, and John P. Gilbert, "A Computer Method for Psychotherapy: Preliminary Communication," *Journal of Nervous and Mental Diseases* 142, no. 2 (1966): 148.

Colby acknowledged that his program and Weizenbaum's were "conceptually equivalent from a computer standpoint" but said that his goal, unlike Weizenbaum's, was to provide therapy. In Colby's view, his program was therapeutic because it communicated an "intent to help, as a psychotherapist does, and to respond as he does by questioning, clarifying, focusing, rephrasing, and occasionally interpreting." Colby et al., "A Computer Method for Psychotherapy": 149.

For Colby, the stumbling blocks to SHRINK's usefulness were its difficulties with language, its "failure to develop an internal cognitive model of the person during on-line communication," and the fact that it had no way of going beyond the getting-to-know-you first stage of a therapy. SHRINK could hold a (limited) conversation but it did not have a model of the patient or a therapeutic plan. Colby considered these to be technical problems in artificial intelligence and announced that the focus of his future work would be to solve them. He would devote his career to creating a computer psychotherapy program that had a model of its patient, a theory of illness, and a theory of how to make things better—in other words, a therapeutic agenda.

Colby's next effort was a computer program, PARRY, that simulated a paranoid patient. See Kenneth Mark Colby, S. Weber, and F. D. Hilf, "Artificial Paranoia," *Artificial Intelligence* 2 (1971): 1–25.

9. Colby et al., "A Computer Method for Psychotherapy": 149.
10. Cited in George Alexander, "Terminal Therapy," *Psychology Today,* September 1978: 56.
11. Cited in Alexander, "Terminal Therapy": 53.
12. Cited in Alexander, "Terminal Therapy": 53.
13. Colby et al., "A Computer Method for Psychotherapy": 151.
14. Reporting on progress in what the computer can do well in clinical settings, Benjamin Kleinmutz, a psychologist at the University of Illinois, made it clear that his argument for computers as clinicians was based on a model that saw the clinician as an information processor. "I do not enter the controversy by asserting that statistics are as good as clinicians but rather by stating that the clinician himself is simply another variant of 'statistical' predictor." Benjamin

Kleinmutz, "The Computer As Clinician," *American Psychologist* 30 (March 1975): 379.

A 1978 paper by Moshe H. Spero, written from a traditional clinical perspective, objected that "if a machine were going to be called a 'therapist' it would have to be able to imitate what human therapists do in all ways considered relevant to successful psychotherapy." Of course, the machine would be falling far short in many respects. It would not have caring, empathy, and intuition. This was the standard argument that traditional therapists used against the idea of computer psychotherapists, much as the students had done. But such objections did not fully acknowledge the gulf that separated their ideas about psychotherapy from those who were trying to computerize it. Moshe H. Spero, "Thoughts on Computerized Psychotherapy," *Psychiatry* 41 (August 1978): 281–82.

15. Stanley Lesse, "The Preventive Psychiatry of the Future," *The Futurist*, October 1976: 232.

16. Jerry O'Dell and James Dickson, "Eliza As a Therapeutic Tool," *Journal of Clinical Psychology* 40 (July 1984): 944.

17. Personal communication, 20 June 1976.

18. "No one would seriously maintain that to understand *King Lear* means no more than to be able to reduce the play to one of those boiled-down outlines used by college students just before a big exam." Weizenbaum, cited in Alexander, "Terminal Therapy": 56.

19. Weizenbaum, *Computer Power and Human Reason,* p. 201.

20. Cited in Alexander, "Terminal Therapy": 56.

21. Cited in Alexander, "Terminal Therapy": 56.

22. ELIZA communicated using only uppercase letters. This typographic convention both aped the teletype machine and reassuringly signaled that a computer, not a person, was "speaking." This convention is now only sometimes used. As we saw in the conversation of Julia, the bot, typographically, computers and people are no longer distinguished.

23. Children, too, worked to make the computer seem more alive. They, of course, are more familiar than adults with the pleasures of animating the world, as is shown in this vignette:

Lucy, five, was the youngest child in the after school day care group. She was plump, small for her age, teased by the other children. She badly needed a friend. On the first day I came to work with the children in her group, Lucy discovered my "Speak and Spell," one of the first electronic toys and games put on the market in the late 1970s. Lucy and the "Speak and Spell" became inseparable. It was her constant companion. Soon, she worked out a special way of keeping it "alive."

"Speak and Spell" has a speaker but no microphone. The only input it can receive is letters typed on its keyboard. But in fantasy, Lucy modified her toy. She used its speaker as her microphone. She called it "Speak and Spell's" "ear," and talked to it. First she spoke softly, "What do I have to spell to you?" And then, more insistently, "What should I spell?" Now screaming, "Tell me!" At this point and always at this point (for this is a sequence I watched many times) Lucy pressed the "Spell" button and the toy spoke: "SPELL . . . GIVE." Lucy settled back, obviously content: she had gotten the toy to address her. Her favorite way of interacting with the toy was to put it in "Say it" mode and to go into "Say it" mode herself, injecting her own "Say it" in the few seconds between the machine's "SAY IT" and its pronouncing the word it "wishes" to have said. So a typical dialogue between Lucy and "Speak and Spell" went like this:

SPEAK AND SPELL: SAY IT . . .
LUCY: Say it . . .
SPEAK AND SPELL: . . . HOUSE
LUCY: That's right, you're very good.

[Sherry Turkle, *The Second Self: Computers and the Human Spirit* (New York: Simon & Schuster, 1984), pp. 40–41]

Lucy said that Speak and Spell is "a little alive." She wanted it to be. Like Gary who coaxed ELIZA into seeming more intelligent than it was, Lucy worked with her machine in a way that made it seem more alive. The issue of computer animism is discussed further in Turkle, *The Second Self,* pp. 29–63.

24. The research literature on ELIZA contains a paper that provides interesting evidence on this point, although it was not a point the paper's authors were trying to make. Seventy normal students in a psychology course were given a version of ELIZA to work with for 44 minutes. Their interactions were grouped in categories and divided into a first and second half of the session. In the second half of the session, the frequency of certain categories of user input increased significantly. These were inquiries about ELIZA, references to ELIZA, and comments of the form, "Earlier you said that. . . ." Clearly, one of the most compelling things about talking to a machine is quite simply that you are talking to a machine. You want to test its limits, you want to check it out. See O'Dell and Dickson, "Eliza as a Therapeutic Tool."

25. These exercises have been a window onto changing attitudes in the years between ELIZA and DEPRESSION 2.0.

In a 1979 MIT class of seventeen undergraduates, three thought a computer could be a psychotherapist. They argued that it would allow people to get things off their chest. With this ventilation model of therapy, even a "dumb" program such as ELIZA could have a positive therapeutic effect. The other students argued that people were uniquely suited to the job of psychotherapist because of their emotions, their mortality, their physicality. In a 1984 class of seventeen undergraduate students, six thought a computer psychotherapist would have merit, a position they justified by pointing out that a computer program could do cognitive therapy, something they now knew quite a bit about.

26. Philosophers argued that lack of embodiment would also interfere with a computer's intelligence because intelligence is constructed from our sense of physical embodiment and connection with our surroundings. The writer best known for this position is Hubert Dreyfus. See, for example, Hubert Dreyfus, *What Computers Still Can't Do* (Cambridge, Mass.: MIT Press, 1992), pp. 235–55, and, for an earlier statement of similar ideas, Hubert Dreyfus, "Why Computers Need Bodies in Order to Be Intelligent," *Review of Metaphysics* 21 (1967): 13–32.

27. Dreyfus, "Why Computers Need Bodies in Order to Be Intelligent."
28. My studies of people's attitudes toward computers in roles previously assigned to humans were not confined to computers as psychotherapists. Another area I investigated was attitudes toward computers used as judges. This was an idea people looked upon with some favor if they focused on human frailty. For example, they imagined that a computer judge might be less racist than a human one.

Just as images of psychotherapy were reflected in opinions about a hypothetical computer psychotherapist, images of the legal system were reflected in opinions about a hypothetical computer judge. In the case of the computer psychotherapist, the computer looked better when therapy was seen as being about information or reprogramming. In the case of the computer judge, the computer looked better when the legal system was seen as racist or unfair in some other way.

In the 1970s and early 1980s, white, middle-class college students usually said that a computer judge was a very bad idea. "Judges weigh precedents and the specifics of the case. A computer could never boil this down to an algorithm." "Judges have to have compassion for the particular circumstances of the people before them. Computers could never develop this quality." "Judges have to be wise. Computers can be smart, only people can be wise." Confronted with a hypothetical computer ready to usurp a human role, people were defined in idealized terms as all the things computers could not do or be. With few exceptions, the dissenters at that time were black, from inner city neighborhoods, and streetwise. Their positive interest in computer judges was not based on having different ideas about computers and their abilities but on having different images of judges. They considered the hypothetical computer judge and said things like, "This is a pretty good idea. He is not going to see a black face and assume guilt. He is not going to see a black face and give a harsher sentence." One student, Ray, interviewed in 1983, put it this way:

I grew up in Chicago. I have tried to mind my own business. But all my life when I walk down the street, white people try to cross to the other side. And all my life, when there has been trouble, I know that it is my friends who are going to be brought in for questioning. I know that if I ever had to go before some judge, there is a good chance that he is going to see my face and he is going to think "nigger."

The computer judge would have a set of rules. He would apply the rules. It might be safer.

By the early 1990s, nonminority students, too, were more sympathetic toward the idea of the computer judge. In 1992 and 1993, conversations about the idea of a computer judge made frequent reference to the Rodney King trial. Many of my students felt that in that case, the legal system ignored the videotaped evidence of police brutality when its object was a black man. When people were distressed by signs of prejudice within the legal system, the idea of a computer judge was appealing. Attitudes about computers often reflect fears about people.

When in 1983 Ray referred to the computer judge and its rules, he was thinking of an expert system whose predictability and reliance on rules reassured him. But during the mid- to late 1980s, this image of artificial intelligence was in the process of eroding. While information processing taught a computer what to do, newer connectionist or neural net models described machines that were said to learn for themselves on the basis of their experiences.

An information-processing model of a computer judge would work from a set of algorithms that embodied the set of rules that real judges claimed they followed when making a decision. Presumably, prejudging a situation based on race would not be among them. But MIT computer science students explained to me that a connectionist computer judge would be fed hundreds of thousands of cases and decisions so that its decision-making powers could emerge from training. By the late 1980s, this new way of thinking about artificial intelligence was sufficiently enmeshed in the MIT student culture to influence reactions to both hypothetical computer psychotherapists and judges. I have said that the idea of neural nets and emergent intelligence seemed to make the notion of computer psychotherapy easier to take because it made the therapist seem less mechanistic, but it did not reassure students who feared prejudices within the legal system. Walt, a Chicano student from Los Angeles who had taken a computer science course that had briefly covered neural nets, put it this way in a 1990 conversation:

If you were training a computer to be a judge and it looked at who had been found guilty and how they were sentenced, it would "learn" that minority people commit more crimes and get harsher sentences. The computer would build a model that minorities were more likely to be guilty, and it would, like the human judges, start to be harder on them. The computer judge would carry all of the terrible prejudices of the real ones. It would not improve the situation but it would freeze the bad one. Of course, you would never be able to find a "rule" within the system that said you should discriminate against minorities. The computer would just do it because in the past, people had done it.

29. Jackie, a graduate student in English, has heard that a physician at a local university counseling center has been dismissed for sexual impropriety. She comments:

I would feel safer with an expert system [psychotherapist]. First of all, if I had it on my own personal computer, I could erase my stuff and wouldn't have to worry about the computer telling people about me. My sessions could be completely confidential. [In contrast] I see that whenever a therapist gets into trouble, the details of the patients he was treating start to get talked about too. And therapists can blab to each other about their cases. Also, no computer could make advances on you.

The more someone's ideas about psychotherapy were dominated by images of human failings, the better a computer looked. When human beings disappoint, there are still machines to count on.

30. For example, 1985 entrants in the computer self-help sweepstakes included The Running Program, Be Your Own Coach, The Original Boston Computer Diet, 28 Day Dieter, Managing Stress, Problem Solving, Stress and Conflict, and Coping with Stress.
31. Peter Garrison, "Technotherapy," *Omni,* December 1987: 162.
32. Among the organizations most cited in the popular press were Harvard Community Health Plan, the Western Institute of Neuropsychiatry, Johnson and Johnson, Arco, and Kaiser Permanente.

   One of the most frequently used programs was The Learning Program (TLP) developed by Roger Gould. TLP has a large database of information that helps it make a diagnosis. It works according to a tree metaphor; as the patient answers questions, he or she moves from large limbs to small branches. The answer to one question determines what question will be asked next. Ultimately, patients should wind up on a twig that pinpoints a problem, at which point the computer will suggest a remedy. TLP is meant to be used in conjunction with brief meetings with a human counselor. Gould's first clinic in a Los Angeles suburb featured eight TLP computers and ten human counselors. For about $70 a session, a mildly troubled Californian could walk in off the street and begin therapy. Gould was widely quoted as saying that most "midlife issues" can be addressed in ten sessions.
33. Paulette M. Selmi et al., "Computer-Administered Cognitive Behavioral Therapy for Depression," *American Journal of Psychiatry* 147 (January 1990): 51–56.
34. In Seligmann, "User Friendly Psychotherapy": 61.
35. Stacey Colino, "Software Shrinks: Freud at Your Fingertips," *Elle,* May 1993: 110.
36. Cited in Kathleen McAuliffe, "Computer Shrinks," *Self,* July 1991: 103.
37. Meaning "less" by intelligence has not been a simple downgrading. It has gone along with a general movement to see intelligence as a multiplicity of specialized intelligences, all functioning in parallel. See Howard Gardner, *Frames of Mind: The Theory of Multiple Intelligences* (New York: Basic Books, 1983).

CHAPTER 5   THE QUALITY OF EMERGENCE

1. For a more detailed discussion of the dichotomy between emergent and information processing AI, see Seymour Papert, "One AI or Many?"; Hubert Dreyfus and Stuart Dreyfus, "Making a Mind Versus Modeling the Brain: Artificial Intelligence Back at a Branchpoint"; and Sherry Turkle, "Artificial Intelligence and Psychoanalysis: A New Alliance"; all contained in *The Artificial Intelligence Debate: False Starts, Real Foundations,* ed. Stephen Graubard (Cambridge, Mass.: MIT Press, 1988). This edited collection was originally published as a special issue on artificial intelligence in *Daedalus* 117, no. 1 (Winter 1988).
2. Marvin Minsky and Seymour Papert, *Perceptrons: An Introduction to Computational Geometry,* rev. ed. (Cambridge, Mass.: MIT Press, 1988 [1969]). The book played an important role in discouraging emergent AI research during the 1970s or, at the very least, discouraging the funders of this research.

Some went so far as to blame Minsky and Papert for the demise of emergent AI. See Dreyfus and Dreyfus, "Making a Mind Versus Modeling the Brain."

3. Douglas R. Hofstadter, *Metamagical Themas: Questing for the Essence of Mind and Pattern* (New York: Basic Books, 1985), p. 631.

4. George Boole, *The Laws of Thought* (La Salle, Ill.: Open Court Publishing Company, 1952).

5. Cited in Jonathan Miller, *States of Mind* (New York: Pantheon, 1983), p. 23.

6. Miller, *States of Mind,* p. 23.

7. See Sherry Turkle, *The Second Self: Computers and the Human Spirit* (New York: Simon & Schuster, 1984), especially Chapter 7, "The New Philosophers of Artificial Intelligence," pp. 239–68. During the 1970s, the media touted the prowess of expert systems, large computer programs embodying the information and rules that professionals ostensibly use to make everyday decisions. To build an expert system, the AI researcher extracted decision rules from a virtuoso in a selected field (medical diagnosis, for example) and embedded them in a machine that then did the diagnosis "for itself." The publicity given to these systems meant that in the popular culture, there was a widespread belief that rule-based, information-rich computers could do amazing things.

8. Hubert Dreyfus, *What Computers Still Can't Do: A Critique of Artificial Reason* (Cambridge, Mass.: MIT Press, 1992).

9. See John Searle, "Minds, Brains, and Programs," *The Behavioral and Brain Sciences* 3 (1980): 417–24, and John Searle, "The Myth of the Computer," *The New York Review of Books,* 29 April 1982: 5.

10. Connectionism had its origin in work on neural nets as early as the 1940s. Warren McCulloch and Walter Pitts, "A Logical Calculus of the Ideas Immanent in Nervous Activity," *Bulletin of Mathematical Biophysics* 5 (1943): 115–33. There seems to be broad consensus that the term "connectionism" was first coined in 1982 by Jerome Feldman, a computer scientist at Berkeley.

11. David Rumelhart, one of the most influential connectionist researchers, wrote:

Our strategy has thus become one of offering a general and abstract model of the computational architecture of brains, to develop algorithms and procedures well suited to this architecture, to simulate these procedures and architecture on a computer, and to explore them as hypotheses about the nature of the human information-processing system. We say that such models are *neurally inspired,* and we call computation on such a system, brain-style computation. Our goal in short is to replace the computer metaphor with the brain metaphor. [David Rumelhart, "The Architecture of Mind: A Connectionist Approach," in *The Foundations of Cognitive Science,* ed. Michael I. Posner (Cambridge, Mass.: MIT Press, 1989), p. 134.]

12. I want to thank Mitchel Resnick for clarifying this point.

13. For example:

In most models, knowledge is stored as a static copy of a pattern. . . . In PDP [connectionist] models, though, this is not the case. In these models, the patterns themselves are not stored. Rather, what is stored is the *strengths* between units that allow these patterns to be recreated. . . . Learning must be a matter of finding the right connection strengths so that the right patterns of activation will be produced under the right circumstances. This is an extremely important property of this class of models, for it opens up the possibility that an information processing mechanism could learn, as a result of tuning its connections, to capture the interdependencies between activations that it is exposed to in the course of processing. [David E. Rumelhart, James L. McClelland, and the PDP Research Group, *Parallel Distributed Processing: Explorations in the Microstructure of Cognition,* vol. 1 (Cambridge, Mass.: Bradford Books/MIT Press, 1986), pp. 31–32.]

14. Computer scientists had long strained against the limitations of serial computers, in which one active processor that could only do one thing at a time, manipulating the passive data in a million cells of memory. It had long been known that, in principle, this bottleneck could be solved by eliminating the distinction between processor and memory so as to make every cell in the computer an active processor. Doing so, however, had always been prohibitively expensive.

15. Alongside the promise of new hardware, new ideas about programming were emerging. A standard computer program in the early 1980s consisted of instructions to do something to data and then move it around: "Add these numbers, put the result in memory, get the content of that memory location." In the mid-1980s researchers felt the need to deal with a different kind of action—not the moving of information but the making of objects. By a coincidence that associated computation with the object-relations tradition in psychoanalytic thought, computer scientists called their new programming methodology object-oriented programming. The coincidence was to be serendipitous.

The contrast between object-oriented and traditional programming is sharp. If one wanted to simulate a line of customers at a post office counter (in order to know, for example, how much longer the average wait would be if the number of clerks were to be reduced), a traditional FORTRAN programmer would assign X's and Y's to properties of the customers and write computer code to manipulate the variables. In contrast, the object-oriented programmer would write a program that created internal objects that behaved like people in a line at the post office. Each simulated inner object would be programmed to advance when the person ahead in the line advanced. Each would know when it reached the counter and then would carry out its transaction. Once created, the objects would be set free to interact. The programmer would not specify what the objects would actually do, but rather "who they would be." The end result would emerge from their local interactions.

If something of the feel of an information processing program is captured by the flow chart, something of the feel of object-oriented programming is captured by the pictures of file folders, documents, and a wastebasket that appear on the screen of a Macintosh. The screen icons are a surface reflection of a programming philosophy in which computers are thought of as electronic puppet shows. On the metaphor of computer systems as electronic puppet shows see Alan Kay, "Software's Second Act," *Science* 85 (November

1985): 122. Writing in 1985, Kay specified that "there are no important limitations to the kind of plays that can be enacted on their screens, nor to the range of costumes or roles that the actors can assume." Kay used this language to describe the philosophy he brought both to the design of Macintosh-style interfaces and to the design of object-oriented programming languages. Kay developed an object-oriented language, Smalltalk, during his time at Xerox PARC and was a major force in the development of the Alto computer, in many ways the prototype of the Macintosh. For a description of the "daylight raid" during which the developers of the Macintosh visited Xerox PARC and "walked away" with the desktop metaphor for the Macintosh, see Steven Levy, *Insanely Great: The Life and Times of the Macintosh, The Computer That Changed Everything* (New York: Vintage, 1994).

16. Cited in Dreyfus and Dreyfus, "Making a Mind Versus Modeling the Brain," p. 35.

17. This language was picked up by emergent AI's popularizers, as for example, in this description of how emergent AIs "learn":

> To train NETalk to read aloud, Sejnowski had given his machine a thousand word transcription of a child's conversation to practice on. The machine was reading the text over and over, experimenting with different ways of matching the written text to the sound of spoken word. If it got a syllable right, NETalk would remember that. If it was wrong, NETalk would adjust the connections between its artificial neurons, trying new combinations to make a better fit. . . . NETalk rambles on, talking nonsense. Its voice is still incoherent, but now the rhythm is somewhat familiar: short and long bursts of vowels packed inside consonants. It's not English, but it sounds something like it, a crude version of the nonsense poem "The Jabberwocky." Sejnowski stops the tape. NETalk was a good student. Learning more and more with each pass through the training text, the voice evolved from a wailing banshee to a mechanical Lewis Carroll. [William F. Allman, *Apprentices of Wonder: Inside the Neural Network Revolution* (New York: Bantam, 1989) p. 2.]

18. Allman, *Apprentices of Wonder*, p. 1.

19. Of course, large information processing programs are not predictable in any simple sense. But the aesthetic of information processing presents a computer whose "rules" are ultimately transparent.

20. W. Daniel Hillis, "Intelligence as an Emergent Behavior; or, The Songs of Eden," in *The Artificial Intelligence Debate*, ed. Graubard, pp. 175–76.

21. Hillis, "Intelligence as an Emergent Behavior," p. 176.

22. Cited in William F. Allman, "Mindworks," *Science* 86 (May 1986): 27.

23. Thomas Kuhn, who wrote the classic statement of how paradigm shifts occur within scientific communities, has cited Max Planck's view that a new scientific truth becomes dominant not because it wins converts but because the proponents of the older system gradually die off. Planck remarked that "a new scientific truth does not triumph by convincing its opponents and making them see the light, but rather because its opponents eventually die, and a new generation grows up that is familiar with it." Cited in Thomas Kuhn, *The Structure of Scientific Revolutions,* 2nd rev. ed. (Chicago: The University of Chicago Press, 1972 [1962]), p. 151.

24. Hofstadter, *Metamagical Themas,* p. 631.
25. In fact, Searle did try to subject neural net models to a similar treatment in a January 1990 interchange with neurophilosophers Paul and Patricia Churchland. The tactic worked to much less effect. See John R. Searle, "Is the Brain's Mind a Computer Program?" *Scientific American,* January 1990: 26–31; and Paul M. Churchland and Patricia Smith Churchland, "Could a Machine Think?" *Scientific American,* January 1990: 32–37.
26. Neurobiologists and connectionists could thus begin to share insights about the part that associative networks of neurons play in thinking. In this way, what seemed in some respects to be a black box theory was in others a decidedly nonbehaviorist attempt to unlock the biggest black box of all, the human brain. For a discussion of brain imaging that demonstrates these connectionist affiliations, see Marcus E. Raichle, "Visualizing the Mind," *Scientific American,* April 1994: 58–64. Raichle characterizes brain imaging—including techniques such as positron-emission tomography (PET) and magnetic resonance imaging (MRI)—as seeking to achieve the mapping of thought as it emerges from the activity of networks of neurons (p. 58). In other words, brain imaging spatializes thought and emergent processes in a way that speaks directly to the connectionist hypothesis that thought is accomplished through network associations in a highly parallel system. The author describes "a network of brain areas as a group of individuals in the midst of a conference call," who needed to know "who was speaking" and "who was in charge" in order to organize their conversation (p. 64).
27. See Lucy A. Suchman, *Plans and Situated Action: The Problem of Human-Machine Communication* (New York: Cambridge University Press, 1987).
28. Dreyfus and Dreyfus, "Making a Mind Versus Modeling the Brain," p. 26.
29. Dreyfus and Dreyfus, "Making a Mind Versus Modeling the Brain," p. 35.
30. Leo Marx, "Is Liberal Knowledge Worth Having?" Address to the Conference of the Association of Graduate Liberal Studies Programs, DePaul University, Chicago, 7 October 1988.
31. See Marvin Minsky, *The Society of Mind* (New York: Simon & Schuster, 1987), and Seymour Papert, *Mindstorms: Children, Computers, and Powerful Ideas,* 2nd rev. ed. (New York: Basic Books, 1993), pp. 167–70. If the purpose of this chapter were to describe the different levels at which one can posit emergence, it would have contrasted connectionism and society models. Connectionism or neural net approaches emphasize hardware; society theory emphasizes software or programming. However, since my central purpose is to contrast information processing with emergent approaches, it is appropriate to discuss society and connectionist models in terms of their similarities rather than their differences.
32. Mitchel Resnick, *Turtles, Termites, and Traffic Jams: Explorations in Massively Parallel Microworlds* (Cambridge, Mass.: MIT Press, 1994).
33. In *Mourning and Melancholia,* Freud argued that the sufferings of a melancholic arise from the ego's introjection of lost love objects toward which the self had ambivalence and, in particular, from mutual reproaches between the self and an internalized father with whom the self identifies. In this paper Freud described the "taking in" of "people" (in psychoanalytic parlance,

objects, and in this case the father) as part of a pathology, but he later came to the conclusion that this process is part of normal development. See Sigmund Freud, *The Standard Edition of the Complete Psychological Works of Sigmund Freud*, vol. 14, ed. and trans. James Strachey (London: Hogarth Press, 1960), pp. 237–58.

34. The language that psychoanalysts need for talking about objects—how they are formed, how they interact—is very different from the language they need for talking about drives. As in computer science, so in psychoanalysis. When one talks about objects, the natural metaphors have to do with making something, not carrying something.

35. Thomas H. Ogden, "The Concept of Internal Object Relations," *The International Journal of Psycho-Analysis* 64 (1983): 227. See Jay R. Greenberg and Stephen A. Mitchell, *Object Relations in Psychoanalytic Theory* (Cambridge, Mass.: Harvard University Press, 1983).

36. Fairbairn reframed the basic Freudian motor for personality development in object-relations terms. For Fairbairn, the human organism is not moved forward by Freud's pleasure principle, the desire to reduce the tensions to which the drive gives rise, but rather by its need to form relationships and internalize them to constitute a self.

37. My class on the psychology of technology was studying psychoanalytic approaches to mind. As an introduction to Freudian ideas I had assigned selections from Freud's *The Psychopathology of Everyday Life*. In this book Freud discusses parapraxis, or slips of the tongue. Freud takes as one of his examples a chairman of a parliamentary session who opens the session by declaring it closed. Freud's interpretation of this slip focuses on the complex feelings that may lie behind it. Is the chairman anxious about the session? Does he have reason to believe that nothing good will come of it? Would he rather be at home? The slip is presumed to tell us about the chairman's wishes. Its analysis lays bare Freud's notion of ambivalence—in this case, the chairman's mixed emotions about attending the session at all.

In the class discussion, one of my students, an MIT computer science major, objected to Freud's example. The chairman's slip was a simple technical matter. In this student's intellectual world, it was natural to code opposite concepts within a computer as the same root with a different "sign bit" attached (hot = − cold, dry = − wet, open = − closed). So if you think of the human mind as storing information in a computer-like memory, substituting "closed" for "open" is easily justified. Or as the student succinctly put it, "A bit was dropped—the sign bit. There might have been a power surge. No problem." For him, understanding the slip required no recourse to ideas of ambivalence or hidden wishes. What Freud had interpreted in terms of emotionally charged feelings, as a window onto conflicts, history, and significant relationships, was for him a bit of information lost or a program derailed. A Freudian slip had become a simple information processing error. What psychoanalysis would interpret broadly in terms of meaning, this student was able to see narrowly in terms of mechanism. For an example of an information processing perspective on the Freudian, see Donald Norman, "Post-Freudian Slips," *Psychology Today*, April 1980: 41–44ff.; Norman, *Slips of the Mind and an Outline of a Theory of Action* (San Diego: Center for Human Information Processing, University of California, November 1979); and Norman, "Categorization of Action Slips," *Psychological Review* 88 (January 1981): 1–15.

At the time of my student's clash with Freudian thinking, his model of computational intelligence was information processing AI. What makes the student's idea about the slip ("a power surge") conflict radically with psychoanalysis is not so much that power surging is alien to psychoanalytic thinking but that any single factor as simple as a sign bit could explain an act of language. Information processing AI supported a view that could be called narrow determination—one bit dropped *could* make a difference. In contrast, psychoanalysis has a logic that calls the whole person into play to explain all his or her actions. It supports an outlook of broad determination. This is why an individual can use something as small as a verbal slip to get in touch with the deepest levels of personality. The slip is seen as a window onto complexity.

In traditional logic, when you say, "All men are mortal; Socrates is a man; therefore Socrates is mortal," your conclusion is determined by two premises. Change one, and you get a new conclusion. Similarly, with an information processing computer model, you drop one bit, one piece of information, and you get a new output. The determination is narrow, like a highway with one lane. Psychoanalysis uses broad determination. It is based on another kind of logic, more like the logic that leads you to say that Shakespeare is a great poet. Coming across a bad poem by Shakespeare does not_call the statement into question. Nor would the discovery that several of Shakespeare's best poems were written by someone else. So even if you learned that the chairman who announced the meeting was closed had a wife at home who was ill, her illness and his desire to be at home would not determine his slip in any simple sense. Psychoanalytic phenomena are as "overdetermined" as judgments of literary merit. Although popular images of a psychoanalytic dream book abound—along with a history of popularizers attempting to write one—there is no such thing as a dictionary of Freudian symbols. The meaning of a dream can only be deciphered from the complex fabric of a particular dreamer's associations to it.

Emergent AI, however, disrupted this easy opposition between computational and psychoanalytic models. While information processing gave concepts like "closed" and "open" actual symbolic representation in a computer, the building blocks of emergent AI do not have that kind of one-to-one relationship with such high-level concepts. What is stored in the computer is data about the relationships among agents who are expected to recreate given patterns. Emergent AI, like psychoanalysis, relies on wide determination. In this kind of system it is not possible for one bit dropped or one rule changed to make a difference to an outcome. In emergent systems, probabilities take the place of algorithms and statistics take the place of rules.

The transition from information processing to emergent AI thus brought about an important change in ideas about determinism. They went from narrow to wide. And with this change, a crucial piece of the wall between psychoanalysis and artificial intelligence was broken down. What is more, the specific way that emergent AI achieved its models of broad determination brought another and even more decisive point of contact between the two fields. Psychoanalysis uses a language of multiple objects and agents to describe our inner landscape. Emergent AI does so as well.

38. The object-relations school (arguably closer to Freudian aspirations to transparency than the Lacanian) is resonant with that theory of emergent AI that offers the clearest view of the entities behind emergence; this is Minsky's society of mind. Society theory shares strengths and weaknesses with the object-relations tradition. The strength is a conceptual framework that offers rich possibilities for interactive process. The weakness is that the framework may be too rich. The postulated objects may be too powerful if they simply explain the mind by saying that there are many minds within it. Psychoanalytic object-relations theorists struggle with this thorny and unresolved issue. Within the field much of the criticism of overpowerful inner objects has been directed at Melanie Klein. For example, Roy Schafer has argued that Klein and the English School of object relations have carried the reification implicit in Freudian metapsychology to a "grotesque extreme": "A multitude of minds is introduced into a single psychic apparatus. . . . The person is being envisaged as a container of innumerable, independent microorganizations that are also microdynamisms." See Roy Schafer, *A New Language for Psychoanalysis* (New Haven, Conn.: Yale University Press, 1976), p. 3; and Schafer, *Aspects of Internalization* (New York: International University Press, 1968), p. 62. Essentially, Klein's critics feel that her idea of "inner idealized figures protecting the ego against terrifying ones is tantamount to proposing that there are internal friendly and hostile 'demons' operating within the mind." See Ogden, "The Concept of Internal Object Relations": 229.

But the problem of infinite regress in accounting for the entities that are then to account for thought has a very different cast in the history of the psychoanalytic movement than it does in the computer science tradition. In psychoanalytic circles the problem of the conveniently appearing objects carries the negative connotation of fuzzy thinking and circular reasoning. But computer scientists are used to relying on a controlled form of circular reasoning—recursion—as a powerful technical tool. Computer science has built a mathematical culture that relies heavily on defining things in terms of themselves. Most of us learned at school to define x^n as x multiplied by itself n times. Power is defined in terms of multiplication. Computer scientists prefer to implement the power function defining x^n as $(x^{n-1})x$. Power is defined in terms of power. The computational aesthetic of recursive thought has been expressed in a poetic and accessible form by Douglas R. Hofstadter, who presents recursive phenomena as a source of power in Bach's music and Escher's art as well as in Gödel's mathematics. See *Gödel, Escher, Bach: An Eternal Golden Braid* (New York: Basic Books, 1979). To put it more sharply, computational models suggest that a way out of the problem of infinite regress is to redefine the problem as a source of power.

One could imagine computationally oriented psychoanalytic theorists finding, in the recursive idea that thoughts might be capable of thinking, an aesthetically pleasing virtue rather than a devastating vice. One could imagine their seeing a reliance on recursion as a source of legitimation rather than as a sign of weakness. In the spirit of George Miller's account of computer memory and behaviorism, psychoanalysts might find it embarrassing to deny human thoughts the ability to think, when in the next generation of society models, "computer thoughts" are presumed to do so.

Of course, computational models have not solved the problem of accounting for objects—what they are and how they come into being. There is something deeply unsatisfying in a theory that cannot go beyond assuming a homunculus within the human, for how then do we explain the homunculus without postulating yet another one within it, and so on? The same is true for a theory that waves off what W. Daniel Hillis called inscrutability by confidently asserting that emergence will take care of things—unless of course, it succeeds in doing that.

39. Minsky, *The Society of Mind,* p. 33.
40. Terry Winograd has accused Minsky of an intellectual "sleight of hand" because, "in a few sentences Minsky has moved from describing the agents as subroutines to speaking of them as entities with consciousness and intention." See Terry Winograd, "Thinking Machines: Can There Be? Are We?" in *The Boundaries of Humanity: Humans, Animals, Machines,* eds. James J. Sheehan and Morton Sosna (Berkeley: University of California Press, 1991), pp. 198–221.
41. David D. Olds, "Connectionism and Psychoanalysis," *Journal of the American Psychoanalytic Association* 42 (1994): 581–612.
42. See Russell Jacoby, *Social Amnesia: A Critique of Conformist Psychology from Adler to Laing* (Boston: Beacon Press, 1975).
43. Olds, "Connectionism and Psychoanalysis": 600. Olds notes a historical similarity between psychoanalysis and artificial intelligence. Information processing AI, like early psychoanalysis, put its emphasis on inner cognitive processes, while connectionism and current trends in psychoanalysis look toward perception, the connections between self and world. The computer scientists have moved away from programming an inner system, and the analysts have moved away from stressing inborn tendencies and the enactment of conflicts established at an early age—things that Olds compares to inner programs. "In the past decade," he remarks, "analysis has begun to place more importance on the interpersonal field where perception is more important than program." (p. 603)
44. Olds, "Connectionism and Psychoanalysis": 590.
45. The connectionist neuroscientist David Rumelhart spoke at the 1992 meeting of the American Psychoanalytic Association to an enthusiastic response. And psychoanalyst Stanley Palombo uses connectionist theory to account for how the dream can rapidly match past and present images. He believes that dreams serve the function of storing short-term memories (day residues) in long-term memory; the manifest dream is an epiphenomenon of this process. (See Stanley Palombo, "Connectivity and Condensation in Dreaming," *Journal of the American Psychoanalytic Association* 40 [1992]: 1139–59.) The psychoanalyst Lee Brauer believes that the connectionist model "can account

for some aspects of 'working through,' by changing images which are generalized without explicit conscious thought." He sees an analogue to object-oriented programs "in which a graph linked to a paper in a word processor may be altered by rerunning the graph in the statistics program." (Personal communication, 3 May 1993.)

46. Bruno Latour, *The Pasteurization of France*, trans. by Alan Sheridan and John Law (Cambridge, Mass.: Harvard University Press, 1988).

47. James L. McClelland and David E. Rumelhart, *Explorations in Parallel Distributed Processing: A Handbook of Models, Programs and Exercises* (Cambridge, Mass.: MIT Press, 1988), p. 10.

48. McClelland and Rumelhart, *Explorations in Parallel Distributed Processing,* p. 10.

49. Papert, "One AI or Many?", p. 14.

50. Cited in Dreyfus and Dreyfus, "Making a Mind Versus Modeling the Brain," pp. 24–25.

51. Kenneth J. Gergen, *The Saturated Self: Dilemmas of Identity in Contemporary Life* (New York: Basic Books, 1992).

52. See Turkle, *The Second Self,* especially Chapter 8, "Thinking of Yourself as a Machine," pp. 271–305.

53. The student who thought of his mind as just a lot of little processors used a hardware metaphor to describe his consciousness. He said it was like a printer. However, when he talked about how he had developed his theory, he spoke about introspection in a language that reintroduced a thinking subject for which his theory, when rigorously stated, had no room. When talking about himself, he reintroduced a self despite himself.

54. Minsky, *The Society of Mind,* p. 182.

55. Minsky, *The Society of Mind,* p. 182.

56. Marvin Minsky, "The Emotion Machine" (Lecture delivered at Massachusetts Institute of Technology, Cambridge, Mass., 16 October 1993).

57. Fredric Jameson, "Postmodernism, or The Cultural Logic of Late Capitalism," *New Left Review* 146 (July–August 1984): 43–44.

CHAPTER 6   ARTIFICIAL LIFE AS THE NEW FRONTIER

1. Richard Dawkins, "Accumulating Small Change," in *The Blind Watchmaker* (New York: W. W. Norton and Co., 1986), pp. 43–76.

2. Dawkins, *The Blind Watchmaker,* pp. 59–60.

3. Dawkins, *The Blind Watchmaker,* p. 60.

4. Norbert Wiener, *God and Golem, Inc.: A Comment on Certain Points Where Cybernetics Impinges on Religion* (Cambridge, Mass.: MIT Press, 1964), p. 17.

5. Wiener, *God and Golem, Inc.*, p. 5.

6. *Simple Rules Yield Complex Behavior: A 13-Minute Introduction,* produced and directed by Linda Feferman. 13 minutes. The Santa Fe Institute, 1993. Videocassette.

7. J. Doyne Farmer and Alletta d'A. Belin, "Artificial Life: The Coming Evolution," in *Artificial Life II: The Proceedings of the Workshop on Artificial Life* ed.

Christopher G. Langton et al., Santa Fe Institute Studies in the Science of Complexity, vol. 10 (Redwood City, Calif.: Addison-Wesley, 1992), p. 815. [Hereafter, this volume will be referred to as *Artificial Life II.*]

8. Christopher G. Langton, "Artificial Life" in *Artificial Life: The Proceedings of an Interdisciplinary Workshop on the Synthesis and Simulation of Living Systems,* ed. Christopher G. Langton, Santa Fe Institute Studies in the Science of Complexity, vol. 6 (Redwood City, Calif.: Addison-Wesley, 1989), pp. 43, 44. [This volume will be hereafter referred to as *Artificial Life I.*]

9. Steven Levy, *Artificial Life: The Quest for a New Frontier* (New York: Pantheon, 1992), pp. 7–8. My narrative of the origins of A-Life life draws on Levy's book, on Mitchell Waldrop's *Complexity: The Emerging Science at the End of Order and Chaos* (New York: Simon & Schuster, 1992), and on Langton, "Artificial Life," in Langton, ed., *Artificial Life I.*

10. Langton, "Artificial Life," in Langton, ed., *Artificial Life I,* p. 2.

11. My thanks to Seymour Papert and Mitchel Resnick for helping me think through this issue.

12. Shortly before his death in 1954, von Neumann had investigated the prospects for a machine that could self-replicate. In doing so, he was inspired by Alan Turing's idea about finite state machines. A finite state machine changes its state in discrete time-steps, taking account of its inner state, incoming information, and a set of simple rules that contain instructions for what to do in every possible case with which the machine could be presented. Turing proved that such a machine, if provided with an infinitely long tape onto which the finite state machine can write/read, would be able to emulate the actions of any other machine. And with the philosopher Alonzo Church, he boldly hypothesized that not only mathematical machines but any aspect of nature can be represented by such a "universal Turing machine." In the 1940s, Warren McCulloch and Walter Pitts suggested that the nervous systems of living organisms could be emulated by a neural network, itself emulatable by a Turing machine (Warren McCulloch and Walter Pitts, "A Logical Calculus of the Ideas Immanent in Nervous Activity," *Bulletin of Mathematical Biophysics* 5 [1943]: 115–33). Using these ideas as points of departure, von Neumann sketched out a fantasy of an artificial creature with a genetic blueprint. He imagined a robot "factory" that lived on a virtual lake. It could gather up components from the lake and use them to fashion copies of itself. When released into the lake, these copies would repeat the process of self-replication. Arbitrary alterations in the system would be analogous to mutations in living systems. They would most often be lethal, but in some cases the mutations would cause the factory to continue to reproduce "but with a modification of traits."

Von Neumann's ideas anticipated the discovery of DNA; his imagined factory kept its reproductive code on an "information tape." And he anticipated one of the fundamental tenets of artificial life research; life was grounded not only in information but in complexity. Von Neumann sensed that once a certain critical mass of complexity was reached, objects would reproduce in an open-ended fashion, evolving by parenting more complicated objects than themselves. In this way, at least on a local level, life defies the second law of thermodynamics, which says that as time passes, energy dissipates and order deteriorates. Life, in contrast, seems to create order over time (John von Neumann, "General and Logical Theory of Automata," in *John von Neumann: Collected Works*, ed. A. H. Taub [Elmsford, N.Y.: Pergamon, 1963], p. 318).

The mathematician Stanislaw Ulam suggested how von Neumann could radically simplify his model. Von Neumann took Ulam's advice and reworked the robot factory into what came to be known as the first cellular automaton. It contained a cell structure embedded on a grid and a rule table that told each individual cell how to change its state in reference to its state and those of its neighbors. As time went on, the original cell structure transformed its neighboring cells into the states that had made it up originally. In other words, the original cell structure (or organism) eventually duplicated itself on the grid. In a further development of the idea, small random changes, analogous to mutations, were allowed to occur. These changes might be passed on to future generations. If a given change increased the organism's fitness, that mutation would flourish and something like evolution would have begun.

Fascinated by von Neumann's work, Conway set out to simplify it, to build the simplest cellular automaton he could.

13. Quoted in Levy, *Artificial Life*, p. 57.
14. Gaston Bachelard, *The Psychoanalysis of Fire* (New York: Quartet, 1987 [1964]).
15. Douglas R. Hofstadter, *Metamagical Themas: Questing for the Essence of Mind and Pattern* (New York: Basic Books, 1985), pp. 492–525.
16. Hofstadter, *Metamagical Themas*, p. 506.
17. Hofstadter, *Metamagical Themas*, p. 506.
18. Richard Doyle, "On Beyond Living: Rhetorics of Vitality and Post Vitality in Molecular Biology" (Ph.D. diss., University of California at Berkeley, 1993), p. 22. Doyle points out that in writing about the air pump, the historians of science Simon Schaffer and Steven Shapin divide the literary/descriptive side of technology from the material technology itself. In their work, says Doyle, ". . . narrative serves as a kind of supplement to the material technology of the air pump, framing it in a coherent and persuasive fashion so that others might be convinced of Boyle's finding at a distance, in the absence of the pump or of Boyle" (p. 5). But Doyle argues that narration often exists prior to material technology and to scientific experiments. In these cases, the narration provides both the organizing metaphors and the history for an experimental project. In Doyle's view, such narration was "a part of the network of power and thinking that made Boyle's project possible" in the first place. It is in this sense that "the rhetorics of code, instruction, and program materialized beliefs into sciences and technologies," including those of artificial life (p. 5).
19. François Jacob, *The Logic of Life: A History of Heredity*, trans. Betty E. Spillmann (New York: Pantheon, 1973), p. 306. Cited in Doyle, "On Beyond Living," p. 23.
20. On this, the work of Jacques Derrida is of foundational importance. See, for example, *Dissemination*, trans. Barbara Johnson (Chicago: University of

Chicago Press, 1981), *Of Grammatology,* trans. Gayatri Chakravorty Spivak (Baltimore: Johns Hopkins University Press, 1976), and *Writing and Difference,* trans. Alan Bass (Chicago: University of Chicago Press, 1978).

21. Cited in Levy, *Artificial Life,* p. 95.

22. Cited in Levy, *Artificial Life,* p. 95.

23. See Ed Regis, *Great Mambo Chicken and the Transhuman Condition: Science Slightly over the Edge* (Reading, Mass.: Addison-Wesley, 1990), p. 193. Cited in Doyle, "On Beyond Living," p. 217. See also Levy, *Artificial Life,* p. 96.

24. Thomas Kuhn, *The Structure of Scientific Revolutions,* 2nd rev. ed. (Chicago: University of Chicago Press, 1970 [1962]).

25. The evolution of the biomorphs dramatized how efficiently evolutionary methods can search "genetic space." The evolutionary methods were excellent for finding the possible outcomes of evolutionary processes. Genetic algorithms could use evolution by unnatural selection to search "problem space," the possible solutions to any given task.

26. After the process of chromosome selection, the user of the program (analogous to the breeder of Dawkins's biomorphs) is left with a new, smaller chromosome population, typically 10 percent of the original number, with an overrepresentation of those chromosomes that were best able to sort numbers. These chromosomes are multiplied by 10 in order to keep the overall population size the same as before. So, if one began with 100 chromosomes, 10 would be selected for reproduction and these 10 would then be copied 10 times to get back to 100. At this point, the 100 chromosomes are mated. Each chromosome is randomly paired with another, and, as in nature, some of the genetic material on each crosses over and becomes attached to the chromosome of its partner. (In crossover, fragments of genetic material —in the case of the algorithm, 0's and 1's—are torn off one chromosome and put on another.) Finally, as in nature, a small degree of mutation is added to the system. A small percentage of the 0's are changed to 1's and 1's to 0's. Now, there is a new population of 100 chromosomes, and this new population is asked to do the original number-sorting problem. The winner chromosomes are once again given pride of place when it is time to select for the next generation. Some loser chromosomes are let in, too. The selected 10 percent are copied, they are mated, there is crossover, there are random mutations, and so on.

27. The differences between researchers in artificial life and researchers in natural biology in their views of the importance of crossover in evolution illustrate a significant difference between artificial life research and an earlier tradition of modeling biological systems on computers. When biology had provided the accepted doctrine and computer scientists had simply been asked to model it, the computer scientists had omitted crossover because the biologists hadn't felt it was critical. Only now, with artificial life setting an independent research agenda in the context of its own experimental settings (virtual ecologies within computers), has the power of crossover been laid on the table.

28. For descriptions of the work on ramps, see W. Daniel Hillis, "Co-Evolving

Parasites Improve Simulated Evolution as an Optimization Procedure," in Langton et al., eds., *Artificial Life II*, pp. 313–24; and Hillis, "Punctuated Equilibrium Due to Epistasis in Simulated Populations," unpub. ms., 1991.

29. Hillis's work on evolution became well known in the late 1980s. Its implications for the future of programming provoked strong reaction. For some young programmers, contemplating programs that evolve into expert programmers felt too humbling. "My whole identity is as a programmer," said Jamal, an MIT sophomore majoring in Electrical Engineering and Computer Science. "It is one thing for a machine to take over on an assembly line or even to be better than people at chess. Reading Hillis makes me feel that I am going to be obsolete the way robots make guys on factory floors obsolete."

While one reaction was to feel devalued as a programmer, another was to devalue programming itself. This reaction was analogous to those that followed the appearance of chess-playing computer programs. When only people could play chess, chess was seen as an activity that called upon the most exalted of human intellectual powers. When computers began to play chess, the first reaction was to insist that they could never be as good as people because chess required a kind of creativity to which only human players had access. But as the chess programs became better than all but a handful of people, chess itself became devalued. If computers could do it, it couldn't be all that special. Now, as computers showed promise as programmers, programming, once valued as esoteric and creative, was recast as something close to undignified or, at best, too much of a bother for the really brilliant people to trouble with. "I'm in computer science so that I can be the master of the machines that make things happen," said one of Jamal's classmates. "So I don't care personally in terms of ego if I am not the actual programmer. But there is an ethical problem there, because we may not understand the programs that the machines are evolving."

30. Cited in Levy, *Artificial Life*, p. 210.

31. Tierran organisms have blocks of computer instructions called electronic templates. Reproduction required that templates be matched to complementary (opposite) templates. Because Tierra was well stocked with potential matching templates, even mutated organisms with altered templates could easily reproduce.

32. Actually, the computer was a virtual computer. When Ray first showed his work to Christopher Langton and James Doyne Farmer at the Santa Fe Institute, they told him that he ran the risk of inadvertently creating a dangerous computer virus. They convinced him to place his creatures in a virtual "cage" by building a simulated computer within his computer. Then he could create organisms that could only function inside this virtual machine. Ray's "close call" evokes fantastic images of computer-generated lifeforms out of control. The kind of mythic fears and concerns formerly associated with AI may be beginning to be displaced onto artificial life. Rodney Brooks, for one, believes that Ray's organisms, even in a virtual machine, are potentially dangerous.

33. When Ray switched off the background cosmic ray mutations, diversity emerged as surely as before. What seemed to be driving evolution was the

way Tierran parasites tampered with their hosts' genomes. This discovery depended on Ray's being able to look inside the genome of successive generations of Tierrans, just as Hillis had been able to look inside successive generations of ramps. Artificial life was once again providing evidence with which traditional biology could evaluate the idea that the evolution of organic complexity owed much to the emergence of parasites.

34. Thomas Ray, Presentation at Artificial Life III: Third Interdisciplinary Workshop on the Synthesis and Simulation of Living Systems, Santa Fe, New Mexico, June 1992.

35. Herbert A. Simon, *The Sciences of the Artificial* (Cambridge, Mass.: MIT Press, 1969), p. 24.

36. Craig Reynolds, "Flocks, Herds, and Schools: A Distributed Behavioral Model," *Computer Graphics* 21 (July 1987): 27.

37. This phrase is drawn from Mitchel Resnick's work. See his *Termites, Turtles, and Traffic Jams: Explorations in Massively Parallel Microworlds* (Cambridge, Mass.: MIT Press, 1994).

38. Reynolds, "Flocks, Herds, and Schools," p. 32.

39. The idea of "simulated natural life," with its complex chaining of real to artificial and back, had been central to the method of the botanist Aristid Lindenmayer. Since the 1970s Lindenmayer had been using mathematical formulas to describe the physical development of plants. Lindenmayer developed a system of algorithms, called L-systems, which were able to generate the growth pattern of particular algae. Other biologists added a graphical dimension to Lindenmayer's work, so that L-systems could generate pictures of botanical forms. When L-systems were used to simulate the development of fern embryos, the embryo grew and divided cells depending on which combinations of cell neighbors pushed against the wall of a given cell. The result of these simple local rules was a virtual fern embryo that was similar to photographs of the real thing taken under a microscope. These successes indicated that the self-organizing principles in nature could at least sometimes be embodied in the pure realm of mathematics and computation. (Again, however, we need to remember that there is no indication that ferns use the same self-organizing principles as L-systems.)

Like Reynolds's boids, the images produced by L-systems were strangely compelling. The aphorism about what pictures are worth held true. L-systems began to interest some natural scientists in an artificial life perspective. Additionally, L-systems opened up a branch of A-Life that began to explore how a single artificial cell might be grown into an actual being, that is, the field began to explore an embryological rather than an evolutionary path to life.

40. Robert Malone, *The Robot Book* (New York: Push Pin Press/Harcourt Brace Jovanovich, 1978), pp. 12–13.

41. Some of these figures—such as Jacques de Vaucanson's 1738 mechanical duck (it walked, swam, flapped its wings, ate, digested, and excreted)—are still present enough in our mythology that they appear as cover illustrations on modern books about artificial life. See, for example, Jean-Arcady Meyer and Stewart W. Wilson, eds., *From Animals to Animats: Proceedings of the*

First International Conference on Simulation of Adaptive Behavior (Cambridge, Mass.: MIT Press, 1991).

42. The Logo language had originally been designed in the 1960s to command a "floor turtle." Papert encouraged children to see floor turtles as creatures and to think about programming as providing them with instructions. When children had questions about what instructions to give, they were encouraged to play turtle themselves. So, for example, if a child asked, "How do you get a turtle to draw a circle?" the answer would be, "Play turtle. You be the turtle." Then, the child, taking the role of the turtle, would walk around in a circle. In the ideal case, the child would figure out that from the point of view of the turtle, a circle emerges when you walk forward a little, turn a little, walk forward a little, turn a little, and so forth. In this way, breaking down the boundaries between the self and a computational object became part of a method for teaching people how to program and to learn about geometry. Papert called such methods body syntonic learning and believed that much of Logo's appeal depended on the way it inspired a physical and emotional identification with the robotic turtle.

    By the late 1980s, however, technological advances made it possible for children to use "screen turtles," which could accept more complicated instructions than the floor turtles. Although children could identify with the screen representations, the Lego-Logo robots returned Logo research to more direct exploration of children's body-to-body identification with cybernetic creatures. See Seymour Papert, *Mindstorms: Children, Computers, and Powerful Ideas,* 2nd rev. ed. (New York: Basic Books, 1993 [1980]). Also see, Seymour Papert, *The Children's Machine: Rethinking School in the Age of the Computer* (New York: Basic Books, 1993).

43. Cited in Levy, *Artificial Life,* p. 286.

44. Langton, "Artificial Life," in Langton, ed., *Artificial Life I,* p. 13.

45. Resnick, *Turtles, Termites, and Traffic Jams,* pp. 131–32.

46. Consider a program in StarLogo that uses local and emergent methods to get a population of termites (screen turtles under a different name) to stockpile woodchips. A first rule, sent out to all termites: If you're not carrying anything and you bump into a woodchip, pick it up. A second rule, also sent out to all termites at the same time as the first: If you're carrying a woodchip and you bump into another woodchip, put down the chip you're carrying. These two rules when played simultaneously on a population of termites wandering around a neighborhood strewn with wood chips will cause the termites to make woodchip piles. (Resnick, *Turtles, Termites, and Traffic Jams,* p. 76.)

    Yet when exposed to the results of this program, Resnick, found people will go to great lengths to back away from the idea that the seemingly purposive behavior of the termites is the result of decentralized decisions. People either assume that there is a leader who has stepped in to direct the process or that there was an asymmetry in the world that gave rise to a pattern, for example, a food source near the final location of a stockpile. (See Resnick, *Turtles, Termites, and Traffic Jams,* pp. 137ff.)

    In the Lego-Logo context, Resnick had developed a programming language called MultiLogo based on agents that communicated with one another. "In using the language," he reports, "children invariably put one of the agents 'in charge' of the others. One student explicitly referred to the agent in charge as 'the teacher.' Another referred to it as 'the mother.' " See Mitchel Resnick, "MultiLogo: A Study of Children and Concurrent Programming," *Interactive Learning Environments* 1, no. 3 (1990): 153–70.

47. Cited in Resnick, *Termites, Turtles, and Traffic Jams*, p. 122.

48. Michael Bremer, *SimAnt User Manual* (Orinda, Calif.: Maxis, 1991), p. 163.

49. Bremer, *SimAnt User Manual*, p. 164. The game manual talks about Douglas Hofstadter's *Gödel, Escher, Bach*, in which an ant colony is explicitly analogized to the emergent properties of the brain. See Douglas R. Hofstadter, *Gödel, Escher, Bach: An Eternal Golden Braid* (New York: Basic Books, 1979). SimAnt wants to make the same point by giving people something concrete to play with.

50. Michael Bremer, *SimLife User Manual* (Orinda, Calif.: Maxis, 1992), p. 2.

51. Bremer, *SimLife User Manual*, p. 6.

52. Bremer, *SimLife User Manual*, pp. 6–7.

53. In Philip K. Dick's science fiction classic, *Do Androids Dream of Electric Sheep?*—the novel on which the film *Blade Runner* was based—a devastating war destroys most biological life. In the aftermath highly developed androids do much of the work. "Natural" biology takes on new importance. In this world, a real insect or frog or fish becomes a valuable and cherished pet. The police are occupied with preserving the boundary between real and artificial life.

54. This term refers not just to Power Rangers but to a computer program that metamorphosed one pictorial image, usually a face, into another.

55. Mitchel Resnick, "Lego, Logo, and Life," in Langton, ed., *Artificial Life I*, p. 402.

56. Levy, *Artificial Life*, pp. 6–7.

57. Peter Kramer, *Listening to Prozac: A Psychiatrist Explores Antidepressant Drugs and the Remaking of the Self* (New York: Viking, 1993), p. xii.

58. Kramer, *Listening to Prozac*, p. xiii.

CHAPTER 7   ASPECTS OF THE SELF

1. Marshall McLuhan, *The Gutenberg Galaxy* (Toronto: University of Toronto Press, 1962).

2. David Riesman, Nathan Glazer, and Reuel Denney, *The Lonely Crowd* (New York: Doubleday Anchor, 1950).

3. Kenneth J. Gergen, *The Saturated Self: Dilemmas of Identity in Contemporary Life* (New York: Basic Books, 1991); Robert Jay Lifton, *The Protean Self: Human Resilience in an Age of Fragmentation* (New York: Basic Books, 1993); Emily Martin, *Flexible Bodies: Tracking Immunity in American Culture from the Days of Polio to the Age of AIDS* (Boston: Beacon Press, 1994).

4. In a March 24, 1992, posting to a WELL (Whole Earth 'Lectronic Link computer bulletin board system) conference called "On Line Personae: Boon or Bête Noire?" F. Randall Farmer noted that in a group of about 50 Habitat users (Habitat is a MUD with a graphical interface), about a quarter experienced

their online persona as a separate creature that acted in ways they do in real life, a quarter experienced their online persona as a separate creature that acted in ways they do not in real life. A quarter experienced their online persona not as a separate creature but as one that acted like them; and another quarter experienced their online persona not as a separate creature but as one which acted in ways unlike them. In other words, there were four distinct and nonoverlapping groups.

5. Katie Hafner, "Get in the MOOd," *Newsweek,* 7 November 1994.

6. Frank Herbert, *Dune* (Philadelphia: Chilton Books, 1965).

7. Erik Erikson, *Childhood and Society,* 2nd rev. ed. (New York: Norton, 1963 [1950]), p. 52.

8. This communication was signed "The Picklingly herbatious one."

9. For more material making the contrast with traditional role playing see Gary Alan Fine, *Shared Fantasy: Role Playing Games as Social Worlds* (Chicago: The University of Chicago Press, 1983). Henry Jenkins's study of fan culture, *Textual Poachers: Television Fans and Participatory Culture* (New York: Routledge, 1992), illuminates the general question of how individuals appropriate fantasy materials in the construction of identity.

10. Clifford Geertz, *Local Knowledge: Further Essays in Interpretive Anthropology* (New York: Basic Books, 1983), cited in Gergen, *The Saturated Self,* p. 113.

11. Players may take out Internet accounts under false names in order to protect their confidentiality. One male player explains that he has created a female persona for the character who "owns" the Internet account from which he MUDs. This way, when he plays a female character and someone finds out the name of the person behind it, that person is a woman. He feels that given the intimacy of the relationships he forms on the MUDs with male-presenting characters, the deception is necessary. "Let's say my MUD lover [a man playing a man] finds out I am really a man. We might still continue our relationship but it would never be the same."

12. Defenses are usually born of situations where they were badly needed, but they can take on a life of their own. Our situations may change but the defense remains. For example, while Stewart was growing up it may have been functional for him to avoid dwelling on his difficulties, but the wholesale denial of his feelings is not serving him well. Therapy aims to give patients greater flexibility in dealing with the world as it is and a sense of distance, even a sense of humor about their particular ways of distorting things. If one can look at one's behavior and say, "There I go again," adjustments and compensations become possible, and one can see one's limitations but maintain one's self-esteem.

13. "The adolescent mind is essentially a mind of the *moratorium. . . .*" Erikson, *Childhood and Society,* p. 262. For the working out of the notion of moratorium in individual lives, see Erikson's psychobiographical works, *Young Man Luther: A Study in Psychoanalysis and History* (New York: Norton, 1958) and *Gandhi's Truth: On the Origins of Militant Nonviolence* (New York: Norton, 1969).

14. Erikson, *Childhood and Society,* p. 222.

15. This point about the advantage of keeping RL and virtual personae close

together has been made by many people I spoke with. It illustrates the diversity of ways people use online personae to "play" other selves or aspects of themselves. In a WELL conference called "Online Personae: Boon or Bête Noire?" Paul Belserene noted, both for himself and for others,

that this conscious persona-building tends to be cheap fuel. After a few months, some things happen that tend to moderate these personae toward your "real" personality. For one thing, the psychic energy required gets expensive. And for another, if you're like most people, you begin to care about the community and people here, and you tend to want to express yourself, to be seen.

But this author felt somewhat differently about what he called "personae-shading from conference-to-conference." About that he said, "We are all many people, and like to express parts of our wholes when we can. But the creation of extreme, artificial personae is, I think, very hard to maintain on a system like the WELL. (Maybe other systems are easier to fool—harder to care about.)"

For some people, however, it is precisely because they care that they want to be very different from what they feel themselves to be. They may not, however, have as much success with integrating their online personae with their sense of an off-line self. Belserene went on to note, "The persistence of my online persona's behavior makes it easier to learn from experience. I've learned a lot about manners, diplomacy, and so on from the WELL. Much of that has spilled over into my 'real-life' persona, which is good." (paulbel [Paul Belserene], The WELL, conference on virtual communities [vc.20.10], 6 April 1992).

16. jstraw (Michael Newman), The WELL, conference on virtual communities (vc.20.26), 25 May 1992.

17. Keeping logs of conversations in which you participate for your own purposes is socially acceptable on MUDs. Sharing them publicly without permission is considered bad manners.

18. In his influential essay "Thick Description," Clifford Geertz argued that the anthropologist never reports raw social discourse. Geertz comments that even reporting a wink implies interpretation, because it means one has already decided that a given contraction of the eyelid was a wink rather than a twitch. And since any wink can be ironic or suggestive, the way in which one reports the wink constitutes an interpretation. In MUDs, the exact form of every communication can be captured in a log file. But the elusiveness of social discourse to which Geertz referred is not pinned down by this technological possibility. Clifford Geertz, "Thick Description: Toward an Interpretive Theory of Culture;" in *The Interpretation of Cultures* (New York: Basic Books, 1973), pp. 3–30.

CHAPTER 8   TINYSEX AND GENDER TROUBLE

1. At the time, I noted that I felt panicky when female or female-presenting characters approached the gender-neutral "me" on the MUD and "waved seductively." And I noted this with considerable irritation. Surely, I thought,

my many years of psychoanalysis should see me through this experience with greater equanimity. They did not.

2. Pavel Curtis, "Mudding: Social Phenomena in Text-Based Virtual Realities," available via anonymous ftp://parcftp.xerox.com/pub/MOO/papers/DIAC92.*. Cited in Amy Bruckman, "Gender Swapping on the Internet," available via anonymous ftp://media.mit.edu/pub/asb/paper/gender-swapping.*.

3. Allucquere Rosanne Stone, Presentation at "Doing Gender on the 'Net Conference," Massachusetts Institute of Technology, Cambridge, Mass., 7 April 1995.

4. The term "gender trouble" is borrowed from Judith Butler, whose classic work on the problematics of gender informs this chapter. See Judith P. Butler, *Gender Trouble: Feminism and the Subversion of Identity* (New York: Routledge, 1990).

5. My thanks to Ilona Issacson Bell for pointing me to this rich example.

6. William Shakespeare, *As You Like It*. Act I, Scene 3. Lines 107–18.

7. Zoe does not MUD any more. She gave me two reasons. First, her MUDding succeeded in making her more assertive at work. Second, she doesn't want her MUDding to succeed in making her "too much" more assertive at home.

I guess I got what I wanted out of MUDs. When I go to work I try to act like my MUD character, but that character is really a part of me now. Well, more like a role model that I've had as a roommate. Not just as a teacher, but [someone] I actually lived with. For two years I did Ulysses for thirty hours a week, so it isn't so hard to do it for a few hours a week during meetings at work or on the phone with clients. But I didn't go all the way with Ulysses. It started to feel dangerous to me. My marriage is still pretty traditional. I am better at talking about my feelings and I think my husband respects me, but he still is Southern. He still likes the feeling of being superior. We need the money so my husband doesn't mind my working. But I do treat my husband more or less the way my father would have wanted me to. I want to have children. If I brought Ulysses home, it would upset my marriage. I don't want that to happen. I'm not ready for that now. Maybe someday, but not now.

8. With the increasing popularity of MUDding, this group has split up into many different groups, each looking at different aspects of MUDding: administrative, technical, social.

9. People feel different degrees of "safety." Most MUDders know responsibility involves not logging sexual encounters and then posting them to public bulletin boards.

On an Internet bulletin board dedicated to MUDding, a posting of "Frequently Asked Questions" described TinySex as "speed-writing interactive erotica" and warned players to participate with caution both because there might be some deception in play and because there might be the virtual equivalent of a photographer in the motel room:

Realize that the other party is not obligated to be anything like he/she says, and in fact may be playing a joke on you (see 'log' below).

"What is a log?"

Certain client programs allow logs to be kept of the screen. A time-worn and somewhat unfriendly trick is to entice someone into having TinySex with you, log the proceedings, and post them to rec.games.mud and have a good laugh at the other person's expense. Logs are useful for recording interesting or useful information or conversations, as well. [Jennifer "Moira" Smith, MUDFAQ, 1 December 1992. This document posted regularly on rec.games.mud.tiny.]

This last response refers to a client program. This is one of a class of programs that facilitate MUDding. A client program stands between a user's computer and the MUD, performing helpful housekeeping functions such as keeping MUD interchanges on different lines. Without a client program, a user's screen can look like a tangle of MUD instructions and player comments. With a client program a user's screen is relatively easy to read.

10. One of the things that has come out of people having virtual experiences as different genders is that many have acquired a new sense of gender as a continuum. In an online discussion the media theorist Brenda Laurel noted that media such as film, radio, and television advertised the idea that sex and gender were identical and that the universe was bi-gendered. Brenda Laurel, The WELL, conference on virtual reality (vr.47.255), 14 January 1993.

11. Since many more men adopt a female persona than vice versa, some have suggested that gender-bending is yet another way in which men assert domination over female bodies. I thank my student Adrian Banard for his insights on this question. The point was also made by Allucquere Rosanne Stone, Presentation at "Doing Gender on the 'Net Conference," Massachusetts Institute of Technology, Cambridge, Mass., 7 April 1995.

12. Lindsay Van Gelder, "The Strange Case of the Electronic Lover," in *Computerization and Controversy: Value Conflicts and Social Choices,* eds. Charles Dunlop and Rob Kling (Boston: Academic Press, 1991), pp. 366–67.

13. Allucquere Rosanne Stone, Presentation at "Doing Gender on the 'Net Conference," Massachusetts Institute of Technology, Cambridge, Mass., 7 April 1995.

14. Lindsay Van Gelder, "The Strange Case of the Electronic Lover," p. 372.

15. John Schwartz of *The Washington Post* reported that:

In a telephone conversation, Mr. X (who spoke on the condition of anonymity) again tried to put events in perspective. "The cycle of fury and resentment and anger instantaneously transmitted, created this kind of independent entity. . . . These people went after me with virtual torches and strung me up. The emotional response is entirely out of proportion to what actually happened. It involved distortions and lies about what I did or did not do." "I was wrong," he said. "The cyber world is the same as the real world. . . . I should have realized that the exact same standards should have applied." Mr. X later announced that he would be leaving the WELL. He had already been shunned. [John Schwartz, "On-line Lothario's Antics Prompt Debate on Cyber-Age Ethics," *The Washington Post,* 11 July 1993: A1.]

I thank Tina Taylor of Brandeis University for pointing out to me in this case, as in others, the complex position of the virtual body. The virtual body is not always the same. It, too, is constructed by context. A virtual body in a MUD is not the same as a virtual body on IRC or on the WELL.

16. Steve Lohr, "The Meaning of Digital Life," *The New York Times,* 24 April 1995.

17. Nicholas Negroponte, *Being Digital* (New York: Knopf, 1995).

CHAPTER 9   VIRTUALITY AND ITS DISCONTENTS

1. Ray Oldenberg, *The Great Good Place: Cafes, Coffee Shops, Community Centers, Beauty Parlors, General Stores, Bars, Hangouts, and How They Get You Through the Day* (New York: Paragon House, 1989).

2. The dance sequence on LambdaMOO proceeds as follows:

Tony hands you a rose which you place between your teeth. Then Tony leads you through a rhythmic tango, stepping across the floor and ending with Tony holding you in a low dip.
Tony smiles.
You say, "I love the tango."
Tony says, "Type @addfeature #5490"
[I do so].
Tony holds his arm out to you, taking you by the hand, and leads you through a graceful waltz with all the style of Fred Astaire and Ginger Rogers.

3. I thank my student Jennifer Light for this helpful analogy.

4. Jean Baudrillard, *Selected Writings,* ed. Mark Poster (Stanford, Calif.: Stanford University Press, 1988), pp. 171–72.

5. Jerry Mander, *Four Arguments for the Elimination of Television* (New York: William Morrow, 1978), p. 24.

6. This critical comment appeared in a discussion group on The WELL:

On the one hand, like most everyone else here, my life has been very positively impacted by my on-line experiences (I've been on-line in one form or another since 1983—remember 300 baud?). I think of this technology as a kind of mental and social amplifier, giving me access to a far wider range of people, viewpoints, and knowledge than I would have otherwise. Yet, on many, perhaps most, days, I feel that the costs are greater than the benefits.

For example, virtuality seems to me to represent the culmination of a several-thousand-year-old trend in Western culture to separate the mind from the body, thought from physicality, man from nature. This trend lies behind the environmental problems we're facing right now, IMO [online slang for "in my opinion"]. Virtuality seems to portend an even greater disregard for the physical environment which nevertheless still does sustain us, and I don't think that's good at all.

Virtuality also implies to me a privileging of the global at the expense of the local. Yes, it's great to be able to get to know people from all over the planet, without regard for their actual geographic location. I really do think that's good. But it seems to me that in the process of creating virtual "neighborhoods" we are withdrawing from our own very real localities. To me this is a continuation of a several-decades-long trend in American society toward the withdrawal of the upper and middle classes from the public sphere, i.e. the streets and parks of our cities and towns. At the same time the on-line community is growing, real communities are collapsing. Most people don't even know their neighbors. The streets are controlled by thugs. Municipalities become more and more dependent upon, and powerless to control, multinational corporations, because local self-reliance, which originates in real-world interactions and organization among local residents, is atrophying. This is not good for democracy or the people of this country as a whole, IMO.

Nor do I think that this medium, while it is great as a *supplement* to f2f [face-to-face] interactions, would be a very healthy, or emotionally satisfying way to conduct *most* of our interactions—which seems to be a goal of at least some of the more rabid VR [virtual reality] advocates. I mean, I don't want to see my friends over a real-time video system, I want to be with them personally.

Virtual sex? How repugnant—even the most intimate of human experiences now mediated through a machine? Not for me, thanks. The ultimate in alienation. . . .

If anyone's up to it I would like to see some discussion about the "dark side" of information technology—and perhaps in the process we can develop some insight into how we might avoid such pitfalls, while still deriving the very real benefits which it potentially provides. [nao, The WELL, conference on virtual communities (vc.121.1), 29 May 1993.]

7. With a growing sensitivity to the importance of "Main Street" to community life, there is some movement to build new housing that plans for Main Streets and front stoops. See, for example, "Bye-Bye Suburban Dream," the cover story of *Newsweek,* 15 May 1995. These are not conceived of as postmodern "appropriations" but as using architecture that once supported community to help create community.

8. Stephen L. Talbott, *The Future Does Not Compute: Warnings from the Internet* (Sebastopol, Calif.: O'Reilly & Associates, 1995), pp. 127–28.

9. Mander, *Four Arguments for the Elimination of Television,* p. 304.

10. E. M. Forster, "The Machine Stops," in *The Science Fiction Hall of Fame,* ed. Ben Bova, Vol. IIB (New York: Avon, 1971). Originally in E. M. Forster, *The Eternal Moment and Other Stories* (New York: Harcourt Brace Jovanovich, 1928). Cited in Pavel Curtis, "Mudding: Social Phenomena in Text-Based Virtual Realities," available via anonymous ftp://parcftp.xerox.com/pub/MOO/papers/DIAC92.*.

11. Peter Kramer, *Listening to Prozac: A Psychiatrist Explores Antidepressant Drugs and the Remaking of the Self* (New York: Viking, 1993).

12. Janice A. Radaway, *Reading the Romance: Women, Patriarchy, and Popular Literature* (Chapel Hill: The University of North Carolina Press, 1991).

13. T. J. Burnside Clap, 1987, Fesarius Publications, quoted in Henry Jenkins, *Textual Poachers: Television Fans and Participatory Culture* (New York: Routledge, 1992), p. 277.

14. Jenkins, pp. 280–81.

15. From CyberMind electronic mailing list, 29 March 1993.

16. Sherry Turkle, "The Subjective Computer: A Study in the Psychology of Personal Computation," *Social Studies of Science* 12 (1982): 201.

17. Turkle, "The Subjective Computer": 201.

18. The sense that virtual is better and safer and more interesting has extended even to those usually most concerned about how we *look.* The editor of *Mademoiselle* magazine, a publication chiefly concerned with fashion and

beauty advice, introduces a special section on electronic communication by declaring that if she "could live anywhere, it would be in Cyberia," i.e. cyberspace. (Gabé Doppelt, *Mademoiselle,* October 1993: 141.)

19. A Spring 1995 special issue of *Time* magazine devoted to cyberspace reported:

The fact is that access to the new technology generally breaks down along traditional class lines. Wealthy and upper-middle-class families form the bulk of the 30% of American households that own computers. Similarly, wealthier school districts naturally tend to have equipment that is unavailable to poorer ones, and schools in the more affluent suburbs have twice as many computers per student as their less-well-funded urban counterparts. [p. 25]

20. See, for example, the work of Alan Shaw of MIT's Media Laboratory. Alan Clinton Shaw, "Social Construction in the Inner City: Design Environments for Social Development and Urban Renewal" (Ph.D. diss., Massachusetts Institute of Technology, Media Laboratory, Epistemology and Learning Group, 1995).

21. *Time,* Spring 1995 (special issue): 24; and Howard Rheingold, *The Virtual Community: Homesteading on the Electronic Frontier* (Reading, Mass.: Addison-Wesley, 1993), pp. 17–37. SeniorNet, founded in 1986 by Mary Furlong, is designed to be permeable to the real. SeniorNet offers its members practical tips on home repair and advice on problems such as how to handle bouts of depression associated with aging. Members say it has given them a sense that their "world is still expanding." The organization sponsors regional face-to-face meetings, and its members regularly visit each other in person. See John F. Dickerson, "Never Too Old," *Time,* Spring 1995 (special issue): 41.

22. Daniel Akst and James Weissman, "At Your Service," *NetGuide,* August 1995: 35–38.

23. Michel Foucault, *Discipline and Punish: The Birth of the Prison,* trans. Alan Sheridan (New York: Pantheon, 1977). See also Mark Poster, *The Mode of Information: Poststructuralism and Social Context* (Chicago: University of Chicago Press, 1990), pp. 69–98.

24. kort (Barry Kort), The WELL, conference on virtual communities (vc.52.28), 18 April 1993.

25. After a brief test period in the United States, Habitat was bought by the Fujitsu Corporation and became a successful commercial venture in Japan, with over one-and-a-half million paid subscribers.

26. Chip Morningstar and F. Randall Farmer, "The Lessons of Lucasfilm's Habitat," in *Cyberspace: First Steps,* ed. Michael Benedikt (Cambridge, Mass.: MIT Press, 1991), p. 289.

27. Morningstar and Farmer, "The Lessons of Lucasfilm's Habitat," p. 289.

28. Julian Dibbell, "Rape in Cyberspace," *The Village Voice,* 21 December 1993: 38.

29. The message was signed by Wonko the Sane.

30. Dibbell, "Rape in Cyberspace": 42.

CHAPTER 10   IDENTITY CRISIS

1. Emily Martin, *Flexible Bodies* (Boston: Beacon Press, 1994), pp. 161–225.
2. mcdee, The WELL, conference on virtual communities (vc.20.17), 18 April 1992.
3. The sentiment that life online could provide a different experience of self was seconded by a participant who described himself as a man whose conversational abilities as an adult were impaired by having been a stutterer as a child. Online he was able to discover the experience of participating in the flow of a conversation.

> I echo [the previous contributor] in feeling that my online persona differs greatly from my persona offline. And, in many ways, my online persona is more "me." I feel a lot more freedom to speak here. Growing up, I had a severe stuttering problem. I couldn't speak a word without stuttering, so I spoke only when absolutely necessary. I worked through it in my early 20s and you wouldn't even notice it now (except when I'm stressed out), but at 37 I'm still shy to speak. I'm a lot more comfortable with listening than with talking. And when I do speak I usually feel out of sync: I'll inadvertently step on other people's words, or lose people's attention, or talk through instead of to. I didn't learn the dynamic of conversation that most people take for granted, I think. Here, though, it's completely different: I have a feel for the flow of the "conversations," have the time to measure my response, don't have to worry about the balance of conversational space—we all make as much space as we want just by pressing "r" to respond. It's been a wonderfully liberating experience for me. (Anonymous)

4. spoonman, The WELL, conference on virtual communities (vc.20.65), 11 June 1992.
5. Kenneth Gergen, *The Saturated Self: Dilemmas of Identity in Contemporary Life* (New York: Basic Books, 1991).
6. bluefire (Bob Jacobson), The WELL, conference on virtual reality (vr.85.146), 15 August 1993.
7. The WELL, conference on virtual reality (vr.85.148), 17 August 1993.
8. Art Kleiner, The WELL, conference on virtual reality (vr.47.41), 2 October 1990.
9. Gergen, *The Saturated Self*, p. 6.
10. Gergen, *The Saturated Self*, p. 17.
11. hlr (Howard Rheingold), The WELL, conference on virtual reality (vr.47.351), 2 February 1993.
12. McKenzie Wark, The WELL, conference on virtual reality (vr.47.361), 3 February 1993.
13. hlr (Howard Rheingold), The WELL, conference on virtual reality (vr.47.362), 3 February 1993.
14. James M. Glass, *Shattered Selves: Multiple Personality in a Postmodern World* (Ithaca, N.Y.: Cornell University Press, 1993).
15. Robert Jay Lifton, *The Protean Self: Human Resilience in an Age of Fragmentation* (New York: Basic Books, 1993), p. 192.
16. Lifton, *The Protean Self*, pp. 229–32.

17. See, for example, "Aion: Phenomenology of the Self," in *The Portable Jung,* ed. Joseph Campbell, trans. R. F. C. Hull (New York: Penguin, 1971).

18. See, for example, Marvin Minsky, *The Society of Mind* (New York: Simon & Schuster, 1985).

19. See, for example, Colin Ross, *Multiple Personality Disorder: Diagnosis, Clinical Features, and Treatment* (New York: John Wiley & Sons, 1989).

20. Claude Lévi-Strauss, *The Savage Mind* (Chicago: University of Chicago Press, 1960).

21. Ian Hacking, *Rewriting the Soul: Multiple Personality and the Sciences of Memory* (Princeton, N.J.: Princeton University Press, 1995), p. 21.

22. Hacking, *Rewriting the Soul,* p. 29.

23. See Hacking, *Rewriting the Soul,* pp. 96ff.

24. Daniel C. Dennett, *Consciousness Explained* (Boston: Little, Brown and Company, 1991).

25. Donna Haraway, "The Actors Are Cyborg, Nature Is Coyote, and the Geography Is Elsewhere: Postscript to 'Cyborgs at Large' " in *Technoculture,* eds. Constance Penley and Andrew Ross (Minneapolis: University of Minnesota Press, 1991), p. 22.

26. Leslie Harris, "The Psychodynamic Effects of Virtual Reality," *The Arachnet Electronic Journal on Virtual Culture* 2, 1 (February 1994), abstract.

27. Harris, "The Psychodynamic Effects of Virtual Reality," section 14.

28. Allucquere Rosanne Stone has referred to our time in history as the close of the mechanical age, underscoring that we no longer look to clockwork or engines to build our images of self and society. See *The War of Desire and Technology at the Close of the Mechanical Age* (Cambridge, Mass.: MIT Press, 1995).

29. *Newsweek,* 17 April 1995: 70.

30. See, for example, Barbara Johnson, *A World of Difference* (Baltimore: Johns Hopkins University Press, 1987); Donald P. Spence, *Narrative Truth and Historical Truth: Meaning and Interpretation in Psychoanalysis* (New York: W. W. Norton & Company, 1982; and Humphrey Morris, ed., *Telling Facts: History and Narration in Psychoanalysis* (Baltimore: Johns Hopkins University Press, 1992).

31. N. Katherine Hayles, *Chaos Bound: Orderly Disorder in Contemporary Literature and Science* (Ithaca, N.Y.: Cornell University Press, 1990), p. 3.

32. W. Daniel Hillis quoted in Steven Levy, *Artificial Life: The Quest for a New Creation* (New York: Pantheon Books, 1992), p. 344.

33. Norbert Wiener, *God and Golem, Inc.: A Comment on Certain Points Where Cybernetics Impinges on Religion* (Cambridge, Mass.: MIT Press, 1964), p. 36.

34. "A Cyborg Manifesto: Science, Technology, and Socialist-Feminism in the Late Twentieth Century" in Donna Haraway, *Simians, Cyborgs, and Women: The Reinvention of Nature* (New York: Routledge, 1991), pp. 149–81.

35. Colin Greenland, "A Nod to the Apocalypse: An Interview with William Gibson," *Foundation* 36 (Summer): 5–9.

36. In *Chaos Bound,* Katherine Hayles refers to Donna Haraway's language of cyborgs when describing the positive and negative sides of how information technology has contributed to reconceptualizing the human:

Haraway argues that information technology has made it possible for us to think of entities (including human beings) as conglomerations that can be taken apart, combined with new elements, and put together again in ways that violate traditional boundaries.

From one perspective this violation is liberating, for it allows historically oppressive constructs to be deconstructed and replaced by new kinds of entities more open to the expression of difference. The problem, of course, is that these new constructs may also be oppressive, albeit in different ways. For example, much feminist thought and writing since the 1940s has been directed toward deconstructing the idea of "man" as a norm by which human experience can be judged. To achieve this goal, another construction has been erected, "woman." Yet as it has been defined in the writing of white, affluent, heterosexual, Western women, this construct has tended to exclude the experiences of black women, Third World women, poor women, lesbian women, and so on. [p. 283]

37. Joel Porte, ed., *Emerson in His Journals* (Cambridge, Mass.: Belknap Press, 1982), p. 81.
38. Fredric Jameson calls these ideas our new "cultural dominant." See Fredric Jameson, "Postmodernism, or the Cultural Logic of Late Capitalism," *New Left Review* 146 (July–August 1984): 53–92. I thank Jim Berkley for directing my attention to the parallels between Jameson's position and consequent developments in personal computation.
39. Victor Turner, *The Ritual Process: Structure and Antistructure* (Chicago: Aldine, 1966).
40. Donna Haraway, "A Cyborg Manifesto," p. 148.

at psychoanalytic ideas as they were "read" at a time of intense social and political turmoil.

In the late 1970s to early 1980s, greatly influenced by the culture of MIT, I turned to the study of the appropriation of ideas surrounding technical artifacts. Like written texts, objects can be "read" in different ways by members of different interpretive communities. Specifically, I explored how ideas borrowed from the computer culture were having their own impact on how people saw their minds, as well as their sense of what it meant to be alive and to be human. I noted signs of a shift from a psychoanalytic to a computer culture: to take a small example, errors once nervously described in terms of their meaning as "Freudian slips" were now being treated in neutral mechanistic terms as "information processing errors."

The mid-1980s was a turning point in the history of the computer culture. The Macintosh was introduced. *Neuromancer* was published. There was a new interest in introducing computing into elementary and secondary education, as well as into general pedagogy at the university level. Networked computing was becoming increasingly important. Simulation software was becoming increasingly available. Indeed, MIT, my own institution, was embarking on a $70 million experiment, Project Athena, whose aim was to use state-of-the-art computer workstations and "courseware" throughout the undergraduate curriculum. On a smaller scale, similar experiments were taking place in classrooms all over the country.

Shortly after the publication of *The Second Self,* I embarked on a series of studies through which I tried to capture different aspects of these changes. These are the studies that contributed to this book. They have tapped the experience of roughly a thousand informants, nearly three hundred of them children.

From 1984 to 1987, I studied the MIT Athena project. I worked on this ethnographic investigation with my faculty colleague, Donald Schön, and three research assistants: Brenda Nielsen, M. Stella Orsini, and Wim Overmeer. We focused on four specific educational settings: the School of Architecture and Planning, the Department of Physics, the Department of Chemistry, and the Department of Civil Engineering.

At the same time as I observed the MIT setting, I also investigated the nascent computer culture at an inner-city elementary school. This project on computers and elementary-school-aged children, which involved observations, interviewing, and psychological testing, was done with Seymour Papert and a team of graduate students at the MIT Media Laboratory. This work helped me refine my ideas on gender and computing and on the psychology of different styles of computer use.

In 1987 to 1989, during a leave of absence from MIT, I began cross-

A NOTE ON METHOD:
THE INNER HISTORY OF TECHNOLOGY

This is a very personal book. It is based on ethnographic and clinical observation, where the researcher, her sensibilities and taste, constitute a principal instrument of investigation. My method is drawn from the various formal disciplines I have studied and grown up in; the work itself is motivated by a desire to convey what I have found most compelling and significant in the evolving relationship between computers and people. My studies of the computer culture since publishing *The Second Self* in 1984 convinced me that information technology is doing more than providing an evocative object for our self-reflection, the essential message of that book. Now it is the basis for a new culture of simulation and a fundamental reconsideration of human identity, the story I tell in *Life on the Screen*.

My training as a social scientist included graduate work in sociology, anthropology, and personality psychology; I am a licensed clinical psychologist and a graduate of the Boston Psychoanalytic Institute. For nearly twenty years I have taught social sciences at MIT and have practiced as a psychotherapist for more than fifteen. In this book I try to use the perspectives of these disciplines to capture something of the inner history of computer technology: how the computer has profoundly shaped our ways of thinking and feeling. This effort has led me to explore the way people appropriate ideas carried by technology, reshaping them for their own complex purposes.

In the early to mid-1970s, I studied the impact of psychoanalytic ideas on people's everyday lives. My case study was the French infatuation with Freud in the years after the student revolution of 1968. In literary studies, a tradition of "reader response" criticism studies texts not as they are written but as they are read. My work on psychoanalysis in France looked cultural investigations: I studied English children's attitudes about computers and the question of what is alive, a question I had earlier studied in the United States. I also did field research in the Soviet Union. There I investigated children's experiences with computers in schools, computer clubs, and computer camps. I also had an opportunity to observe the politics of computing and computer-mediated communication in the rapidly changing Soviet political landscape. In 1989, I returned to the question of the computer as an evocative object in the lives of American children. This time, my investigations included children's use of online communication.

Beginning in the late 1970s, I had studied the computer "as Rorschach," focusing on different styles of programming. In the late 1980s, I pursued the question of styles and the psychology of computing by looking at the way people related to different operating systems, interfaces, and specific software. By 1992, my research turned almost exclusively to the question of identity and the Internet. In July of that year I held a series of weekly pizza parties for MUDders in the Boston area with my research assistant Amy Bruckman of the MIT Media Laboratory. There the conversation quickly turned to multiple personae, romance, and what can be counted on as real in virtual space. The following fall, I continued interviewing people intensively involved in MUDding and expanded my investigations to the world of Internet Relay Chat (IRC), newsgroups, bulletin boards, and commercial online services and continued investigating the online lives of children and teenagers.

My research has two distinct parts: a field research component and a clinical component. In the field research I put myself in some of the places where people and computers meet; I observe interactions and take detailed field notes about humor, conflict, collaboration, and styles of use. I try to understand the social lives and cultural meanings that people are constructing as their lives become increasingly entwined with computer technology. As I do this work in such places as computer laboratories, programming classes, grade-school classrooms, and personal-computer user groups, I have informal conversations with those around me. These conversations tend to be relatively brief, a half-hour to an hour in length. Depending on the age and situation of my informant, the conversations may take place over coffee, during school snacks of milk and cookies, or in a corner of a computer laboratory. Since the late 1970s, I have taught courses about the computer culture and the psychology of computation, and some of my material comes from the give-and-take of the classroom. This is particularly true in my discussion of changing attitudes toward computer psychotherapy, in which I have been able to draw from nearly two decades of talking to students about this question.

What I call the clinical component of my work has a rather different tone. There, I pursue more detailed interviews, usually in my office, sometimes in the homes of the people I am interviewing. In these studies, there will typically be several hours of talk. I have spoken with some of my informants for six to eight hours over several years. In these more lengthy interviews I am better able to explore an individual's life history and tease out the roles technology has played. I call these studies clinical, but my role in them is as a researcher, not a therapist. This is an important distinction, because talking to people for many hours about their lives brings up many of the same issues that would come up in therapy sessions. Although it is beyond the scope of this note to discuss the strategies one uses to maintain this distinction, I want to signal that I consider it the researcher's responsibility to guide people out of conversations where this distinction will be hard to maintain.

In this book I follow a consistent policy of disguising the identities of all my informants. I have invented names, places (virtual and real), and some elements of personal background. Of course, I try to invent a disguise that captures what seem to me to be the crucial variables of life history. In reporting cases of people who have part of their identities on the Internet, I follow the same policy as for other informants: I protect confidentiality by disguising identities. This means that among other things, I change MUD names, character names, and city names. In the case of the WELL, there is a clear community norm that "You Own Your Own Words." I have asked contributors to WELL discussions how they wish to be identified. Different people have made different choices. When I use materials from publicly archived online sources I simply indicate the source.

Part of my field research has consisted of going to scientific conferences, for example in the fields of artificial intelligence, virtual reality, artificial life, and computer-human interaction. In these settings, I also talk to many people, and here, too, I protect confidentiality, although I do identify people by name when I draw from the public record or when they specifically ask to be so cited.

Virtual reality poses a new methodological challenge for the researcher: what to make of online interviews and, indeed, whether and how to use them. I have chosen not to report on my own findings unless I have met the Internet user in person rather than simply in persona. I made this decision because of the focus of my research: how experiences in virtual reality affect real life and, more generally, on the relationship between the virtual and the real. In this way, my work on cyberspace to this point is conservative because of its distinctly real-life bias. Researchers with different interests and theoretical perspectives will surely think about this decision differently.

ACKNOWLEDGMENTS

Writing a book provides a welcome opportunity to thank the many who have helped. My first debt is of course to the subjects of this book who shared their experiences of the computer culture with me. Their generosity and patience made my work a pleasure. I want to thank Ralph Willard for the many times his critical reading of drafts and insightful editorial suggestions provided timely and much appreciated encouragement. John Berlow, in this project as in others, has been a deeply appreciated dialogue partner. His critical readings have been invaluable. Mary-Kay Wilmers, Jennifer June Cobb and John Fox allowed me to be warmed by their friendship and to profit from their editorial insight. At Simon & Schuster, Mary Ann Naples and Laurie Chittenden provided the right mix of practical support and patience.

My research assistants Jim Berkley, Amy Bruckman, and Tina Taylor each embody the new culture of computing, humanistic as well as technical, about which I am trying to write. This book owes much to their special combination of gifts. Conversations with Jim helped give shape to ideas about computers as the carriers for postmodern ideas. Amy worked with me in my early explorations of both MUDding and Macintosh computing and challenged me to analyze my attraction and resistance to each. Tina contributed challenging critical readings and meticulous reference work. All three read drafts of my manuscript and provided suggestions and corrections.

For nearly twenty years, MIT has provided a stimulating environment for my studies of the computer culture. Michael Dertouzos, Director of the Laboratory for Computer Science, and former provost Walter Rosenblith provided early support. Hal Abelson introduced me to MUDs. I shall always be grateful. Much of my work on children and computers was done in collaboration with the Epistemology and Learning Group at the MIT Media Laboratory. For many years, Seymour Papert and I worked together on the way computers encourage pluralistic ways of knowing. Seymour read an early draft of this manuscript and offered a tonic mixture of criticism and good ideas. Mitchel Resnick taught me as we taught together and improved this work through his careful reading of chapter drafts. I thank them and all the students who worked on Project Headlight —a program that introduced the best of the computer culture to an inner-city school. Donald Schön and I worked on an ethnographic study of MIT's Project Athena, which brought computer-intensive learning to the MIT undergraduate community. I thank him and our research assistants, Brenda Nielsen, M. Stella Orsini, and Wim Overmeer for their observations and insights. I am also grateful to the MacArthur Foundation, the National Science Foundation, the MIT Provost's Office, and the Office of the Dean of the School of Humanities and Social Sciences for funding aspects of the work reported in this book.

At MIT, my students have always been my teachers. I thank the students in Psychology and Technology; Gender, Technology, and Computer Culture; Computers and People; Systems and the Self; and most recently, Identity and the Internet. And I thank my colleagues, both faculty and graduate students, in MIT's Program in Science, Technology, and Society. From this talented group, I must single out Michael Fischer whose close reading of an early draft of this book provided worthy guideposts for revision.

My work owes much to several teachers at Harvard, Daniel Bell, Stanley Hoffman, George Homans, Martin Peretz, David Riesman, and Laurence Wylie who gave me the confidence to pursue my interests even when they were off the beaten track, and to Sandra Siler who helped me define the style of intellectual contribution to which I feel best suited. During the years of writing this book, the friendship of Jacqueline Bolles, Candace Karu, and Nancy Rosenblum has been sustaining. Each read and commented on sections of the manuscript in progress. If I have succeeded in talking about technology in a way that makes humanists want to listen, it is because I tried to write to them. Henry Jenkins provided the *Star Trek* tapes that I watched for inspiration. Barbara Masi kept me up to date on attitudes about computer psychotherapy. Stacey Whiteman's patience, dedication, and good humor helped me through the power surges, system freezes, and lost files that are part of the life of digitally dependent authors. Cynthia Col and Sean Doyle generously contributed technical support, help with galleys, and wit for my Web home page. John Brockman, Katie Hafner, Jeremy Harding, Larry Husten, Wendy Kaminer, Mitch Kapor, Joyce Kingdon, Jennifer Light, Susan Marsten, Katinka Matson, Ted Metcalfe, Lynn Rhenisch, Gail Ross, Sharon Traweek, and Gérard Vishniac provided much-appreciated suggestions and encouragement.

Susan Turkle sent flowers when it mattered most. Robert and Edith Bonowitz and their daughters Harriet and Mildred were always present. The memories they left me have made everything possible.

Finally, I give thanks not only to but for my daughter, Rebecca. I write this book in the hope that she will be proud of me.

Stockbridge, Massachusetts
June 1995

INDEX